Current Topics in Microbiology and Immunology

Volume 374

For further volumes:
http://www.springer.com/series/82

Jean Pieters · John D. McKinney
Editors

Pathogenesis of *Mycobacterium tuberculosis* and its Interaction with the Host Organism

Responsible Series Editor: Fritz Melchers

Springer

Editors
Jean Pieters
Biozentrum
University of Basel
Basel
Switzerland

John D. McKinney
Laboratory of Microbiology and Microsystems
Swiss Federal Institute of Technology Lausanne (EPFL)
Lausanne
Switzerland

ISSN 0070-217X ISSN 2196-9965 (electronic)
ISBN 978-3-642-40231-9 ISBN 978-3-642-40232-6 (eBook)
DOI 10.1007/978-3-642-40232-6
Springer Heidelberg New York Dordrecht London

Library of Congress Control Number: 2013949198

Printed on acid-free paper

Springer is part of Springer Science+Business Media (www.springer.com)

Preface

Tuberculosis has been with mankind ever since the dawn of civilization, with evidence of human infection as early as 9,000 years ago. Although the causative agent, *Mycobacterium tuberculosis*, has been identified more than a century ago, tuberculosis remains one of the most prevalent diseases on the globe, with currently one third of the population being latently infected with *M. tuberculosis*. Despite concerted efforts from scientists with diverse backgrounds, including medical doctors, microbiologists, cell biologists and immunologists, global control of tuberculosis continues to be highly challenging. However, research over the past few years has highlighted several issues that may contribute to overcoming some of the difficulties of eradicating *M. tuberculosis*. In this volume, a number of experts discuss various topics related to the virulence of *M. tuberculosis*.

Mycobacterium tuberculosis is an obligate intracellular pathogen that has gained the capacity to survive within those cells of the immune system that are in fact designed to eradicate microbial pathogens, the macrophages. As a result, pathogenic mycobacteria are able to cope with a plethora of antimicrobial activities, against which these bacilli have evolved numerous mechanisms of resistance. In addition, one emerging problem is the appearance of strains of *M. tuberculosis* that show various degrees of drug resistance. This problem has become of increasing concern also because of the recent discovery of extremely drug-resistant (XDR) as well as totally drug resistant (TDR) strains of *M. tuberculosis*. The emergence of these highly problematic strains may be a result of the slow generation time of *Mycobacterium spp.*, their exposure to a battery of drugs for prolonged time periods, the sometimes poor compliance of patients with anti-tuberculosis therapy or most likely a combination of these factors. One realization however has been that there is a need for a better understanding of the genetic diversity in different strains of *M. tuberculosis*, as outlined by Gagneux (Chap. 1). Also, recent work suggests that *M. tuberculosis*, unlike many other bacterial pathogens, relies heavily on chromosomal mutagenesis to drive its evolution within the human host. The intricate balance between maintenance of chromosome integrity and evolutionary progress is discussed in detail by Warner and colleagues in Chap. 2.

The emerging problem of multi-drug resistance requires a far better understanding of the mechanisms involved in drug resistance, and Smith and colleagues (Chap. 3) review the current understanding of the molecular basis of the development of resistance mechanisms. Furthermore, despite the fact that *M. tuberculosis* is extensively exposed to diverse antimycobacterial drugs as well as host antimicrobials, the bacilli can survive exposure to many of these compounds through the upregulation of antimicrobial efflux pumps. In Chap. 4, Szumowski and colleagues propose that drug tolerance has mainly evolved as a result of the need of intracellularly residing mycobacteria to employ efflux pumps to protect against environmental toxins.

To successfully survive under challenging conditions, *M. tuberculosis* not only needs to be able to fight off attacks from toxic components through efflux pumps and other drug resistance mechanisms, but it also requires a diverse array of secretion systems in order to communicate with and modulate its environment, the macrophage. Van der Woude and colleagues (Chap. 5) review the different secretion systems that are employed by *M. tuberculosis*, which have recently attracted much interest as determining factors for intracellular survival.

Mycobacterium tuberculosis is notoriously able to survive in a so-called "dormant" or non-growing state within host macrophages. The capacity to survive without growth also creates challenges in terms of finding anti-mycobacterial drugs that are able to interfere with survival of non-proliferating bacteria rather than controlling growth per se. Manina and McKinney (Chap. 6) highlight the importance of analyzing bacteria that are non-growing but metabolically active, knowledge that might be important for designing better strategies to eliminate non-replicating mycobacteria. Metabolism becomes also important when considering the natural niche of *M. tuberculosis*, the macrophage phagosome. In Chap. 7, Ehrt and Rhee discuss the importance of the metabolic network of intracellularly residing *M. tuberculosis* for survival within its host.

Once within macrophage phagosomes, *M. tuberculosis* can rely on several distinct mechanisms to subvert the antimicrobial activities of the macrophages. Many of these mechanisms allow *M. tuberculosis* to prevent intraphagosomal destruction, and in Chap. 8 Jayachandran and colleagues discuss work that has deciphered several of these mechanisms. Besides intraphagosomal survival, *M. tuberculosis* can also extensively interact with other host components, and Stanley and Cox (Chap. 9) review recent work that has added several layers of complexity to the interaction of *M. tuberculosis* with its host.

Together, these Chapters provide a topical overview on the current understanding of diverse mechanisms that are involved in the virulence of *M. tuberculosis*, ranging from their genetic, metabolic and molecular makeup as well as the exquisite strategies these bacteria utilize to circumvent host innate immune responses. The work described in this volume may therefore provide a stimulus for further exploration of the intricate biology of *M. tuberculosis* and its interaction with their hosts.

Jean Pieters
John McKinney

Contents

Genetic Diversity in *Mycobacterium tuberculosis*

Sebastien Gagneux

Abstract Recent years have witnessed an increased appreciation of the extent and relevance of strain-to-strain variation in *Mycobacterium tuberculosis*. This paradigm shift can largely be attributed to an improved understanding of the global population structure of this organism, and to the realisation that the various members of the *M. tuberculosis* complex (MTBC) harbour more genetic diversity than previously realised. Moreover, many studies using experimental models of infection have demonstrated that MTBC diversity translates into significant differences in immunogenecity and virulence. However, linking these experimental phenotypes to relevant clinical phenotypes has been difficult, and to date, largely unsuccessful. Nevertheless, emerging high-throughput technologies, in particular next-generation sequencing, offer new opportunities, and have already lead to important new insights. Given the complexity of the host-pathogen interaction in tuberculosis, systems approaches will be key to define the role of MTBC diversity in the fight against one of humankind's most important pathogens.

Contents

S. Gagneux (✉)
Swiss Tropical and Public Health Institute (Swiss TPH),
University of Basel, Basel, Switzerland
e-mail: Sebastien.Gagneux@unibas.ch

Current Topics in Microbiology and Immunology (2013) 374: 1–25

DOI: 10.1007/82_2013_329

Published Online: 16 May 2013

1 Introduction

The outcome of tuberculosis (TB) infection and disease is extraordinarily diverse, ranging from lifelong asymptomatic infection to pulmonary TB and disseminated disease. In the past, this variable outcome has largely been attributed to host and environmental factors, but there is mounting evidence suggesting variation in the causative agent might also play a role. Indeed, the fact that this book, which is primarily focusing on the pathogenesis and host-pathogen interaction in TB, includes a chapter on "genetic diversity in *M. tuberculosis*" is a testimony to the recent paradigm shift with respect to the importance of studying strain variation (Comas and Gagneux 2009). Early sequencing studies revealed little DNA diversity across clinical isolates (Musser et al. 2000; Sreevatsan et al. 1997) leading to the dogma that strain variation in *M. tuberculosis* was "negligible" and therefore clinically irrelevant. Even though genetic diversity in *M. tuberculosis* is limited compared to other bacteria (Achtman 2008), this does not necessarily mean that the existing variability, as minor it may be, will not translate into any relevant phenotypes. Already in the 1960s, Mitchison and colleagues observed differences in virulence in clinical strains of *M. tuberculosis* when infecting guinea-pigs (Mitchison et al. 1960). At the time, no molecular tools were available to classify these strains into phylogenetically meaningful groupings. When the first genotyping tools became available in the early 1990s (van Embden et al. 1993), epidemiologist noticed very soon that *M. tuberculosis* strains also differed in their propensity to spread between individuals, and that some of these strains were associated with prolonged outbreaks (Rajakumar et al. 2004; Valway et al. 1998; Zhang et al. 1999). In some cases, experimental studies identified molecular features that might contribute to the success of outbreak strains (Newton et al. 2006; Reed et al. 2004). However, the relevance of these characteristics relative to non-bacterial factors remains unclear.

Over the last 10 years, several reviews have been published on the extent and relevance of strain diversity in *M. tuberculosis* (Gagneux and Small 2007; Kato-Maeda et al. 2001a; Malik and Godfrey-Faussett 2005; Nicol and Wilkinson 2008; Parwati et al. 2010). One of the most comprehensive articles published recently reviewed 100 papers on the topic, and concluded that even though the phenotype of clinical strains of *M. tuberculosis* undoubtedly differs significantly in vitro and in animal models of infection, if and how these differences are reflected in clinical settings is less clear (Coscolla and Gagneux 2010). Two years later, this view largely still holds. Nevertheless, recent technological advances, in particular large-scale DNA sequencing and other–omics platforms, are now able to approach the subject at an unprecedented scale, with the potential to generate important novel insights. Given the biological complexities in TB, it is too simplistic to think that variable strain behaviour will be the product of just one or few genomic differences. It is much more likely that the various genomic features of a given strain will have combined effects. Such epistatic interactions will be difficult to detect and link to relevant phenotypes, but systems biology approaches might offer a way forward (Comas and Gagneux 2011).

This chapter summarises some of the recent advances in our understanding of the nature and consequence of strain variation in *M. tuberculosis,* with a special focus on comparative whole-genome sequencing. I will start by reviewing the current definition of the *M. tuberculosis* complex (MTBC), and briefly discuss the main genotyping methods for MTBC as well as their limitations. I will then review our current understanding of the global phylogeography of human-associated MTBC based on the available whole-genome data, discuss some of the new forms of genomic variation that have become evident through comparative genome sequencing, and review recent examples where large-scale DNA sequence data was used to predict putative phenotypic effects. Moving from the genome diversity to phenotypic variation, I will review some recent insights into strain-specific transcriptomes and strain-specific differences in innate immune recognition. Finally, I will discuss recent findings on the microevolution of drug-resistant MTBC, and end with a few thoughts on possible future research directions.

2 Genomic Diversity Among Human-Associated MTBC

2.1 The Current Definition of MTBC

MTBC consists of several closely related species and sub-species of acid-fast bacteria also referred to as 'ecotypes' (Smith et al. 2005). Even though these ecotypes share identical 16S ribosomal RNA sequences and 99.9 % nucleotide identity at the whole-genome level (with the exception of *Mycobacterium canettii* and the other so-called "smooth tubercle bacilli" further discussed below), they appear to be adapted to different host species. *M. tuberculosis* sensu stricto and *Mycobacterium africanum* are the main agents of TB in humans. *M. africanum* is limited to West Africa for reasons that are unknown, but causes up to 50 % of human TB in parts of that region (de Jong et al. 2010). In addition to these human pathogens, several animal-associated members of MTBC are found in various domestic and wild animals. These include *Mycobacterium bovis*, the agent of bovine TB (Garnier et al. 2003), *Mycobacterium caprae* (sheep and goats) (Niemann et al. 2002), *Mycobacterium microti* (voles) (Frota et al. 2004), *Mycobacterium pinnipedii* (seals and see lions) (Cousins et al. 2003), *Mycobacterium mungi* (mangoose) (Alexander et al. 2010),

Mycobacterium orygis (antelope) (van Ingen et al. 2012) and the "dassie bacillus" (rock hyrax) (Mostowy et al. 2004a).

M. canettii and the other smooth TB bacilli (also referred to as "*Mycobacterium prototuberculosis*") are regarded by some as human-adapted pathogens (Gutierrez et al. 2005), but several factors point to a likely environmental reservoir for these microbes (Koeck et al. 2010). First, only about 60 isolates have been described in the literature since the original discovery of *M. canettii* in 1969 (van Soolingen et al. 1997). Second, the majority of these organisms have been isolated from patients in Djibouti (Fabre et al. 2010). Third, no human-to-human transmission has been documented thus far (Koeck et al. 2010). And forth, these bacteria show clear evidence of extensive ongoing horizontal gene transfer (HGT) (Gutierrez et al. 2005), which stands in contrast to the other members of MTBC (the role of HGT and recombination in MTBC is further discussed below).

For the remainder of this chapter, I will focus on *M. tuberculosis* sensu stricto and *M. africanum* , which are the most important agents of human TB. Based on the various strain genotyping techniques, *M. tuberculosis* and *M. africanum* have been further subdivided into strain lineages and families. For example, the "Beijing" family of strains was originally defined based on characteristic IS*6110* RFLP patterns (van Soolingen et al. 1995). This strain family has gained considerable attention because of its association with drug resistance (reviewed in Borrell and Gagneux 2009) and hypervirulence in some experimental models (reviewed in Parwati et al. 2010). Other human-associated MTBC lineages have been studied less, partially because of the lack of a standardised and phylogenetically robust classification system and associated nomenclature (Gagneux and Small 2007). Ultimately, such a system should be based on whole-genome sequencing of a large and representative collection of MTBC clinical strains. However, until genome-sequencing becomes more readily available, the different genotyping methodologies and the corresponding nomenclatures are likely to continue to be used in parallel (for a comparison of the various nomenclatures for human-associated MTBC refer to Coscolla and Gagneux 2010). Let me thus briefly review these different genotyping techniques, and discuss some of their advantages and disadvantages.

2.2 Current Genotyping Methodologies for MTBC

Strain typing in MTBC is usually performed for one of three main purposes: (i) "classical" molecular epidemiology , (ii) phylogenetic and evolutionary studies and (iii) strain classification . However, the existing genotyping techniques are not equally suited for all of these applications. Classical molecular epidemiological studies in TB usually involve measuring ongoing transmission, differentiating between relapse and re-infection, or detecting laboratory cross-contamination (Kato-Maeda et al. 2011). Such studies require highly discriminatory genotyping tools such as IS*6110* RFLP analysis (van Embden et al. 1993) or MIRU-VNTR typing (Supply et al. 2006). These techniques rely on mobile and repetitive DNA elements, respectively, which exhibit a rapid rate of change (i.e. a fast "molecular clock"), providing a high discriminatory power for differentiating between patient isolates. However, whilst useful for genotyping closely related strains, these high rates of change lead to convergent evolution and homoplasy (i.e., emergence of identical patterns in phylogenetically unrelated strains), which complicates phylogenetic inference and strain classification (Comas et al. 2009). Spoligotyping is another genotyping technique which has been widely used for both molecular epidemiology and evolutionary studies. It usually shows a lower discriminatory power than IS*6110* RFLP or MIRU-VNTR typing, and is therefore often used as a complementary genotyping method in epidemiological studies (Kamerbeek et al. 1997). Spoligotyping is based on the CRISPR (Clustered Regularly Interspaced Short Palindromic Repeats) region, also known as the Direct Repeat (DR) region of MTBC. Unique so-called spacer sequences are interspersed between repetitive sequences and are variably present or absent in a given MTBC isolate. Interrogation of 43 of these unique spacer sequences results in a strain-specific fingerprint which is easily digitalised and compared across laboratories. Several large international databases have been compiled containing spoligotypes of thousands of clinical isolates from many countries (Allix-Beguec et al. 2008; Demay et al. 2012). Based on some characteristic spoligopatterns, several MTBC lineages and families have been defined. For example, the Beijing family shows a typical spoligopattern in which the first 34 unique spacers are deleted (Bifani et al. 2002). Unfortunately, spoligotyping patterns, too, are prone to homoplasy as individual spacers can be deleted independently in phylogenetically unrelated strains (Comas et al. 2009). For example, a recent study reported clinical strains exhibiting a "pseudo-Beijing" spoligotype, referring to the fact that these strains showed a spoligotype pattern characteristic of Beijing strains but belonged to a different phylogenetic lineage (Fenner et al. 2011). Moreover, some spoligotyping patterns do not exhibit any of the recognised signatures and can therefore not be classified into any of the known strain families (Flores et al. 2007).

Two additional genotyping approaches have been developed primarily for phylogenetic studies and strain classification in MTBC; one is based on genomic deletions (also referred to as large sequence polymorphisms (LSPs) or regions of difference (RDs)), and the other on single nucleotide polymorphisms (SNPs). Starting about 12 years ago, scientists have used bacterial artificial chromosome libraries and DNA microarrays to interrogate the genome of MTBC strains (Behr et al. 1999; Gordon et al. 1999; Kato-Maeda et al. 2001b; Tsolaki et al. 2004). These studies identified many genomic regions that were absent compared to the H37Rv reference genome, which at the time was the only whole-genome sequence of MTBC available (Cole et al. 1998). Many of these genomic deletions can be used as robust phylogenetic markers for assigning MTBC strain into meaningful groupings. This is because due to the virtual absence of large-scale HGT (further discussed below), once a particular genomic region is lost in MTBC, this region cannot be reacquired, and all progeny of the strain experiencing the loss initially will inherit this deletion. The observation that phylogenetic trees constructed using such genomic deletion markers showed no homoplasy was per se strong support for the already established notion that MTBC exhibits no ongoing HGT and a highly clonal population structure. Several groups used such an approach to revisit the evolutionary history of MTBC and define sets of discrete strain lineages (Brosch et al. 2002; Gagneux et al. 2006b; Hirsh et al. 2004; Mostowy et al. 2002, 2004b; Tsolaki et al. 2005). From a strict phylogenetic point of view; however, these evolutionary scenarios do not represent actual phylogenetic trees but mere cladograms, as their branches do not reflect actual evolutionary distances (i.e., genomic deletions are stochastic events for which no evolutionary models are available). Moreover, phylogenetic inference using genomic deletions is limited by the fact that these markers are usually defined through a one-way comparison with the H37Rv reference genome.

SNP typing is rapidly gaining importance for MTBC genotyping, primarily because of available technologies originally developed for other organisms. It is important to note however that, similarly to genomic deletion analysis, SNP typing does not provide the necessary resolution to be applicable in classical molecular epidemiological studies of TB transmission (Kato-Maeda et al. 2011). Nevertheless, SNP typing is ideal for classifying MTBC strains into lineages as it exhibits negligible levels of homoplasy (Comas et al. 2009). Moreover, various typing methodologies have already been proposed (Dos Vultos et al. 2008; Filliol et al. 2006; Gutacker et al. 2006; Mestre et al. 2011). Importantly, in contrast to de novo SNP discovery through DNA sequencing (Baker et al. 2004; Hershberg et al. 2008), mere SNP typing is of limited use in phylogenetic studies because of the problem known as 'phylogenetic discovery bias' (Pearson et al. 2004). This bias refers to the fact that if only a few reference genomes are used to identify the SNPs used for subsequent genotyping, the resulting phylogeny will be biased, because the genetic diversity among strains not included in the initial SNP discovery will not be detected. As a consequence, such strains will automatically fall in intermediate positions on the phylogenetic tree, leading to the problems known as 'branch collapse' and 'linear phylogeny' (Achtman 2008; Alland et al. 2003; Smith et al. 2009). Recently, two new complementary SNP-typing assays have been developed to screen for the main lineages of MTBC (Stucki et al. 2012). In contrast to most other methods available, the SNPs used in these new assays have been identified by comparing many genomes representative of the known global MTBC diversity (Fig. 1).

2.3 From Genotypes to Whole Genomes

Whole-genome sequencing offers many advantages over the current methods used for genotyping MTBC. The main disadvantage remains the relatively high cost and the bioinformatics capacity required to analyse the data. However, both of these

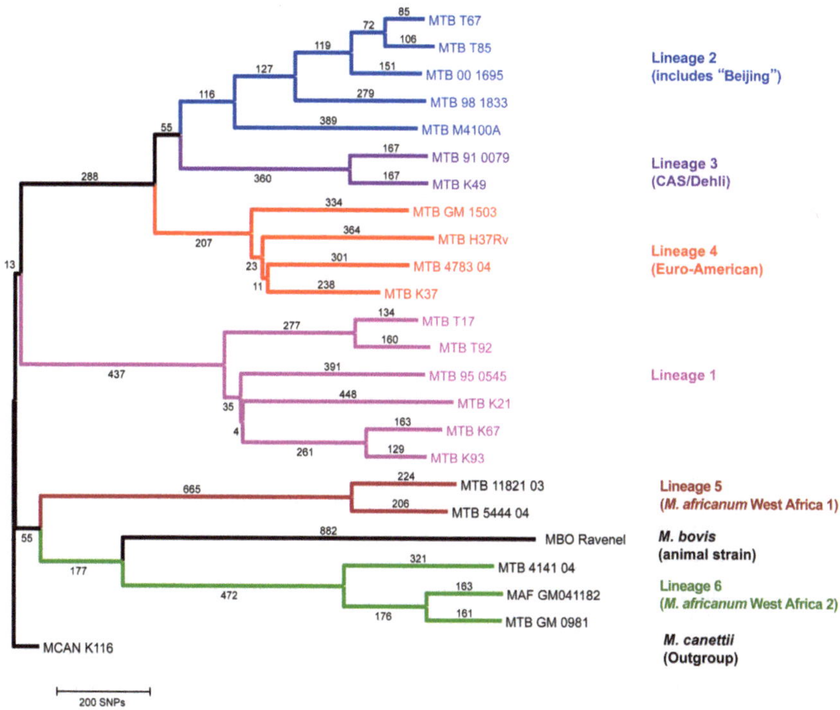

Fig. 1 Global phylogeny of MTBC based on 24 whole-genome sequences. *M. canettii* is used as outgroup. Coloured branches indicate the six main human-associated lineages. Numbers on branches indicate the number of SNPs (adapted from Bentley et al. 2012; Comas et al. 2010)

obstacles will likely be surmounted in the near future thanks to further advances in DNA sequencing technologies and the development of more user-friendly analytical tools. Several recent publications support the notion that genome sequencing is already on its way to become the new gold standard for molecular epidemiological studies of MTBC. Specifically, these studies have shown that genome sequencing has a much higher discriminatory power than the standard genotyping tools, and that clinical strains exhibiting identical genotyping profiles can sometimes harbour extensive genetic diversity, with obvious implications for the interpretation of transmission patterns (Casali et al. 2012; Gardy et al. 2011; Niemann et al. 2009). Comparative genome sequencing has also been used to study closely related strains isolated sequentially from a single patient (Comas et al. 2012; Saunders et al. 2011), from chains of ongoing transmission (Sandegren et al. 2011; Schurch et al. 2009, 2010), or from experimentally infected macaques (Ford et al. 2011), in the hope of learning more about the mechanisms driving the micro-evolution of MTBC. Intriguingly, the study in macaques found that the mutation rate of MTBC in latently infected animals was not significantly different from that of bacteria isolated from macaques with active TB. This suggests that

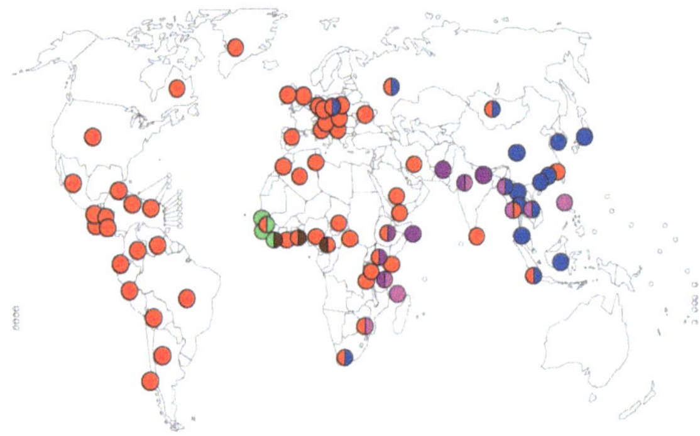

Fig. 2 Global phylogeography of human-associated MTBC. The geographic distribution of the six main human-associated MTBC liemnages is indicated. Each dot corresponds to a country and the dominant MTBC lineage(s) are indicated by colours corresponding to Fig. 1 (adapted from Gagneux et al. 2006b)

dormant and/or persister cells are also prone to mutational damage, and therefore drug resistance could also be acquired during latent infections (Ford et al. 2011).

With respect to the global diversity of MTBC, comparative genome sequencing is revealing a degree of heterogeneity which was previously unrecognised. Specifically, because DNA sequence data (as opposed to e.g. genome deletions) can be used to infer genetic distances, this heterogeneity can be measured quantitatively and appreciated in the context of a robust phylogenetic framework. Based on recently published genome sequences (Bentley et al. 2012; Comas et al. 2010), six main phylogenetic lineages can be distinguished among the human-associated MTBC (in addition to all the animal-associated variants not further discussed here; Fig. 1). This includes two lineages traditionally referred to as *M. africanum* (de Jong et al. 2010). These six lineages have already been described using various genotyping methods and shown to be distributed non-randomly around the world (Fig. 2) (Brudey et al. 2006; Gagneux et al. 2006b). However, as mentioned above, DNA sequence data can be used to compute evolutionary distances and construct robust phylogenetic trees. As shown in Fig. 1, the genetic distance between two human-associated MTBC strains can reach close to 2,000 SNPs, which is equivalent to the evolutionary distance between H37Rv and *M. bovis* (Garnier et al. 2003). Moreover, about two-thirds of amino acid coding SNPs in MTBC are non-synonymous, and a large proportion of these nSNPs have been predicted to affect gene function (Hershberg et al. 2008). Many more genomes will be necessary to better define the global diversity of MTBC (Fig. 2). In particular, based on the currently available genomes, geographic coverage is limited, and potential phylogenetic substructures within the main lineages are not well understood (Fig. 1). Moreover, we know from genotyping studies that much more MTBC diversity exists that has not yet been characterised by whole-genome sequencing (Demay et al. 2012; Gagneux et al. 2006b).

In addition to genomic deletions and functional SNPs, recent data suggest that gene duplications might also be an important source of genome plasticity in MTBC. In 2000, Brosch et al. reported, two duplicated regions in *M. bovis* BCG Pasteur (Brosch et al. 2000). Until recently, genomic duplications were considered rare events limited to a few BCG sub-strains (Leung et al. 2008), possibly reflecting the result of long-term in vitro evolution (Brosch et al. 2007). However, two recent studies reported gene duplications in clinical isolates of *M. tuberculosis*, including one large-scale duplication of more than 500 Kb (Domenech et al. 2010; Weiner et al. 2012). Intriguingly, these duplications occurred multiple times independently and across different MTBC lineages, suggesting that they were positively selected. Moreover, some of these duplications are instable when strains are grown in vitro, arguing against a phenomenon linked to prolonged in vitro growth, but rather in support of in vivo selection and adaptation. Future studies will help define the extent and significance of genomic duplications in MTBC.

2.4 What is the Role of HGT and Recombination in MTBC?

There is strong evidence that past instances of HGT have contributed to the emergence of MTBC as a successful pathogen (Behr and Gagneux 2011). Specifically, genomic comparisons of MTBC and other mycobacteria have identified many genes that the common ancestor of all MTBC appears to have acquired horizontally (Becq et al. 2007; Rosas-Magallanes et al. 2006; Veyrier et al. 2009). Some of these genes have been implicated in the pathogenesis of TB. Similarly, analysis of housekeeping genes in *M. canettii* and other smooth TB bacilli demonstrated mosaic structures, suggesting ongoing HGT among these rare and extant members of MTBC (Gutierrez et al. 2005). However, whether ongoing (as opposed to past) HGT plays any role in the evolution of MTBC is controversial (except in the case of *M. canettii* and the other smooth TB bacilli discussed above). For a long time, the common view was that MTBC exhibited a strictly clonal population structure with little evidence for ongoing HGT. With a few exceptions largely considered anecdotal (Hughes et al. 2002; Liu et al. 2006), most of the available evidence supported this view until recently. This included the observation of a rather stable G + C content across most of the MTBC genome (Cole et al. 1998), strong linkage disequilibrium between minisatellite loci (Supply et al. 2003), congruence of phylogenies derived from many different molecular parkers including genomic deletions (Baker et al. 2004; Filliol et al. 2006; Gagneux et al. 2006b; Gagneux and Small 2007; Gutacker et al. 2006; Hirsh et al. 2004; Wirth et al. 2008), and negligible homoplasy in DNA sequence data (Comas et al. 2009; Hershberg et al. 2008). In particular, the fact that genomic deletions in MTBC can be used as robust phylogenetic markers has been demonstrated multiple times (Brosch et al. 2002; Hirsh et al. 2004; Mostowy et al. 2002), and suggests that once a particular genomic region has been lost in MTBC, the corresponding genes cannot be reacquired via HGT. Moreover, all known drug resistance determinants in MTBC represent *de novo* acquired chromosomal mutations (Sandgren et al. 2009), indicating that HGT plays no significant role in the emergence of drug resistance in this organism.

A recent study has now challenged the dogma of strict clonality in MTBC (Namouchi et al. 2012). The authors used 24 MTBC genomes and performed a series of analyses designed to detect ongoing HGT. Based on their results, the authors concluded that MTBC does in fact show evidence of recombination, mostly involving short DNA fragments of around 50 bp. This observation is striking, given that in other bacteria, recombination usually involves much longer stretches of DNA (Didelot et al. 2010). This short size of recombination tracts might explain why the genomic deletions used to genotype MTBC, which comprise much larger DNA segments, show no evidence of homoplasy linked to HGT or recombination. The authors found that 150 kb (i.e. about 3 %) of the MTBC genomes showed significant recombination tracts. However, the nature of the putative donor organisms could not be determined, except for some instances were the recombining tracks matched regions in the genome of *M. canettii*. More work is needed, including larger collections of high-quality genomes, to define the role and relevance of HGT and recombination in MTBC, and to identify the sources and mechanisms of acquisition of foreign DNA in this organism.

2.5 Using DNA Sequence Data to Predict the Impact of Genome Diversity

DNA sequences can also be used to analyse the relative strength and direction of selection acting on these sequences. For example, a recent study used whole-genome sequences from 21 clinical strains covering the global diversity of MTBC to study the genetic diversity of 491 experimentally confirmed human T cell epitopes (Comas et al. 2010). In contrast to other pathogens were immune pressure drives increased antigenic diversity (Deitsch et al. 2009), the authors found that in MTBC, T cell epitopes were evolutionarily hyperconserved, showing the lowest dN/dS values (i.e. ratio of non-synonymous to synonymous mutations) in the genome (Fig. 3), with more than 95 % of these epitopes harbouring no amino acid substitution at all. The authors concluded from their findings that the host immune responses directed towards these hyperconserved T cell epitopes might offer a net benefit to the bacteria rather than to the host, by promoting tissue damage, ultimately leading to enhanced transmission of MTBC. This notion is supported by the fact that patient with cavitary disease are particularly likely to generate secondary cases (Rodrigo et al. 1997). Also, in TB patients co-infected with HIV, CD4 T cell counts are inversely correlated with the likelihood of developing cavitations (Kwan and Ernst 2011). The latter suggests that CD4 T cells directly or indirectly contribute to the formation of cavitations, and that TB patients with low CD4 T cell counts might be less likely to transmit MTBC (Brites and Gagneux 2011; Cruciani et al. 2001).

The observation that known human T cell epitopes in MTBC are evolutionarily hyperconserved does of course not exclude the possibility that other, as yet unknown epitopes might be involved in antigenic variation. Indeed, one of the limitations in the study by Comas et al. was that due to technical reasons, the

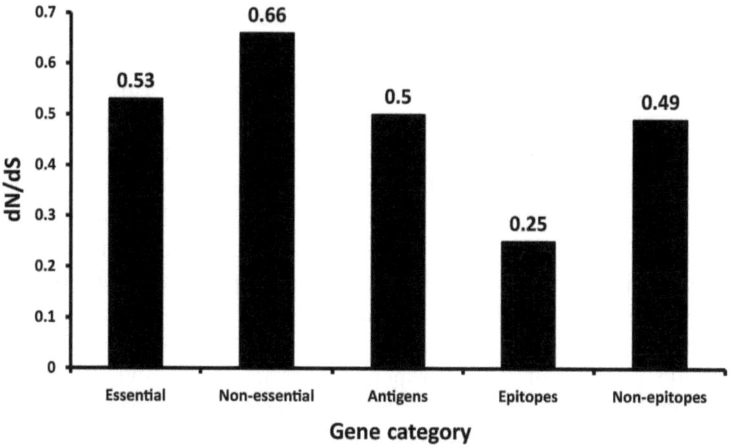

Fig. 3 Human T cell epitopes of MTBC are evolutionarily hyperconserved. The ratios of non-synonymous to synonymous nucleotide substitutions are indicated for each gene category (adapted from Comas et al. 2010)

members of the PE/PPE gene family had to be excluded from the analysis (Brennan and Delogu 2002). Some of these genes have previously been shown to be highly diverse (Talarico et al. 2005, 2008), prompting the view they may be implicated in antigenic variation (Banu et al. 2002). A recent study addressed this possibility by analysing the genetic diversity of various PE/PPE genes across a panel of 40 phylogenetically diverse clinical strains (McEvoy et al. 2012). The authors confirmed that PE/PPE genes were more genetically diverse than other genes, but the observed dN/dS values were in average around 1.0, suggesting these genes evolve neutrally. The latter finding argues against the notion that PE/PPE diversity is a result of immune escape. One of the inherent problems in studying PE/PPE genes is that hardly any information exists as to their specific function(s). Future studies will show whether or not the findings by McEvoy et al. can be extrapolated to other PE/PPE genes.

One of the striking findings of comparative DNA sequencing in MTBC has been the observation that in average about two-thirds of coding SNPs are non-synonymous, thus leading to a change in the encoded amino acid (Comas et al. 2010; Fleischmann et al. 2002). Moreover, in silico analyses have shown that 58 % of nonsynonymous SNPs in MTBC are predicted to affect gene function (Hershberg et al. 2008). Recent data from our laboratory using whole-genome sequences have confirmed that even in the deep phylogenetic branches that define the main lineages of MTBC (Fig. 1), close to 50 % of non-synonymous SNPs are predicted to be functional (unpublished data). Taken together, these data highlight the extent of genetic diversity among MTBC strains and suggest that this diversity is likely to have phenotypic consequences.

2.6 From MTBC Genotype to Phenotype

Some evidence that genetic diversity in MTBC does indeed translate into phenotypic diversity comes from studies of gene expression. One early study used DNA microarrays to study the transcriptome of 10 clinical strains and found about 500 genes differentially expressed (Gao et al. 2005). More recently, Homolka et al. studied a panel of 17 MTBC strains covering several of the main phylogenetic lineages of MTBC (Fig. 1) using DNA microarrays, and determined their transcriptional profile when grown in vitro and in resting or activated mouse macrophages (Homolka et al. 2010b). The authors detected both strain-specific and lineage-specific gene expression patterns. These patterns were consistent with the phylogenetic position of the corresponding strains, supporting the association between phylogenetic diversity discussed above and phenotypic variation observed experimentally. Moreover, the authors were able to define a core transcriptome that was common to all strains when grown intracellularly. Considering recent technological advances, including the development of RNA sequencing (RNA-seq), future studies will likely uncover additional MTBC diversity at the level of the genome-wide transcriptome. RNA-seq offers many advantages over DNA microarrays, as it allows for detection of unknown transcripts, including novel regulatory RNAs in a strand-specific manner (Sorek and Cossart 2010). A recent RNA-seq analysis of H37Rv grown in vitro revealed many 5' and 3' untranslated regions, antisense transcripts and intergenic small RNAs (Arnvig et al. 2011), supporting a role for non-coding RNA in the biology of MTBC (Arnvig and Young 2012).

In addition to differences in gene expression, MTBC phenotypic diversity has also become evident at the level of the host-pathogen interaction, at least in experimental settings. Many studies have reported MTBC strain-specific differences in immunogenicity and virulence using various models of infection; these have been reviewed in detail in (Coscolla and Gagneux 2010). However, only a few studies have been able to link the observed immune- or virulence-phenotypes to particular molecular characteristics of the infecting MTBC strain (Manca et al. 1999; Newton et al. 2006; Reed et al. 2004). In a recent study, Portevin et al. managed to link strain-specific immune-phenotypes with the phylogenetic classification of these strains using a human monocyte-derived macrophage infection model (Portevin et al. 2011). Specifically, the authors infected macrophages from multiple donors with one of 28 strains of MTBC and measured the cytokine profiles at several time points. They found that these strains varied widely in their stimulation of inflammatory cytokines, suggesting that human macrophages perceive and react to these strains very differently (Fig. 4). Despite this wide range of immune responses, the authors detected statistically significant differences between the different MTBC lineages, suggesting that the phylogenetic diversity of MTBC is also reflected at the host-pathogen interface.

For a long time, one of the main challenges in the field has been linking strain phenotypes defined in the laboratory to relevant clinical phenotypes; so far no

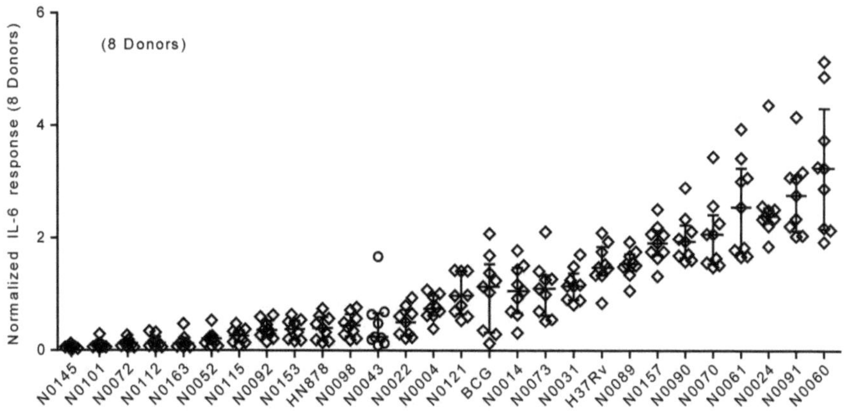

Fig. 4 IL-6 response in human monocyte-derived macrophages infected with 28 different MTBC strains. Each strain was tested in eight different human donors (adapted from Portevin et al. 2011)

consistent picture has emerged (Coscolla and Gagneux 2010). Perhaps one of the most promising avenues has been the study of *M. africanum* West Africa 2 (i.e. MTBC Lineage 6 in Fig. 1), which causes up to 50 % of TB in some countries of West African (de Jong et al. 2010). Recently, it was shown that Lineage 6 was attenuated in mice compared to H37Rv (Bold et al. 2012). Consistent with these laboratory findings, individuals infected with Lineage 6 strains were less likely to progress towards active TB compared to individuals infected with other strains, even though no difference was observed at the level of transmission (de Jong et al. 2008). Moreover, Lineage 6 has been associated with HIV co-infection in the Gambia, suggesting Lineage 6 might behave as an opportunistic pathogen (de Jong et al. 2005). No such association was observed in Ghana (Meyer et al. 2008), suggesting that even within a particular MTBC lineage, additional phylogeographic variation exists that might be clinically relevant.

3 Micro-Evolution of Drug-Resistant MTBC

A particular aspect of genetic diversity in MTBC that is of obvious clinical relevance relates to drug resistance. MTBC strains resistant to an ever increasing number of antibiotics are threatening global TB control (Gandhi et al. 2010). Over the last years, many of the molecular mechanisms leading to resistance have been elucidated for many of the first- and second-line drugs (Zhang and Yew 2009). Similarly, many of the mutations causing resistance have been identified (Sandgren et al. 2009). What is much less understood is how drug-resistant MTBC strains evolve over the longer run (Borrell and Gagneux 2009). In particular, little is known on how different drug resistance-conferring mutations might interact with each other, with pre-existing strain-specific mutations, or other mutations acquired subsequent to the resistance mutations (Borrell and Gagneux 2011). A particular class of mutations represented by the latter are referred to as compensatory mutations.

3.1 Compensatory Evolution

Acquisition of drug resistance determinants in bacteria is often associated with reduced Darwinian fitness in absence of the drug (Andersson and Levin 1999). However, this fitness cost can be mitigated by subsequent mutations at secondary sites, a phenomenon known as compensatory evolution. Compensation has been described in many bacterial species and antibiotics (reviewed in Andersson and Hughes 2010). In MTBC, a widely cited example has been mutations in the promoter of *ahpC* leading to over-expression of this gene, believed to compensate for the loss of the katalase-peroxidase (KatG) activity in isoniazid-resistant strains (Sherman et al. 1996). KatG activates isoniazid but also protects the bacterial cell against host-mediated oxidative stress. Over-expression of *ahpC*, which encodes an alkyl hydroperoxide reductase is thought to partially protect against excess oxidative damage in strains lacking KatG activity. However, the actual role of *ahpC* promotor mutations in isoniazid resistance remains unclear (Heym et al. 1997). Moreover, molecular epidemiological studies have found these mutations to be rare in clinical settings (Gagneux et al. 2006a), suggesting they play a minor role in the epidemiology of drug-resistant MTBC. Similarly, a compensatory mechanism in the 16S ribosomal RNA of MTBC strains resistant to aminoglycosides was reported (Shcherbakov et al. 2010). However, the corresponding mutation is only rarely observed in clinical settings (Georghiou et al. 2012). By contrast, Comas et al. described a set of novel compensatory mutations in *rpoA* and *rpoC*, which encode the alpha- and beta-prime subunits of the RNA polymerase (Comas et al. 2012). These mutations occurred exclusively in rifampicin-resistant strains with resistance-conferring mutations in *rpoB*, and were associated with an increased competitive fitness in vitro. Moreover, they occurred in up to 30 % of clinical MDR strains, and were significantly over-represented in MDR strains from countries known to suffer from a high burden of MDR-TB (Fig. 5). This suggests that these compensatory mutations contribute to the spread of drug-resistant MTBC strains. Subsequently, the same mutations were reported from XDR strains from Russia (Casali et al. 2012), and in experimentally evolved *Salmonella* (Brandis et al. 2012). While the mechanism(s) leading to compensation are still unknown, genetic reconstruction of some these *rpoA* and *rpoC* mutations in *Salmonella* showed that they were directly responsible for the improved growth rate of strains carrying rifampicin resistance-conferring mutations in *rpoB* (Brandis et al. 2012).

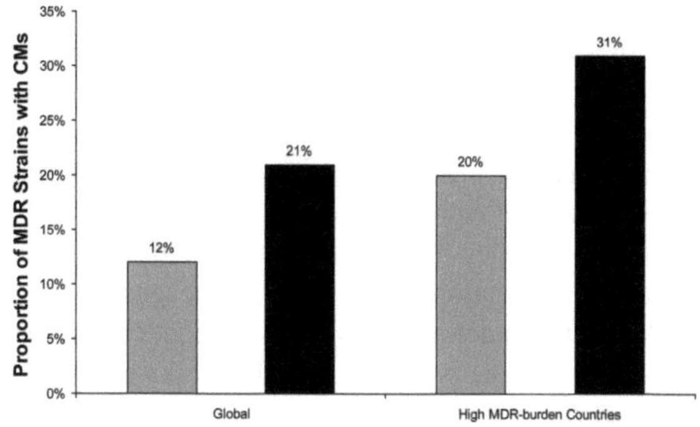

Fig. 5 Compensatory evolution in rifampicin-resistant MTBC. The proportion of MDR strains carrying compensatory mutations (CMs) in *rpoA* or *rpoC* is indicated for strains isolated globally or from countries with a high MDR-TB burden. Light bars indicate mutations very likely to have a compensatory role; black bars indicate all putative compensatory mutations identified (adapted from Comas et al. 2012)

3.2 The Role of the Strain Genetic Background

The interaction between a drug resistance mutation and a compensatory mutation is an example of epistasis (Borrell and Gagneux 2011). Epistatic interaction can generally be defined as a case in which the phenotypic effect of one mutation is modified by the presence or absence of another mutation (Phillips 2008). In addition to epistatic interactions represented by compensatory evolution, interactions between different drug-resistance mutations have also been described. Importantly, work in *Escherichia coli* and *Pseudomonas aeruginosa* has shown that the fitness cost associated with a mutation conferring resistance to one drug can be ameliorated by a mutation conferring resistance to a second drug (Trindade et al. 2009; Ward et al. 2009). This phenomenon is referred to as sign epistasis, as the "sign" of the fitness effect associated with the second mutation switches from "negative" to "positive" (Weinreich et al. 2005). Currently, we do not know whether sign epistasis plays any role in the evolution of MDR-TB, but work is needed to explore this disturbing possibility.

There is increasing evidence that the strain genetic background can influence the evolutionary trajectory of drug-resistant MTBC (Borrell and Gagneux 2011). Presumably, this is due to epistatic interactions between pre-existing strain-specific genomic features, and subsequently acquired drug-resistance conferring mutations and compensatory mutations (Mueller et al. 2012). Experimental work in *E. coli* has shown that such interactions can limit the available mutational pathway(s) towards drug resistance (Toprak et al. 2012). In MTBC, one particular phylogenetic lineage known as "Beijing" has repeatedly been associated with drug resistance in clinical settings (reviewed in Borrell and Gagneux 2009). The underlying basis of this phenomenon remains unknown, but several hypotheses have been proposed (Parwati et al. 2010). For example, based on various non-ysynonymous mutations found in DNA repair genes of Beijing strains, it was hypothesised that these strains might exhibit an intrinsically elevated mutation rate, leading to an enhanced likelihood of acquiring resistance mutations (Dos Vultos et al. 2008). However, fluctuation assays have yielded conflicting results with respect to the in vitro mutation rate of Beijing compared to other MTBC strains (de Steenwinkel et al. 2012; Werngren and Hoffner 2003).

Several studies have reported associations between particular drug resistance-conferring mutations and MTBC lineages, suggesting that the strain genetic background, as defined by MTBC lineage, can influence the mutational pathway to resistance (Gagneux et al. 2006a; Homolka et al. 2010a). For example in isoniazid-resistant MTBC, *inhA* promoter regions, which are one of the main causes of isoniazid resistance in clinical settings (Zhang and Yew 2009), have been associated with Lineage 1 of MTBC (Fig. 1). The fact that this association was observed in three independent studies supports a biological basis for this phenomenon (Baker et al. 2004; Fenner et al. 2012; Gagneux et al. 2006a). Moreover, Fenner et al. showed recently that putative epistatic interactions between specific isoniazid resistance-conferring mutations and different strain genetic backgrounds can influence the level of resistance to isoniazid in vitro (Fenner et al. 2012).

4 Conclusions

The past few years have witnessed much progress in our understanding and appreciation of strain-to-strain diversity in MTBC. This has partially been a consequence of new opportunities arising from access to novel technologies, including DNA microarrays, and more recently next-generation DNA sequencing (Comas and Gagneux 2009). Fresh insights have also been gained by learning from work in other bacteria. Fifteen years ago, human-associated MTBC was essentially regarded as a "clone", and strain variation was generally considered irrelevant by the majority of the basic science community (Kato-Maeda et al. 2001a). Today, we know that MTBC harbours more genetic diversity than previously realised, and experimental studies show very clearly that part of this diversity translates into important phenotypic variation (Coscolla and Gagneux 2010). Yet, the role of this diversity for clinical TB remains largely elusive, partially because of the difficulty of linking MTBC strain genotypes to relevant clinical phenotypes. In addition to differences at the DNA level, MTBC strains have been shown to differ at the transcriptome level (Homolka et al. 2010b). Yet, nothing is known with respect to strain variation at the level of regulatory RNAs, which are highly abundant in this organism (Arnvig et al. 2011). Similarly, epigenetics is a rapidly emerging field that has recently been proposed as an avenue to discover host-derived biomarkers for TB (Esterhuyse et al. 2012). However, little work has been done on epigenetics in MTBC, and it is unknown whether strain variability will be reflected at the epigenetic level as well. Much effort today is dedicated to the development of new diagnostics, drugs and vaccines against TB (Young et al. 2008). Based on the existing evidence from other bacterial pathogens, strain genetic diversity in MTBC should be considered during the development of new tools and strategies to better control TB (Gagneux and Small 2007). Increasingly, systems approaches should be used (Comas and Gagneux 2011), not only to determine if and how diversity in MTBC matters for disease control, but to better understand the biology and epidemiology of one of the most important human diseases.

Acknowledgments I thank Douglas Young, Iñaki Comas and all the members of my group for the stimulating discussions and comments on the manuscript. Work in my laboratory is supported by the Swiss National Science Foundation (grant number PP0033-119205), the National Institutes of Health (AI090928 and HHSN266200700022C) the Leverhulme-Royal Society Africa Award (AA080019) and the European Research Council (309540).

References

Achtman M (2008) Evolution, population structure, and phylogeography of genetically monomorphic bacterial pathogens. Annu Rev Microbiol 62:53–70

Alexander KA, Laver PN, Michel AL, Williams M, van Helden PD, Warren RM, Gey van Pittius NC (2010) Novel *Mycobacterium tuberculosis* complex pathogen, *M. mungi*. Emerg Infect Dis 16:1296–1299

Alland D, Whittam TS, Murray MB, Cave MD, Hazbon MH, Dix K, Kokoris M, Duesterhoeft A, Eisen JA, Fraser CM, Fleischmann RD (2003) Modelling bacterial evolution with comparative-genome-based marker systems: application to *Mycobacterium tuberculosis* evolution and pathogenesis. J Bacteriol 185:3392–3399

Allix-Beguec C, Harmsen D, Weniger T, Supply P, Niemann S (2008) Evaluation and strategy for use of MIRU-VNTRplus, a multifunctional database for online analysis of genotyping data and phylogenetic identification of *Mycobacterium tuberculosis* complex isolates. J Clin Microbiol 46:2692–2699

Andersson DI, Hughes D (2010) Antibiotic resistance and its cost: is it possible to reverse resistance? Nat Rev Microbiol 8:260–271

Andersson DI, Levin BR (1999) The biological cost of antibiotic resistance. Curr Opin Microbiol 2:489–493

Arnvig KB, Young D (2012) Non-coding RNA and its potential role in *Mycobacterium tuberculosis* pathogenesis. RNA Biol 9:427–436

Arnvig KB, Comas I, Thomson NR, Houghton J, Boshoff HI, Croucher NJ, Rose G, Perkins TT, Parkhill J, Dougan G, Young DB (2011) Sequence-based analysis uncovers an abundance of non-coding RNA in the total transcriptome of *Mycobacterium tuberculosis*. PLoS Pathog 7:e1002342

Baker L, Brown T, Maiden MC, Drobniewski F (2004) Silent nucleotide polymorphisms and a phylogeny for *Mycobacterium tuberculosis*. Emerg Infect Dis 10:1568–1577

Banu S, Honore N, Saint-Joanis B, Philpott D, Prevost MC, Cole ST (2002) Are the PE-PGRS proteins of *Mycobacterium tuberculosis* variable surface antigens? Mol Microbiol 44:9–19

Becq J, Gutierrez MC, Rosas-Magallanes V, Rauzier J, Gicquel B, Neyrolles O, Deschavanne P (2007) Contribution of horizontally acquired genomic islands to the evolution of the tubercle bacilli. Mol Biol Evol 24:1861–1871

Behr M, Gagneux S (2011) The rise and fall of the *Mycobacterium tuberculosis* complex. In: Tibayrenc M (ed) Genetics and evolution of infectious diseases. Elsevier, London, pp 651–667

Behr MA, Wilson MA, Gill WP, Salamon H, Schoolnik GK, Rane S, Small PM (1999) Comparative genomics of BCG vaccines by whole-genome DNA microarray. Science 284:1520–1523

Bentley SD, Comas I, Bryant JM, Walker D, Smith NH, Harris SR, Thurston S, Gagneux S, Wood J, Antonio M, Quail MA, Gehre F, Adegbola RA, Parkhill J, de Jong BC (2012) The genome of *Mycobacterium africanum* West African 2 reveals a lineage-specific locus and genome erosion common to the *M. tuberculosis* complex. PLoS Negl Trop Dis 6:e1552

Bifani PJ, Mathema B, Kurepina NE, Kreiswirth BN (2002) Global dissemination of the *Mycobacterium tuberculosis* W-Beijing family strains. Trends Microbiol 10:45–52

Bold TD, Davis DC, Penberthy KK, Cox LM, Ernst JD, de Jong BC (2012) Impaired fitness of *Mycobacterium africanum* despite secretion of ESAT-6. J Infect Dis 205:984–990

Borrell S, Gagneux S (2009) Infectiousness, reproductive fitness and evolution of drug-resistant *Mycobacterium tuberculosis*. Int J Tuberc Lung Dis 13:1456–1466

Borrell S, Gagneux S (2011) Strain diversity, epistasis and the evolution of drug resistance in *Mycobacterium tuberculosis*. Clin Microbiol Infect 17:815–820

Brandis G, Wrande M, Liljas L, Hughes D (2012) Fitness-compensatory mutations in rifampicin-resistant RNA polymerase. Mol Microbiol 85:142–151

Brennan MJ, Delogu G (2002) The PE multigene family: a 'molecular mantra' for mycobacteria. Trends Microbiol 10:246–249

Brites D, Gagneux S (2011) Old and new selective pressures on *Mycobacterium tuberculosis*. Infect Genet Evol 12:678–685

Brosch R, Gordon SV, Buchrieser C, Pym AS, Garnier T, Cole ST (2000) Comparative genomics uncovers large tandem chromosomal duplications in *Mycobacterium bovis* BCG Pasteur. Yeast 17:111–123

Brosch R, Gordon SV, Marmiesse M, Brodin P, Buchrieser C, Eiglmeier K, Garnier T, Gutierrez C, Hewinson G, Kremer K, Parsons LM, Pym AS, Samper S, van Soolingen D, Cole ST (2002) A new evolutionary scenario for the *Mycobacterium tuberculosis* complex. Proc Natl Acad Sci USA 99:3684–3689

Brosch R, Gordon SV, Garnier T, Eiglmeier K, Frigui W, Valenti P, Dos Santos S, Duthoy S, Lacroix C, Garcia-Pelayo C, Inwald JK, Golby P, Nunez Garcia J, Hewinson RG, Behr MA, Quail MA, Churcher C, Barrell BG, Parkhill J, Cole ST (2007) Genome plasticity of BCG and impact on vaccine efficacy. Proc Natl Acad Sci USA

Brudey K, Driscoll JR, Rigouts L, Prodinger WM, Gori A, Al-Hajoj SA, Allix C, Aristimuno L, Arora J, Baumanis V, Binder L, Cafrune P, Cataldi A, Cheong S, Diel R, Ellermeier C, Evans JT, Fauville-Dufaux M, Ferdinand S, Garcia de Viedma D, Garzelli C, Gazzola L, Gomes HM, Gutierrez MC, Hawkey PM, van Helden PD, Kadival GV, Kreiswirth BN, Kremer K, Kubin M, Kulkarni SP, Liens B, Lillebaek T, Ly HM, Martin C, Mokrousov I, Narvskaia O, Ngeow YF, Naumann L, Niemann S, Parwati I, Rahim MZ, Rasolofo-Razanamparany V, Rasolonavalona T, Rossetti ML, Rusch-Gerdes S, Sajduda A, Samper S, Shemyakin I, Singh UB, Somoskovi A, Skuce R, van Soolingen D, Streicher EM, Suffys PN, Tortoli E, Tracevska T, Vincent V, Victor TC, Warren R, Yap SF, Zaman K, Portaels F, Rastogi N, Sola C (2006) *Mycobacterium tuberculosis* complex genetic diversity : mining the fourth international spoligotyping database (SpolDB4) for classification, population genetics and epidemiology. BMC Microbiol 6:23

Casali N, Nikolayevskyy V, Balabanova Y, Ignatyeva O, Kontsevaya I, Harris SR, Bentley SD, Parkhill J, Nejentsev S, Hoffner SE, Horstmann RD, Brown T, Drobniewski F (2012) Microevolution of extensively drug-resistant tuberculosis in Russia. Genome Res 22:735–745

Cole ST, Brosch R, Parkhill J, Garnier T, Churcher C, Harris D, Gordon SV, Eiglmeier K, Gas S, Barry CE 3rd, Tekaia F, Badcock K, Basham D, Brown D, Chillingworth T, Connor R, Davies R, Devlin K, Feltwell T, Gentles S, Hamlin N, Holroyd S, Hornsby T, Jagels K, Barrell BG et al (1998) Deciphering the biology of *Mycobacterium tuberculosis* from the complete genome sequence. Nature 393:537–544

Comas I, Gagneux S (2009) The past and future of tuberculosis research. PLoS Pathog 5:e1000600

Comas I, Gagneux S (2011) A role for systems epidemiology in tuberculosis research. Trends Microbiol 19:492–500

Comas I, Homolka S, Niemann S, Gagneux S (2009) Genotyping of genetically monomorphic bacteria: DNA sequencing in *Mycobacterium tuberculosis* highlights the limitations of current methodologies. PLoS ONE 4:e7815

Comas I, Chakravartti J, Small PM, Galagan J, Niemann S, Kremer K, Ernst JD, Gagneux S (2010) Human T cell epitopes of *Mycobacterium tuberculosis* are evolutionarily hyperconserved. Nat Genet 42:498–503

Comas I, Borrell S, Roetzer A, Rose G, Malla B, Kato-Maeda M, Galagan J, Niemann S, Gagneux S (2012) Whole-genome sequencing of rifampicin-resistant *Mycobacterium tuberculosis* strains identifies compensatory mutations in RNA polymerase genes. Nat Genet 44:106–110

Coscolla M, Gagneux S (2010) Does *M. tuberculosis* genomic diversity explain disease diversity? Drug Discov Today Dis Mech 7:e43–e59

Cousins DV, Bastida R, Cataldi A, Quse V, Redrobe S, Dow S, Duignan P, Murray A, Dupont C, Ahmed N, Collins DM, Butler WR, Dawson D, Rodriguez D, Loureiro J, Romano MI, Alito A, Zumarraga M, Bernardelli A (2003) Tuberculosis in seals caused by a novel member of the *Mycobacterium tuberculosis* complex: *Mycobacterium pinnipedii* sp. nov. Int J Syst Evol Microbiol 53:1305–1314

Cruciani M, Malena M, Bosco O, Gatti G, Serpelloni G (2001) The impact of human immunodeficiency virus type 1 on infectiousness of tuberculosis: a meta-analysis. Clin Infect Dis 33:1922–1930

de Jong BC, Hill PC, Brookes RH, Otu JK, Peterson KL, Small PM, Adegbola RA (2005) *Mycobacterium africanum*: a new opportunistic pathogen in HIV infection? Aids 19:1714–1715

de Jong BC, Hill PC, Aiken A, Awine T, Antonio M, Adetifa IM, Jackson-Sillah DJ, Fox A, Deriemer K, Gagneux S, Borgdorff MW, McAdam KP, Corrah T, Small PM, Adegbola RA (2008) Progression to active tuberculosis, but not transmission, varies by *Mycobacterium tuberculosis* lineage in The Gambia. J Infect Dis 198:1037–1043

de Jong BC, Antonio M, Gagneux S (2010) *Mycobacterium africanum*–review of an important cause of human tuberculosis in West Africa. PLoS Negl Trop Dis 4:e744

de Steenwinkel JE, ten Kate MT, de Knegt GJ, Kremer K, Aarnoutse RE, Boeree MJ, Verbrugh HA, van Soolingen D, Bakker-Woudenberg IA (2012) Drug susceptibility of *Mycobacterium tuberculosis* Beijing genotype and association with MDR TB. Emerg Infect Dis 18:660–663

Deitsch KW, Lukehart SA, Stringer JR (2009) Common strategies for antigenic variation by bacterial, fungal and protozoan pathogens. Nat Rev Microbiol 7:493–503

Demay C, Liens B, Burguiere T, Hill V, Couvin D, Millet J, Mokrousov I, Sola C, Zozio T, Rastogi N (2012) SITVITWEB—a publicly available international multimarker database for studying *Mycobacterium tuberculosis* genetic diversity and molecular epidemiology. Infect Genet Evol 12:755–766

Didelot X, Lawson D, Darling A, Falush D (2010) Inference of homologous recombination in bacteria using whole-genome sequences. Genetics 186:1435–1449

Domenech P, Kolly GS, Leon-Solis L, Fallow A, Reed MB (2010) A massive gene duplication event amongst clinical isolates of the *Mycobacterium tuberculosis* W/Beijing family. J Bacteriol 192:4562–4570

Dos Vultos T, Mestre O, Rauzier J, Golec M, Rastogi N, Rasolofo V, Tonjum T, Sola C, Matic I, Gicquel B (2008) Evolution and diversity of clonal bacteria: the paradigm of *Mycobacterium tuberculosis*. PLoS ONE 3:e1538

Esterhuyse MM, Linhart HG, Kaufmann SH (2012) Can the battle against tuberculosis gain from epigenetic research? Trends Microbiol 20:220–226

Fabre M, Hauck Y, Soler C, Koeck JL, van Ingen J, van Soolingen D, Vergnaud G, Pourcel C (2010) Molecular characteristics of "*Mycobacterium canettii*" the smooth *Mycobacterium tuberculosis* bacilli. Infect Genet Evol 10:1165–1173

Fenner L, Malla B, Ninet B, Dubuis O, Stucki D, Borrell S, Huna T, Bodmer T, Egger M, Gagneux S (2011) "Pseudo-Beijing": evidence for convergent evolution in the direct repeat region of *Mycobacterium tuberculosis*. PLoS One 6:e24737

Fenner L, Egger M, Bodmer T, Altpeter E, Zwahlen M, Jaton K, Pfyffer GE, Borrell S, Dubuis O, Bruderer T, Siegrist HH, Furrer H, Calmy A, Fehr J, Stalder JM, Ninet B, Bottger EC, Gagneux S (2012) Effect of mutation and genetic background on drug resistance in *Mycobacterium tuberculosis*. Antimicrob Agents Chemother 56:3047–3053

Filliol I, Motiwala AS, Cavatore M, Qi W, Hernando Hazbon M, Bobadilla Del Valle M, Fyfe J, Garcia–Garcia L, Rastogi N, Sola C, Zozio T, Guerrero MI, Leon CI, Crabtree J, Angiuoli S, Eisenach KD, Durmaz R, Joloba ML, Rendon A, Sifuentes-Osornio J, Ponce de Leon A, Cave MD, Fleischmann R, Whittam TS, Alland D (2006) Global phylogeny of *Mycobacterium tuberculosis* based on single nucleotide polymorphism (SNP) analysis: insights into tuberculosis evolution, phylogenetic accuracy of oher DNA fingerprinting systems, and recommendations for a minimal standard SNP set. J Bacteriol 188:759–772

Fleischmann RD, Alland D, Eisen JA, Carpenter L, White O, Peterson J, DeBoy R, Dodson R, Gwinn M, Haft D, Hickey E, Kolonay JF, Nelson WC, Umayam LA, Ermolaeva M, Salzberg SL, Delcher A, Utterback T, Weidman J, Khouri H, Gill J, Mikula A, Bishai W, Jacobs Jr WR, Jr., Venter JC, Fraser CM (2002) Whole-genome comparison of *Mycobacterium tuberculosis* clinical and laboratory strains. J Bacteriol 184:5479–5490

Flores L, Van T, Narayanan S, Deriemer K, Kato-Maeda M, Gagneux S (2007) Large sequence polymorphisms classify *Mycobacterium tuberculosis* with ancestral spoligotyping patterns. J Clin Microbiol 45:3393–3395

Ford CB, Lin PL, Chase MR, Shah RR, Iartchouk O, Galagan J, Mohaideen N, Ioerger TR, Sacchettini JC, Lipsitch M, Flynn JL, Fortune SM (2011) Use of whole genome sequencing to estimate the mutation rate of *Mycobacterium tuberculosis* during latent infection. Nat Genet 43:482–486

Frota CC, Hunt DM, Buxton RS, Rickman L, Hinds J, Kremer K, van Soolingen D, Colston MJ (2004) Genome structure in the vole bacillus, *Mycobacterium microti*, a member of the *Mycobacterium tuberculosis* complex with a low virulence for humans. Microbiology 150:1519–1527

Gagneux S, Small PM (2007) Global phylogeography of *Mycobacterium tuberculosis* and implications for tuberculosis product development. Lancet Infect Dis 7:328–337

Gagneux S, Burgos MV, DeRiemer K, Encisco A, Munoz S, Hopewell PC, Small PM, Pym AS (2006a) Impact of bacterial genetics on the transmission of isoniazid-resistant *Mycobacterium tuberculosis*. PLoS Pathog 2:e61

Gagneux S, Deriemer K, Van T, Kato-Maeda M, de Jong BC, Narayanan S, Nicol M, Niemann S, Kremer K, Gutierrez MC, Hilty M, Hopewell PC, Small PM (2006b) Variable host-pathogen compatibility in *Mycobacterium tuberculosis*. Proc Natl Acad Sci USA 103:2869–2873

Gandhi NR, Nunn P, Dheda K, Schaaf HS, Zignol M, van Soolingen D, Jensen P, Bayona J (2010) Multidrug-resistant and extensively drug-resistant tuberculosis: a threat to global control of tuberculosis. Lancet 375:1830–1843

Gao Q, Kripke KE, Saldanha AJ, Yan W, Holmes S, Small PM (2005) Gene expression diversity among *Mycobacterium tuberculosis* clinical isolates. Microbiology 151:5–14

Gardy JL, Johnston JC, Sui SJ, Cook VJ, Shah L, Brodkin E, Rempel S, Moore R, Zhao Y, Holt R, Varhol R, Birol I, Lem M, Sharma MK, Elwood K, Jones SJ, Brinkman FS, Brunham RC, Tang P (2011) Whole-genome sequencing and social-network analysis of a tuberculosis outbreak. N Engl J Med 364:730–739

Garnier T, Eiglmeier K, Camus JC, Medina N, Mansoor H, Pryor M, Duthoy S, Grondin S, Lacroix C, Monsempe C, Simon S, Harris B, Atkin R, Doggett J, Mayes R, Keating L, Wheeler PR, Parkhill J, Barrell BG, Cole ST, Gordon SV, Hewinson RG (2003) The complete genome sequence of *Mycobacterium bovis*. Proc Natl Acad Sci USA 100:7877–7882

Georghiou SB, Magana M, Garfein RS, Catanzaro DG, Catanzaro A, Rodwell TC (2012) Evaluation of genetic mutations associated with *Mycobacterium tuberculosis* resistance to amikacin, kanamycin and capreomycin: a systematic review. PLoS One 7:e33275

Gordon SV, Brosch R, Billault A, Garnier T, Eiglmeier K, Cole ST (1999) Identification of variable regions in the genomes of tubercle bacilli using bacterial artificial chromosome arrays. Mol Microbiol 32:643–655

Gutacker MM, Mathema B, Soini H, Shashkina E, Kreiswirth BN, Graviss EA, Musser JM (2006) Single-nucleotide polymorphism-based population genetic analysis of *Mycobacterium tuberculosis* strains from 4 geographic sites. J Infect Dis 193:121–128

Gutierrez C, Brisse S, Brosch R, Fabre M, Omais B, Marmiesse M, Supply P, Vincent V (2005) Ancient origin and gene mosaicism of the progenitor of *Mycobacterium tuberculosis*. PLoS Pathogens 1:1–7

Hershberg R, Lipatov M, Small PM, Sheffer H, Niemann S, Homolka S, Roach JC, Kremer K, Petrov DA, Feldman MW, Gagneux S (2008) High functional diversity in *Mycobacterium tuberculosis* driven by genetic drift and human demography. PLoS Biol 6:e311

Heym B, Stavropoulos E, Honore N, Domenech P, Saint-Joanis B, Wilson TM, Collins DM, Colston MJ, Cole ST (1997) Effects of overexpression of the alkyl hydroperoxide reductase AhpC on the virulence and isoniazid resistance of *Mycobacterium tuberculosis*. Infect Immun 65:1395–1401

Hirsh AE, Tsolaki AG, DeRiemer K, Feldman MW, Small PM (2004) Stable association between strains of *Mycobacterium tuberculosis* and their human host populations. Proc Natl Acad Sci USA 101:4871–4876

Homolka S, Meyer CG, Hillemann D, Owusu-Dabo E, Adjei O, Horstmann RD, Browne EN, Chinbuah A, Osei I, Gyapong J, Kubica T, Ruesch-Gerdes S, Niemann S (2010a) Unequal distribution of resistance-conferring mutations among *Mycobacterium tuberculosis* and *Mycobacterium africanum* strains from Ghana. Int J Med Microbiol 300:489–495

Homolka S, Niemann S, Russell DG, Rohde KH (2010b) Functional genetic diversity among *Mycobacterium tuberculosis* complex clinical isolates: delineation of conserved core and lineage-specific transcriptomes during intracellular survival. PLoS Pathog 6:e1000988

Hughes AL, Friedman R, Murray M (2002) Genomewide pattern of synonymous nucleotide substitution in two complete genomes of *Mycobacterium tuberculosis*. Emerg Infect Dis 8:1342–1346

Kamerbeek J, Schouls L, Kolk A, van Agterveld M, van Soolingen D, Kuijper S, Bunschoten A, Molhuizen H, Shaw R, Goyal M, van Embden J (1997) Simultaneous detection and strain differentiation of *Mycobacterium tuberculosis* for diagnosis and epidemiology. J Clin Microbiol 35:907–914

Kato-Maeda M, Bifani PJ, Kreiswirth BN, Small PM (2001a) The nature and consequence of genetic variability within *Mycobacterium tuberculosis*. J Clin Invest 107:533–537

Kato-Maeda M, Rhee JT, Gingeras TR, Salamon H, Drenkow J, Smittipat N, Small PM (2001b) Comparing genomes within the species *Mycobacterium tuberculosis*. Genome Res 11:547–554

Kato-Maeda M, Metcalfe JZ, Flores L (2011) Genotyping of *Mycobacterium tuberculosis*: application in epidemiologic studies. Future Microbiol 6:203–216

Koeck JL, Fabre M, Simon F, Daffe M, Garnotel E, Matan AB, Gerome P, Bernatas JJ, Buisson Y, Pourcel C (2010) Clinical characteristics of the smooth tubercle bacilli "*Mycobacterium canettii*" infection suggest the existence of an environmental reservoir. Clin Microbiol Infect 17:1013–1019

Kwan CK, Ernst JD (2011) HIV and tuberculosis: a deadly human syndemic. Clin Microbiol Rev 24:351–376

Leung AS, Tran V, Wu Z, Yu X, Alexander DC, Gao GF, Zhu B, Liu J (2008) Novel genome polymorphisms in BCG vaccine strains and impact on efficacy. BMC Genomics 9:413

Liu X, Gutacker MM, Musser JM, Fu YX (2006) Evidence for recombination in *Mycobacterium tuberculosis*. J Bacteriol 188:8169–8177

Malik AN, Godfrey-Faussett P (2005) Effects of genetic variability of *Mycobacterium tuberculosis* strains on the presentation of disease. Lancet Infect Dis 5:174–183

Manca C, Tsenova L, Barry CE 3rd, Bergtold A, Freeman S, Haslett PA, Musser JM, Freedman VH, Kaplan G (1999) *Mycobacterium tuberculosis* CDC1551 induces a more vigorous host response in vivo and in vitro, but is not more virulent than other clinical isolates. J Immunol 162:6740–6746

McEvoy C, Cloete R, Muller B, Schurch A, van Helden P, Gagneux S, Warren R, Van Pittius NC (2012) Comparative analysis of *Mycobacterium tuberculosis* PE and PPE genes reveals high sequence variation and an apparent absence of selective constraints. PLoS ONE 7:e30593

Mestre O, Luo T, Dos Vultos T, Kremer K, Murray A, Namouchi A, Jackson C, Rauzier J, Bifani P, Warren R, Rasolofo V, Mei J, Gao Q, Gicquel B (2011) Phylogeny of *Mycobacterium tuberculosis* Beijing strains constructed from polymorphisms in genes involved in DNA replication, recombination and repair. PLoS One 6:e16020

Meyer CG, Scarisbrick G, Niemann S, Browne EN, Chinbuah MA, Gyapong J, Osei I, Owusu-Dabo E, Kubica T, Rusch-Gerdes S, Thye T, Horstmann RD (2008) Pulmonary tuberculosis: virulence of *Mycobacterium africanum* and relevance in HIV co-infection. Tuberculosis (Edinb) 88:482–489

Mitchison DA, Wallace JG, Bhatia AL, Selkon JB, Subbaiah TV, Lancaster MC (1960) A comparison of the virulence in guinea-pigs of South Indian and British tubercle bacilli. Tubercle 41:1–22

Mostowy S, Cousins D, Brinkman J, Aranaz A, Behr MA (2002) Genomic deletions suggest a phylogeny for the *Mycobacterium tuberculosis* complex. J Infect Dis 186:74–80

Mostowy S, Cousins D, Behr MA (2004a) Genomic interrogation of the dassie bacillus reveals it as a unique RD1 mutant within the *Mycobacterium tuberculosis* complex. J Bacteriol 186:104–109

Mostowy S, Onipede A, Gagneux S, Niemann S, Kremer K, Desmond EP, Kato-Maeda M, Behr M (2004b) Genomic analysis distinguishes *Mycobacterium africanum*. J Clin Microbiol 42:3594–3599

Mueller B, Borrell S, Rose G, Gagneux S (2012) The heterogeneous evolution of multidrug-resistant *Mycobacterium tuberculosis*. Trends Genet (in press)

Musser JM, Amin A, Ramaswamy S (2000) Negligible genetic diversity of *Mycobacterium tuberculosis* host immune system protein targets: evidence of limited selective pressure. Genetics 155:7–16

Namouchi A, Didelot X, Schock U, Gicquel B, Rocha EP (2012) After the bottleneck: genome-wide diversification of the *Mycobacterium tuberculosis* complex by mutation, recombination, and natural selection. Genome Res 22:721–734

Newton SM, Smith RJ, Wilkinson KA, Nicol MP, Garton NJ, Staples KJ, Stewart GR, Wain JR, Martineau AR, Fandrich S, Smallie T, Foxwell B, Al-Obaidi A, Shafi J, Rajakumar K, Kampmann B, Andrew PW, Ziegler-Heitbrock L, Barer MR, Wilkinson RJ (2006) A deletion defining a common Asian lineage of *Mycobacterium tuberculosis* associates with immune subversion. Proc Natl Acad Sci USA 103:15594–15598

Nicol MP, Wilkinson RJ (2008) The clinical consequences of strain diversity in *Mycobacterium tuberculosis*. Trans R Soc Trop Med Hyg 102:955–965

Niemann S, Richter E, Rusch-Gerdes S (2002) Biochemical and genetic evidence for the transfer of *Mycobacterium tuberculosis* subsp. Caprae Aranaz et al. 1999 to the species *Mycobacterium bovis* Karlson and Lessel 1970 (approved lists 1980) as *Mycobacterium bovis* subsp. *caprae* comb. nov. Int J Syst Evol Microbiol 52:433–436

Niemann S, Koser CU, Gagneux S, Plinke C, Homolka S, Bignell H, Carter RJ, Cheetham RK, Cox A, Gormley NA, Kokko-Gonzales P, Murray LJ, Rigatti R, Smith VP, Arends FP, Cox HS, Smith G, Archer JA (2009) Genomic diversity among drug sensitive and multidrug resistant isolates of *Mycobacterium tuberculosis* with identical DNA fingerprints. PLoS ONE 4:e7407

Parwati I, van Crevel R, van Soolingen D (2010) Possible underlying mechanisms for successful emergence of the *Mycobacterium tuberculosis* Beijing genotype strains. Lancet Infect Dis 10:103–111

Pearson T, Busch JD, Ravel J, Read TD, Rhoton SD, U'Ren JM, Simonson TS, Kachur SM, Leadem RR, Cardon ML, Van Ert MN, Huynh LY, Fraser CM, Keim P (2004) Phylogenetic discovery bias in *Bacillus anthracis* using single-nucleotide polymorphisms from whole-genome sequencing. Proc Natl Acad Sci USA 101:13536–13541

Phillips PC (2008) Epistasis–the essential role of gene interactions in the structure and evolution of genetic systems. Nat Rev Genet 9:855–867

Portevin D, Gagneux S, Comas I, Young D (2011) Human macrophage responses to clinical isolates from the *Mycobacterium tuberculosis* complex discriminate between ancient and modern lineages. PLoS Pathog 7:e1001307

Rajakumar K, Shafi J, Smith RJ, Stabler RA, Andrew PW, Modha D, Bryant G, Monk P, Hinds J, Butcher PD, Barer MR (2004) Use of genome level-informed PCR as a new investigational approach for analysis of outbreak-associated *Mycobacterium tuberculosis* isolates. J Clin Microbiol 42:1890–1896

Reed MB, Domenech P, Manca C, Su H, Barczak AK, Kreiswirth BN, Kaplan G, Barry CE 3rd (2004) A glycolipid of hypervirulent tuberculosis strains that inhibits the innate immune response. Nature 431:84–87

Rodrigo T, Cayla JA, Garcia de Olalla P, Galdos-Tanguis H, Jansa JM, Miranda P, Brugal T (1997) Characteristics of tuberculosis patients who generate secondary cases. Int J Tuberc Lung Dis 1:352–357

Rosas-Magallanes V, Deschavanne P, Quintana-Murci L, Brosch R, Gicquel B, Neyrolles O (2006) Horizontal transfer of a virulence operon to the ancestor of *Mycobacterium tuberculosis*. Mol Biol Evol 23:1129–1135

Sandegren L, Groenheit R, Koivula T, Ghebremichael S, Advani A, Castro E, Pennhag A, Hoffner S, Mazurek J, Pawlowski A, Kan B, Bruchfeld J, Melefors O, Kallenius G (2011) Genomic Stability over 9 Years of an isoniazid resistant *Mycobacterium tuberculosis* outbreak strain in Sweden. PLoS One 6:e16647

Sandgren A, Strong M, Muthukrishnan P, Weiner BK, Church GM, Murray MB (2009) Tuberculosis drug resistance mutation database. PLoS Med 6:e2

Saunders NJ, Trivedi UH, Thomson ML, Doig C, Laurenson IF, Blaxter ML (2011) Deep resequencing of serial sputum isolates of *Mycobacterium tuberculosis* during therapeutic failure due to poor compliance reveals stepwise mutation of key resistance genes on an otherwise stable genetic background. J Infect 62:212–217

Schurch AC, Kremer K, Kiers A, Daviena O, Boeree MJ, Siezen R, Smith NH, van Soolingen D (2009) The tempo and mode of molecular evolution of *Mycobacterium tuberculosis* at patient-to-patient scale. Infect Genet Evol 10:108–114

Schurch AC, Kremer K, Daviena O, Kiers A, Boeree MJ, Siezen RJ, van Soolingen D (2010) High resolution typing by integration of genome sequencing data in a large tuberculosis cluster. J Clin Microbiol 48:3403–3406

Shcherbakov D, Akbergenov R, Matt T, Sander P, Andersson DI, Bottger EC (2010) Directed mutagenesis of *Mycobacterium smegmatis* 16S rRNA to reconstruct the in vivo evolution of aminoglycoside resistance in *Mycobacterium tuberculosis*. Mol Microbiol 77:830–840

Sherman DR, Mdluli K, Hickey MJ, Arain TM, Morris SL, Barry CE 3rd, Stover CK (1996) Compensatory ahpC gene expression in isoniazid-resistant *Mycobacterium tuberculosis*. Science 272:1641–1643

Smith NH, Kremer K, Inwald J, Dale J, Driscoll JR, Gordon SV, van Soolingen D, Glyn Hewinson R, Maynard Smith J (2005) Ecotypes of the *Mycobacterium tuberculosis* complex. J Theor Biol 239:220–225

Smith NH, Hewinson RG, Kremer K, Brosch R, Gordon SV (2009) Myths and misconceptions: the origin and evolution of *Mycobacterium tuberculosis*. Nat Rev Microbiol 7:537–544

Sorek R, Cossart P (2010) Prokaryotic transcriptomics: a new view on regulation, physiology and pathogenicity. Nat Rev Genet 11:9–16

Sreevatsan S, Pan X, Stockbauer KE, Connell ND, Kreiswirth BN, Whittam TS, Musser JM (1997) Restricted structural gene polymorphism in the *Mycobacterium tuberculosis* complex indicates evolutionarily recent global dissemination. Proc Natl Acad Sci USA 94:9869–9874

Stucki D, Malla B, Hostettler S, Huna T, Feldmann J, Yeboah-Manu D, Borrell S, Fenner L, Comas I, Coscolla M, Gagneux S (2012) Two new rapid SNP-typing methods for classifying

Mycobacterium tuberculosis complex into the main phylogenetic lineages. PLoS ONE 7:e41253

Supply P, Warren RM, Banuls AL, Lesjean S, Van Der Spuy GD, Lewis LA, Tibayrenc M, Van Helden PD, Locht C (2003) Linkage disequilibrium between minisatellite loci supports clonal evolution of *Mycobacterium tuberculosis* in a high tuberculosis incidence area. Mol Microbiol 47:529–538

Supply P, Allix C, Lesjean S, Cardoso-Oelemann M, Rusch-Gerdes S, Willery E, Savine E, de Haas P, van Deutekom H, Roring S, Bifani P, Kurepina N, Kreiswirth B, Sola C, Rastogi N, Vatin V, Gutierrez MC, Fauville M, Niemann S, Skuce R, Kremer K, Locht C, van Soolingen D (2006) Proposal for standardization of optimized Mycobacterial Interspersed Repetitive Unit-Variable Number Tandem Repeat typing of *Mycobacterium tuberculosis*. J Clin Microbiol 44:4498–4510

Talarico S, Cave MD, Marrs CF, Foxman B, Zhang L, Yang Z (2005) Variation of the *Mycobacterium tuberculosis* PE_PGRS 33 gene among clinical isolates. J Clin Microbiol 43:4954–4960

Talarico S, Zhang L, Marrs CF, Foxman B, Cave MD, Brennan MJ, Yang Z (2008) *Mycobacterium tuberculosis* PE_PGRS16 and PE_PGRS26 genetic polymorphism among clinical isolates. Tuberculosis (Edinb) 88:283–294

Toprak E, Veres A, Michel JB, Chait R, Hartl DL, Kishony R (2012) Evolutionary paths to antibiotic resistance under dynamically sustained drug selection. Nat Genet 44:101–105

Trindade S, Sousa A, Xavier KB, Dionisio F, Ferreira MG, Gordo I (2009) Positive epistasis drives the acquisition of multidrug resistance. PLoS Genet 5:e1000578

Tsolaki AG, Hirsh AE, DeRiemer K, Enciso JA, Wong MZ, Hannan M, Goguet de la Salmoniere YO, Aman K, Kato-Maeda M, Small PM (2004) Functional and evolutionary genomics of *Mycobacterium tuberculosis*: insights from genomic deletions in 100 strains. Proc Natl Acad Sci USA 101:4865–4870

Tsolaki AG, Gagneux S, Pym AS, Goguet de la Salmoniere YO, Kreiswirth BN, Van Soolingen D, Small PM (2005) Genomic deletions classify the Beijing/W strains as a distinct genetic lineage of *Mycobacterium tuberculosis*. J Clin Microbiol 43:3185–3191

Valway SE, Sanchez MP, Shinnick TF, Orme I, Agerton T, Hoy D, Jones JS, Westmoreland H, Onorato IM (1998) An outbreak involving extensive transmission of a virulent strain of *Mycobacterium tuberculosis*. N Engl J Med 338:633–639

van Embden JD, Cave MD, Crawford JT, Dale JW, Eisenach KD, Gicquel B, Hermans P, Martin C, McAdam R, Shinnick TM et al (1993) Strain identification of *Mycobacterium tuberculosis* by DNA fingerprinting: recommendations for a standardized methodology. J Clin Microbiol 31:406–409

van Ingen J, Rahim Z, Mulder A, Boeree MJ, Simeone R, Brosch R, van Soolingen D (2012) Characterization of *Mycobacterium orygis* as *M. tuberculosis* complex subspecies. Emerg Infect Dis 18:653–655

van Soolingen D, Qian L, de Haas PE, Douglas JT, Traore H, Portaels F, Qing HZ, Enkhsaikan D, Nymadawa P, van Embden JD (1995) Predominance of a single genotype of *Mycobacterium tuberculosis* in countries of east Asia. J Clin Microbiol 33:3234–3238

van Soolingen D, Hoogenboezem T, de Haas PE, Hermans PW, Koedam MA, Teppema KS, Brennan PJ, Besra GS, Portaels F, Top J, Schouls LM, van Embden JD (1997) A novel pathogenic taxon of the *Mycobacterium tuberculosis* complex, Canetti: characterization of an exceptional isolate from Africa. Int J Syst Bacteriol 47:1236–1245

Veyrier F, Pletzer D, Turenne C, Behr MA (2009) Phylogenetic detection of horizontal gene transfer during the step-wise genesis of *Mycobacterium tuberculosis*. BMC Evol Biol 9:196

Ward H, Perron GG, Maclean RC (2009) The cost of multiple drug resistance in *Pseudomonas aeruginosa*. J Evol Biol 22:997–1003

Weiner B, Gomez J, Victor TC, Warren RM, Sloutsky A, Plikaytis BB, Posey JE, van Helden PD, Gey van Pittius NC, Koehrsen M, Sisk P, Stolte C, White J, Gagneux S, Birren B, Hung D, Murray M, Galagan J (2012) Independent large scale duplications in multiple *M. tuberculosis* lineages overlapping the same genomic region. PLoS One 7:e26038

Weinreich DM, Watson RA, Chao L (2005) Perspective: sign epistasis and genetic constraint on evolutionary trajectories. Evolution 59:1165–1174

Werngren J, Hoffner SE (2003) Drug-susceptible *Mycobacterium tuberculosis* Beijing genotype does not develop mutation-conferred resistance to rifampin at an elevated rate. J Clin Microbiol 41:1520–1524

Wirth T, Hildebrand F, Allix-Beguec C, Wolbeling F, Kubica T, Kremer K, van Soolingen D, Rusch-Gerdes S, Locht C, Brisse S, Meyer A, Supply P, Niemann S (2008) Origin, spread and demography of the *Mycobacterium tuberculosis* complex. PLoS Pathog 4:e1000160

Young DB, Perkins MD, Duncan K, Barry CE (2008) Confronting the scientific obstacles to global control of tuberculosis. J Clin Invest 118:1255–1265

Zhang Y, Yew WW (2009) Mechanisms of drug resistance in *Mycobacterium tuberculosis*. Int J Tuberc Lung Dis 13:1320–1330

Zhang M, Gong J, Yang Z, Samten B, Cave MD, Barnes PF (1999) Enhanced capacity of a widespread strain of *Mycobacterium tuberculosis* to grow in human macrophages. J Infect Dis 179:1213–1217

DNA Metabolism in Mycobacterial Pathogenesis

Digby F. Warner, Tone Tønjum and Valerie Mizrahi

Abstract Fundamental aspects of the lifestyle of *Mycobacterium tuberculosis* implicate DNA metabolism in bacillary survival and adaptive evolution. The environments encountered by *M. tuberculosis* during successive cycles of infection and transmission are genotoxic. Moreover, as an obligate pathogen, *M. tuberculosis* has the ability to persist for extended periods in a subclinical state, suggesting that active DNA repair is critical to maintain genome integrity and bacterial viability during prolonged infection. In this chapter, we provide an overview of the major DNA metabolic pathways identified in *M. tuberculosis*, and situate key recent findings within the context of mycobacterial pathogenesis. Unlike many other bacterial pathogens, *M. tuberculosis* is genetically secluded, and appears to rely solely on chromosomal mutagenesis to drive its microevolution within the human host. In turn, this implies that a balance between high versus relaxed fidelity mechanisms of DNA metabolism ensures the maintenance of genome integrity, while accommodating the evolutionary imperative to adapt to hostile and fluctuating environments. The inferred relationship between mycobacterial DNA repair and genome dynamics is considered in the light of emerging data from whole-genome sequencing studies of clinical *M. tuberculosis* isolates which have revealed the potential for considerable heterogeneity within and between different bacterial and host populations.

D. F. Warner · V. Mizrahi (✉)
MRC/NHLS/UCT Molecular Mycobacteriology Research Unit and DST/NRF
Centre of Excellence for Biomedical TB Research,
Institute of Infectious Disease and Molecular Medicine,
University of Cape Town, P/Bag X3, Rondebosch 7700, South Africa
e-mail: valerie.mizrahi@uct.ac.za

D. F. Warner
e-mail: digby.warner@uct.ac.za

T. Tønjum
Centre for Molecular Biology and Neuroscience,
Institute of Microbiology Rikshospitalet-Radiumhospitalet Medical Centre,
University of Oslo, 0027 Oslo, Norway

Current Topics in Microbiology and Immunology (2013) 374: 27–51
DOI: 10.1007/82_2013_328

Published Online: 1 May 2013

Contents

1 Introduction

Mycobacterium tuberculosis is characterized by a number of features that are inextricably linked to DNA replication and repair. First, as a slow-growing intracellular pathogen that preferentially targets host phagocytic cells, *M. tuberculosis* is exposed to host-derived reactive oxygen and nitrogen species which exacerbate the oxidative and nitrosative stresses imposed by normal cellular metabolism (Nathan and Shiloh 2000; Zahrt and Deretic 2002; Russell 2007). Consistent with this idea, insights from animal models of tuberculosis infection (Sassetti and Rubin 2003; Darwin and Nathan 2005; Dutta et al. 2010) suggest that the environments encountered by *M. tuberculosis* during residence in the human host are inherently genotoxic and pose a major threat to the integrity of the genome. Moreover, analyses of *M. tuberculosis* gene expression in various infection models (Talaat et al. 2004), and in clinical samples (Rachman et al. 2006), suggest that DNA repair mechanisms are active throughout the course of infection (Gorna et al. 2010). Second, *M. tuberculosis* bacilli are able to persist for extended periods within the human host in a poorly understood subclinical state, in some cases reactivating decades later to cause post-primary tuberculosis (Lillebaek et al. 2002). This implies a continual role for DNA repair mechanisms in the maintenance of genome integrity—and bacterial viability—throughout the course of prolonged infection. Finally, there is strong evidence that *M. tuberculosis* is secluded genetically, and might even be regarded as a monomorph (Achtman 2012): single nucleotide polymorphisms (SNPs) and large sequence polymorphisms offer the only means to infer phylogenies reliably among circulating strains that can be characterized by limited diversity (Comas et al. 2009). Furthermore, resistance to anti-tubercular drugs arises exclusively from mutations in

Table 1 Notable findings from mycobacterial DNA metabolism research

Discovery	Significance	Reference
Mycobacterial genomes lack identifiable MMR components	Absence of a mutator phenotype suggests alternative mechanism to ensure replication fidelity	(Mizrahi and Andersen 1998)
Novel mechanisms of regulation of RecA-dependent and RecA-independent DNA damage (SOS) responses	RecA processing via splicing of intein Simultaneous operation of alternate SOS responses Regulation of LexA/RecA-independent response by ClpR ClpR implicated in genomic stability	(Davis et al. 1992), (Rand et al. 2003), (Wang et al. 2011)
Mycobacteria employ a novel, dual-polymerase damage tolerance system	TLS by a C-family DNA polymerase, and adaptor role defined for a pseudo Y-family polymerase Implicated in the evolution of rifampicin resistance in vivo	(Boshoff et al. 2003), (Warner et al. 2010)
Mycobacteria possess a NHEJ system for DSB repair	First prokaryote in which NHEJ was identified	(Della et al. 2004), (Gong et al. 2005), (Shuman and Glickman 2007)
Mycobacteria encode novel helicase-nuclease complex	Heterodimeric AdnAB motor-nuclease limited to Actinobacteria and required for DSB repair	(Sinha et al. 2009b)

chromosomal genes or regulatory elements that are associated in some way with drug action (Sandgren et al. 2009; Almeida Da Silva and Palomino 2011), providing compelling support for the idea that chromosomal mutagenesis drives the microevolution of *M. tuberculosis* within the human host. In combination, these three features imply competing roles for DNA repair and damage tolerance pathways in ensuring the maintenance of genome integrity while accommodating the evolutionary imperative to adapt to variable environments and changing selection pressures. Moreover, the ability of *M. tuberculosis* to respond rapidly at the transcriptional level to environmental cues (Rachman et al. 2006; Rohde et al. 2012) suggests that functional transcription-coupled DNA repair could be critical for mycobacterial adaptation and survival (Prabha et al. 2011), particularly under infectious conditions which might require the bacillus to make maximum use of limited transcriptional and translational resources.

Significant advances have been made in the last decade toward elucidating mechanisms of DNA metabolism in *M. tuberculosis* and the related non-pathogen, *M. smegmatis*, and in exploring the role of DNA metabolic enzymes or pathways in mycobacterial pathogenesis. In many cases, these studies have contributed novel insights to the field of DNA metabolism in general (Table 1), and have reinforced

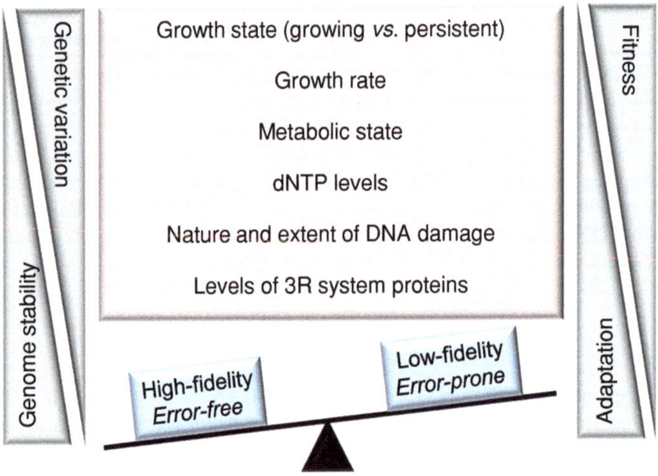

Fig. 1 The balance between high-fidelity genome maintenance and adaptation. Multiple factors are expected to influence the fidelity of replication and repair during host colonization (highlighted in *box*). Adapted from Warner (2010)

the idea that current models of bacterial DNA repair should be expanded to include alternatives to the well characterized *E. coli* system. Different aspects of mycobacterial DNA metabolism have been reviewed in a number of recent articles, and the reader is referred to these for additional mechanistic and biological insight (Mizrahi et al. 2000; Warner and Mizrahi 2006; Davis and Forse 2009; Dos Vultos et al. 2009; Gorna et al. 2010; Warner 2010; Kurthkoti and Varshney 2011; Kurthkoti and Varshney 2012). In this chapter, we provide an overview of DNA metabolism in *M. tuberculosis*, focusing particularly on recent findings in the area of DNA repair, and discuss the role of specific DNA metabolic pathways in pathogenesis. We then consider insights from whole-genome sequencing projects on the inferred relationship between mycobacterial DNA metabolic pathways and genome dynamics, and suggest possible areas for future study.

2 Fidelity of the "3R" System: Striking the Right Balance

The need for *M. tuberculosis* to maintain genome stability in order to remain fit—that is, able to establish an infection, grow and persist within a host, and transmit between hosts—must be balanced against the need for the organism to adapt genetically to the stresses encountered during infection, including those imposed by anti-tubercular drug administration. This balance is set by the fidelity of the so-called "3R" system of DNA replication, repair, and recombination (Fig. 1). The high-fidelity operation of the system favors genome stability, whereas relaxed fidelity—or loss of specific 3R functions (Dos Vultos et al. 2008)—might facilitate

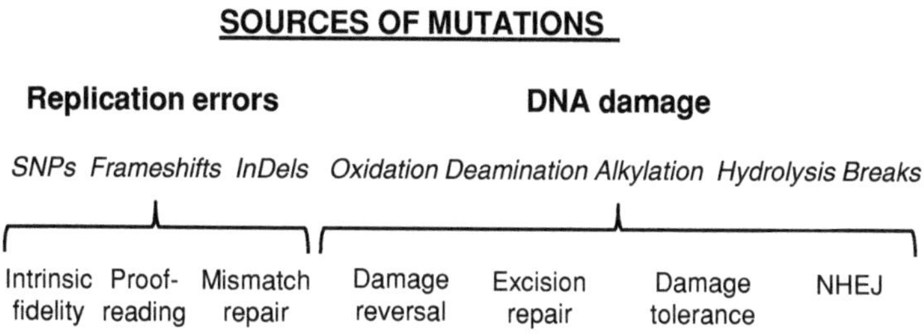

Fig. 2 DNA damage lesions and the repair mechanisms restricting their fixation

genetic adaptation (Warner 2010). In this context, it is interesting to note that genomic analyses have implicated DNA repair in the evolution of *M. tuberculosis* pathogenesis (McGuire et al. 2012). Moreover, comparatively high numbers of polymorphisms are found in genes of the 3R system of *M. tuberculosis* compared to housekeeping genes, identifying strong selection pressure on 3R genes as a common component in the modern evolutionary history of different strain lineages (Dos Vultos et al. 2008). It is tempting, therefore, to speculate that the identified polymorphisms result in a relaxation of 3R fidelity that facilitates the adaptation of this clonal pathogen to conditions of stress. To date, however, in vitro mutation rate analyses by fluctuation assay have failed to associate a hypermutable (or mutator) phenotype with a specific strain genotype (Werngren and Hoffner 2003).

2.1 Replication Fidelity

Mutations can arise through replication errors or as a consequence of DNA damage (Fig. 2). Based on fluctuation analyses that utilized *rpoB* as the target for rifampicin resistance, and which were corrected for the mutational target size, the in vitro mutation rate of *M. tuberculosis* was estimated at $\sim 2 \times 10^{-10}$ per base pair per round of replication (Ford et al. 2011). This is comparable to *E. coli*, in which intrinsic replication fidelity, proof-reading, and post-replicative mismatch repair (MMR) contribute 10^{-5}, $10^{-2,}$ and 10^{-3}, respectively, to the overall error rate of 10^{-10} (Mizrahi et al. 2000). Given the lack of a canonical MMR system in *M. tuberculosis*, this result is notable since it suggests that intrinsic fidelity and/or proofreading makes a proportionally greater contribution to mycobacterial replication fidelity, or that alternative mechanisms exist for the correction of replication errors, perhaps including a non-orthologous system for MMR (Mizrahi and Andersen 1998). While the formal possibility exists that the organism possesses an alternative MMR pathway, multiple lines of evidence instead suggest that *M. tuberculosis* has adapted to the lack of MMR through the activity of alternative repair components (Springer et al. 2004; Machowski et al. 2007; Wanner et al. 2008). This idea has been reinforced by the recent observation (Guthlein et al. 2009) that nucleotide excision repair (NER) might mitigate the absence of MMR via UvrD1-dependent processing of recombination-associated mismatches (see Sect. 2.3). As for replication fidelity and proof-reading, these have not been studied at all in *M. tuberculosis* or other mycobacteria and so represent an important area for future research.

2.2 Excision Repair and DNA Damage Reversal

In contrast to the paucity of information on replication fidelity, considerable progress has been made in characterizing the DNA damage repair and reversal systems that operate in *M. tuberculosis* and the model organism, *M. smegmatis*, and in elucidating the contribution of these systems to growth, survival, mutation avoidance, and mutation induction under conditions of genotoxic stress. Damage repair and reversal in mycobacteria have been investigated using genetic and biochemical approaches; however, there are relatively few examples of studies which have utilized a combination of both (Warner and Mizrahi 2011). Moreover, as a result of the functional redundancy and/or overlap that exists within and between certain DNA repair systems, definitive phenotypes have been difficult to establish for some individual and even multiple gene knockout mutants (Guo et al. 2010). Nevertheless, the phenotypes in animal models of infection of *M. tuberculosis* strains deficient in specific DNA damage repair or reversal components have proved very useful in inferring the nature and extent of genotoxicity of the in vivo environment. These derive in particular from screens of pools of transposon mutants by Transposon Site Hybridization (TraSH) (Sassetti and Rubin 2003) or Designer Arrays for Defined Mutant Analysis (DeADMAn) methodology (Dutta et al. 2010), both of which have elucidated mutants that are comparatively disadvantaged for growth and/or survival as a result of impaired DNA repair or reversal function. So, the base excision repair (BER) components *ung*, *xthA*, and *nfo* are critical for growth of *M. tuberculosis* in mice (Sassetti et al. 2003) while *fpg1* and *nei2* are required for bacillary viability in non-human primates (Dutta et al. 2010); only *uvrB*, the gene encoding the key NER excinuclease, is essential in both models.

In addition to reinforcing the idea that the in vivo environment encountered by *M. tuberculosis* is genotoxic, the genes identified in these studies specifically implicate deaminated cytosine, abasic sites, oxidative base damage, and bulky adducts as the major lesions requiring repair in order for the organism to retain full replicative fitness during infection. These findings are consistent with the type of damage expected for an intracellular organism (Fig. 3) with a genome of high G + C content (O'Sullivan et al. 2005). However, an important caveat is that, of the five genes listed above, *uvrB* is the only repair component which has been independently validated by in vivo phenotyping of a defined allelic exchange mutant (Darwin and Nathan 2005).

The finding that *fpg1* is required for growth and survival of *M. tuberculosis* in the lungs of non-human primates (Dutta et al. 2010) underscores the risk posed to *M. tuberculosis* by persisting oxidative base damage, and is consistent with recent

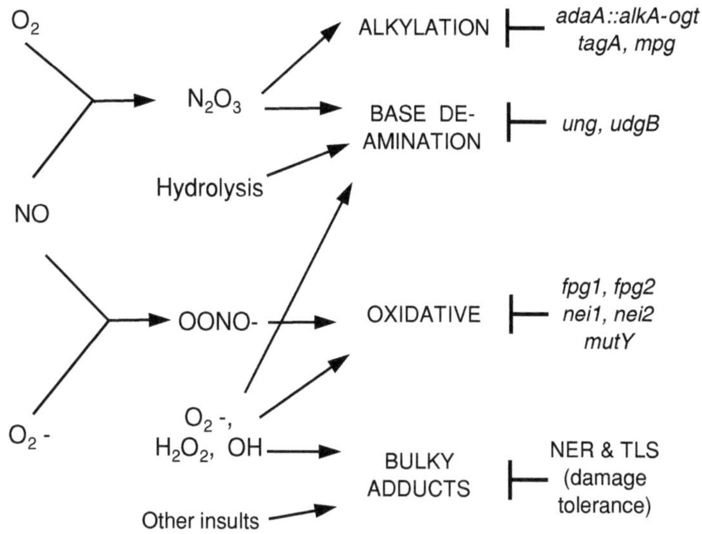

Fig. 3 The type of DNA damage expected to be sustained in *M. tuberculosis* and the genes implicated in damage repair and reversal (deRojas-Walker et al. 1995; Durbach et al. 2003)

evidence from whole-genome sequencing analyses of bacilli recovered from the lungs of non-human primates which identified polymorphisms that are potentially associated with oxidative DNA damage (Ford et al. 2011). At a functional level, the importance of *fpg1* is also consistent with the fact that a second predicted *fpg* homolog in the *M. tuberculosis* genome, *fpg2*, is truncated at its 5' terminus and so non-functional (Olsen et al. 2009; Guo et al. 2010). In contrast, the identification in the same model of *ung*, which encodes a class 1 uracil DNA glycosylase, is unexpected as *M. tuberculosis* also possesses a class 5 uracil DNA glycosylase, UdgB (Srinath et al. 2007; Wanner et al. 2009), which exhibits broader substrate specificity than Ung, but is also able to excise uracil. Similarly intriguing is the classification of *nei2* as essential for growth of *M. tuberculosis* in the non-human primate lung: the *M. tuberculosis* genome contains a second *nei* homolog, *nei1*, which encodes a DNA glycosylase that appears to be specific for oxidized pyrimidines (Guo et al. 2010). It is notable that a recombinant form of Nei2 did not show any biochemical activity in vitro; however, heterologous expression of *M. tuberculosis nei2* in *E. coli* decreased the spontaneous mutation frequency of a triple *fpg mutY nei* mutant as well as a double *nei nth* mutant, confirming that Nei2 is functional in vivo and is able to recognize both oxidized guanine and cytosine products (Guo et al. 2010).

DNA damage also results from exposure to alkylating compounds that are generated from endogenous metabolism as well as the host environment, including sources of nitrosative stress. Alkylation damage is both genotoxic and mutagenic; consistent with this dual threat, the *M. tuberculosis* genome encodes repair systems that are specifically directed to the repair of these lesions. Loss of the alkylation damage repair and reversal operon, *adaA::alkA-ogt*, had no effect on the growth of *M. tuberculosis* in mouse organs even though the mutant strain was hypersensitive in vitro to the cytotoxic and mutagenic effects of the alkylating agent, N-methyl-N'-nitro-N-nitrosoguanidine (Durbach et al. 2003). This result suggests that *M. tuberculosis* does not sustain significant alkylation damage in vivo, or that the alternate DNA glycosylases, TagA and/or Mpg, could substitute for the DNA glycosylase function of AlkA in recognizing and excising bases damaged by alkylation. More recent work has established that the adaptive response to alkylation damage is mediated exclusively by the DNA methyltransferases, AdaA, and Ogt (Yang et al. 2011a), which are restricted to the suppression of alkylation-induced mutagenesis. In light of these observations, the failure to detect an in vivo phenotype in the *adaA::alkA-ogt* knockout mutant (Durbach et al. 2003) appears to be even more significant since it implies that any hypermutability that might arise as a consequence of loss of this response could be tolerated without adversely affecting virulence. Interestingly, several studies have identified polymorphisms in the *adaA::alkA-ogt* operon in both susceptible and multidrug-resistant clinical strains (Ebrahimi-Rad et al. 2003; Nouvel et al. 2007), perhaps indicating a link between impaired alkylation damage repair and mutability. This possibility raises an important question: how might hypermutability be detected utilizing in vivo infection models where colony counts—and, perhaps, histopathology—are the standard measures of bacterial viability and "fitness"?

2.3 Recombination and End-Joining

Double-strand breaks (DSBs) in chromosomal DNA are caused by a variety of genotoxic agents from endogenous and exogenous sources, and constitute a major threat to bacillary viability. It is not surprising, therefore, that multiple mechanisms exist for the detection, processing, and repair of these potentially cytotoxic lesions. Until very recently, non-homologous end-joining (NHEJ) and homologous recombination (HR) were thought to constitute the major systems in *M. tuberculosis* for DSB repair (Shuman and Glickman 2007). However, the application in *M. smegmatis* of an elegant reporter assay that distinguishes DSB repair outcomes has expanded the mycobacterial DSB repair pathway complement to include single-strand annealing (SSA), simultaneously demonstrating that these pathways can be distinguished according to their differential requirements for the DSB-resecting, helicase-nuclease machines, AdnAB, and RecBCD (Gupta et al. 2011). That is, mycobacterial RecBCD is a dedicated SSA nuclease, while AdnAB is required for RecA-dependent HR. This represents a key distinction between mycobacteria and *E. coli*, in which the absence of an AdnAB homolog requires that RecBCD functions as the dominant helicase–nuclease and drives HR. Moreover, as Gupta and colleagues (2011) proposed, the identification of mycobacterial SSA might imply a key role in the repair of an *M. tuberculosis* genome that contains multiple repeat loci (Cole et al. 1998). The observation that alternative DSB pathways operate in *M. tuberculosis* and appear to be regulated asymmetrically provides another example of a mycobacterial function that has required a re-evaluation of prevailing models of bacterial DNA repair. This assessment is further supported by the recent biochemical characterization of a mycobacterial XPB helicase, which raises the additional possibility that *M. tuberculosis* is able to repair DSBs via synthesis-dependent strand annealing (Balasingham et al. 2012).

Although it was originally thought to be limited to eukaryotes, the first evidence for prokaryotic NHEJ was provided by the characterization of the mycobacterial system, which involves the activity of two proteins, Ku and LigD (Della et al. 2004). Mycobacterial Ku is a small protein comprising a core domain which is required for dimerization and enables binding of free DNA ends via the sugar-phosphate backbone. Recognition and binding of DNA ends by Ku is critical since it limits their degradation, and ensures the recruitment of LigD, a much larger protein that is responsible for end-processing and ligation and contains polymerase, nuclease, and ligase activities within a single polypeptide (Della et al. 2004). The *M. tuberculosis* genome also contains a number of other predicted ligases; however, with the exception of the essential replicative ligase, LigA (Gong et al. 2004), the potential role of these other proteins in processes such as NHEJ remains unclear (Aniukwu et al. 2008). Moreover, recent data have shown that a Sir2-like NAD-dependent deacetylase functionally interacts with Ku during NHEJ (Li et al. 2011), suggesting that multiple interacting components might mediate NHEJ activity and function.

Owing to the processing involved, NHEJ is inherently error-prone (Gong et al. 2005). However, the sensitivity of *M. smegmatis ku* and *ligD* mutants to DNA damage during stationary phase and in response to desiccation (Pitcher et al. 2007) suggests that NHEJ might fulfil a key repair function under special circumstances, including long-term survival and during transmission (Davis and Forse 2009). Importantly, the presence of multiple DSB repair mechanisms with different intrinsic fidelities reinforces the need to balance the activity of (error-free) DNA repair mechanisms in maintaining genomic integrity, and the operation of mutagenic pathways that might generate diversity during host infection (Fig. 1). It also suggests the possibility that DNA repair functions might have evolved specifically to cope with the high G + C content of the mycobacterial genome that is likely to influence both the type and frequency of mutational events (O'Sullivan et al. 2005).

DNA unwinding by helicases and gyrases is a prerequisite for repair and replicative polymerases to access the nucleic acid strands during replication, repair, recombination, transcription, and RNA processing (Schmid and Linder 1992; Matson et al. 1994). Like other mycobacteria, *M. tuberculosis* encodes two superfamily I DNA helicases, UvrD1, and UvrD2 (Davis and Forse 2009), which display 3'–5' polarity (Sinha et al. 2007). The differential essentialities of these two helicases—*uvrD2* is predicted to be essential (Sinha et al. 2008; Williams et al. 2011), whereas *uvrD1* is not—suggests non-redundant helicase function. Interestingly, recent evidence has shown that although *uvrD2* is essential, expression of a mutant form of UvrD2 lacking helicase activity was able to rescue the lethal effect of *uvrD2* deletion (Williams et al. 2011). UvrD1 depends on an interaction at its C-terminus with the key DNA-binding NHEJ protein, Ku, to stimulate its DNA unwinding activity (Sinha et al. 2007). In contrast, the in vitro helicase activity of a truncated UvrD2 mutant can be restored in trans by Ku (Sinha et al. 2007). UvrD1 and UvrD2 are structurally distinct: although UvrD1 resembles UvrD-like helicases, UvrD2 possesses an unusual, superfamily II-like C-terminal domain that appears to be limited to *Actinomycetales* and is not required for helicase function (Sinha et al. 2008). Instead, genetic analyses suggest that the inferred essentiality vests in the N-terminally encoded ATPase activity, implying a potential role for UvrD2 in DNA translocation, and/or protein displacement. As noted above, UvrD1 has also been shown to inhibit RecA-mediated strand exchange (Singh et al. 2010) and, together with the other NER component UvrB, has been implicated in ensuring the fidelity of HR in the absence of functional MMR (Guthlein et al. 2009).

In addition to UvrD1/2, the *M. tuberculosis* genome encodes another superfamily I helicase, the heterodimeric helicase-nuclease, AdnAB (discussed above), that is required for HR and appears to be limited to the actinomycetes (Sinha et al. 2009a). Added to the list of helicases in *M. tuberculosis* are a number of superfamily II helicases, including homologs of the eukaryotic basal transcription factor II (TFII) helicases, XPB, and XPD, which together with other TFII complex proteins, are required for NER and proofreading of transcription initiation (Coin et al. 2007). Whereas the function and role of mycobacterial XPD (DinG) is unknown, in vitro work has shown that XPB is an ATP-dependent 3'–5' DNA helicase (Biswas et al. 2009) which, in addition to DNA unwinding, is able to catalyze the ATP-independent annealing of complementary DNA strands in vitro (Balasingham et al. 2012). The catalog of superfamily II helicases has recently been extended with the characterization of a structurally distinct clade of DNA-dependent ATPases which are well-represented in mycobacteria (Yakovleva and Shuman 2012). It is notable that SftH provides yet another example of a DNA repair function that is elucidated in the mycobacterial model; however, additional studies combining biochemical and genetic analyses will be required to elucidate the relative contributions of the different helicases to specific mycobacterial DNA repair functions, as well as transcription-coupled repair (TCR) processes.

2.4 Transcription-Coupled Repair

In addition to the threat posed to genomic integrity, DNA lesions which hinder the function of RNA polymerase can have serious real-time consequences for cellular homeostasis and survival, especially where these occur in genes undergoing active transcription. Stalling of RNA polymerase triggers the TCR pathway, an important subclass of NER which functions to ensure that actively transcribed genes are repaired more rapidly than inactive regions of the genome. In bacterial organisms including *M. tuberculosis*, TCR is mediated by the *mfd* (*m*utation *f*requency *d*ecline)-encoded protein (Prabha et al. 2011), a homolog of the eukaryotic transcription-repair coupling factor (TRCF). As in other bacteria, *M. tuberculosis* Mfd does not recognize DNA damage directly, but through its interaction with the N-terminus of the RNA polymerase β subunit (Westblade et al. 2010). Binding of RNA polymerase by Mfd initiates a cascade of events which ensure the removal of the stalled RNA polymerase from the site of damage, and the recruitment of the NER components to effect the necessary repair. Unlike the *E. coli* protein, which is prevalent as a monomer, *M. tuberculosis* Mfd appears to exist as both monomers and hexamers which form via a self-interaction that requires the C-terminal region (Prabha et al. 2011). Although it is tempting to speculate that this might indicate the adaptation of this repair factor to a specific mycobacterial function(s) or property, that possibility requires further investigation.

Recently, the model for bacterial TCR was expanded with the demonstration that, in *E. coli*, the transcription elongation factor, NusA, directs NER to transcribed genes through a pathway that is independent of Mfd, and so offers an alternative to "classic" TCR (Cohen et al. 2010). NusA was previously shown to interact with the *dinB*-encoded *E. coli* DNA polymerase IV (Cohen et al. 2009), and the recent observations of Cohen and colleagues implicate this interaction in the recruitment of the TLS polymerase for bypass of transcription-blocking DNA lesions in an additional process dubbed "transcription-coupled TLS" (Cohen et al. 2010). Although similar processes have not been described in *M. tuberculosis*, it seems likely that multiple mechanisms exist to ensure the maintenance and operation of cellular information pathways and, in the context of obligate pathogen, to maximize the use of limited transcriptional and translational resources during host infection. This conclusion is supported by several other lines of evidence which indicate a role for DNA repair components in regulating cellular functions that are extensions of—or even distinct from—their "traditional" repair activities: for example, the DNA glycosylase, Tag, appears to modulate mycobacterial growth and morphology through a direct interaction with the chromosome partitioning factor, ParA (Huang and He 2012). Similarly, in *M. smegmatis*, disruption of the *uvrA*-encoded NER subunit results in an impaired ability to survive under nutrient- and oxygen-limited conditions (Cordone et al. 2011). The contribution of these and other DNA repair functions to mycobacterial physiology and pathogenesis therefore represents an important area for future research.

3 DNA Damage Tolerance

Persisting DNA lesions that have escaped detection and repair present a threat to the organism. Uracil, 5-hydroxyuracil, 8-oxo-guanine and O^6-methylguanine have strong miscoding potential, and can result in mutation induction and fixation through normal DNA replication. Because they do not cause distortions in the DNA helix, these lesions are difficult to detect, and so present the major threat to genome integrity. If replication-blocking lesions such as cyclobutane dimers and other bulky adducts persist, they can result in replication fork collapse and cell death. Specialized translesion synthesis (TLS) polymerases offer a solution to this problem: by allowing bypass of replication-blocking lesions, these enzymes provide a mechanism for tolerating DNA damage (Yang and Woodgate 2007). The importance of this function is suggested by the fact that TLS polymerases are conserved across all kingdoms of life (Ohmori et al. 2001). However, while these polymerases can display a high fidelity of replication across cognate lesions, their replication fidelities on non-cognate lesions or on undamaged templates are often reduced (Yang and Woodgate 2007). For this reason, TLS function is often associated with mutagenesis (Andersson et al. 2010).

3.1 The Y-Family of Specialist DNA Polymerases

Although it now seems obvious that specialized DNA polymerases might have evolved to replicate across persisting lesions, the discovery that the replicative flexibility they provide can incur error rates two to four orders of magnitude greater than those of replicative polymerases demanded a redefinition of traditional concepts of DNA synthesis fidelity (Yang and Woodgate 2007). Moreover, the fact that many of these polymerases are upregulated in response to stress suggested a potential role in induced mutagenesis (Andersson et al. 2010). Most TLS polymerases fall into the Y polymerase superfamily comprising a wide range of structurally related proteins present in bacteria, archaea, and eukaryotes (Ohmori et al. 2001). Of these, the DinB subfamily—represented by *E. coli* Pol IV—is the most widespread. Members of the Y-family catalyze low-fidelity synthesis on non-substrate templates such as undamaged DNA or non-cognate lesions, lack intrinsic 3' → 5' proofreading exonuclease activity, adopt a distributive mode of DNA replication, and are able to support TLS across DNA lesions which might block replication by members of the other polymerase families. The structural relation of members of this family to the other main DNA polymerases, the features enabling faithful bypass of DNA lesions but low-fidelity synthesis of undamaged or non-cognate DNA and, finally, the factors that co-ordinate multiple different polymerase classes at a replication fork, have all been subjects of active research (Yang and Woodgate 2007). These studies have revealed that, while Y-family polymerases share little sequence homology with the other polymerase families, they exhibit a similar overall architecture. Critically, the active site of Y-family polymerases is sufficiently flexible to allow bypass of lesions that might distort replicative polymerase geometry, but with a consequent reduction in stringency that has implicated members of this family in mutagenesis. However, while it has been suggested that TLS polymerases might have been selected equally for their ability to ensure continued replication as for their inherent capacity for mutagenesis, it seems increasingly likely that mutagenesis is a secondary consequence of the reduced fidelity that tolerance necessarily demands (reviewed in Andersson et al. (2010)).

3.2 Damage Tolerance and Mutagenesis in Mycobacteria

The *E. coli* genome contains three DNA polymerases which function in TLS—the B-family polymerase Pol II, and the Y-family polymerases Pol IV and Pol V which are encoded by *dinB* and *umuDC,* respectively—all of which are upregulated in the DNA damage or SOS response (Goodman 2002). In contrast to *E. coli*, there is no B-family DNA polymerase in *M. tuberculosis*. Therefore, it was originally assumed that all specialist bypass function in *M. tuberculosis* would depend on the two canonical Pol IV polymerase homologs, originally annotated as DinP and DinX (Mizrahi and Andersen 1998). The DNA damage response in *M. tuberculosis* is unusual (Rand et al. 2003) in that it comprises both a classic RecA-dependent SOS regulon (Smollett et al. 2012) as well an alternate, RecA-independent, damage regulon (Gamulin et al. 2004) that is controlled by a mycobacterial ClpR-like regulator (Wang et al. 2011) (Table 1). Surprisingly, neither Pol IV homolog is included in the mycobacterial DNA damage response (Boshoff et al. 2003), an additional departure from the *E. coli* model. Instead, the *dnaE1* and *dnaE2*-encoded catalytic (α) subunits of DNA polymerase III represent the only *bona fide* DNA polymerases that are induced as part of the mycobacterial SOS regulon (Boshoff et al. 2003; Warner et al. 2010). Furthermore, DnaE2 has been shown to be the central player in damage-induced base substitution mutagenesis in *M. tuberculosis*, identifying the mycobacterial polymerase as the founder member of a novel family of DnaE-type family C polymerases from Gram-positive bacteria that catalyse TLS. Loss of DnaE2 activity renders *M. tuberculosis* hypersensitive to DNA damage and eliminates induced mutagenesis. Moreover, functional inactivation of *dnaE2* attenuates virulence and reduces the frequency of drug resistance in vivo. Coupled with the induction of *dnaE2* during stationary infection, these observations implicate DnaE2-mediated DNA repair in the virulence of *M. tuberculosis* and in the adaptive evolution of drug resistance during persistent infection (Boshoff et al. 2003).

Subsequent efforts to understand the role of TLS-mediated DNA damage tolerance in mutation induction in *M. tuberculosis* led to the functional characterization of a novel damage tolerance system that is widely distributed across other bacterial genera (Boshoff et al. 2003; Warner et al. 2010). The mycobacterial version comprises the C-family DNA polymerase, DnaE2 (Boshoff et al. 2003) [more recently referred to as "ImuC" (McHenry 2011b)], and two "accessory factors", ImuA', and ImuB, (Warner et al. 2010) which are encoded by genes carried on a split, mutagenesis cassette that forms part of the SOS regulon controlled by RecA and LexA (Smollett et al. 2012). ImuB is one of three putative Y-family polymerase homologs in the *M. tuberculosis* genome. However, it is distinguished from the canonical Y-family homologs, DinB1 (DinX), and DinB2 (DinP), by two key features: in contrast to *bona fide* DNA polymerases, ImuB lacks the invariant active site acidic residues that are necessary for catalysis and, in addition, possesses an extended C-terminal domain comprising stretches of predicted structural disorder that are reminiscent of protein–protein interaction sites.

The features of the *imuA'-imuB/dnaE2* system were elucidated by genetic studies in *M. smegmatis* and *M. tuberculosis* using DNA damage survival and induced mutagenesis as phenotypic readouts, and yeast-two hybrid analysis to probe the protein interaction network. Consistent with the predicted inability of ImuB to catalyze nucleotidyl transfer, DnaE2 was identified as the TLS polymerase with ImuB apparently acting as hub protein that interacts with both ImuA' and DnaE2 via the C-terminal domain, and with the β-clamp via a canonical hexapeptide motif. Although the function of ImuA' remains cryptic, homology modeling suggests a RecA-type structure, and so reinforces the notion that the *imuA'-imuB/dnaE2*-encoded system constitutes a non-orthologous replacement of the PolV mutasome, UmuD'$_2$C•RecA•ATP, which has been extensively characterized in *E. coli* (Jiang et al. 2009). As noted elsewhere, the elucidation of this system represents an important example of the contribution of mycobacterial research to the DNA repair field in general (McHenry 2011a, b) (Table 1).

In contrast to the demonstrated role for ImuB in damage tolerance, the functions of the two other Y-family polymerases in *M. tuberculosis* remain poorly understood (Kana et al. 2010). These proteins are homologous to *E. coli* Pol IV which has been implicated in TLS across N^2-dG adducts (Jarosz et al. 2006) and cytotoxic alkylation damage (Bjedov et al. 2007). To date, extensive phenotypic characterization of mutants of *M. tuberculosis* lacking *dinB1* and/or *dinB2* has failed to yield discernible phenotypes in vitro and in vivo (Kana et al. 2010). As noted above, in contrast to the *E. coli* model in which three specialist DNA polymerases, Pol II (B-family), Pol IV, and Pol V (both Y-family) are induced as part of the SOS regulon (reviewed in (Goodman 2002)), the *M. tuberculosis* DNA damage response does not include *dinB1* or *dinB2* and is instead limited to the two α subunits, DnaE1 and DnaE2 (Warner et al. 2010). Elucidating the function of the mycobacterial DinB proteins will therefore require the combined application of biochemical assays as well as phenotypic analyses in DNA repair-defective strain backgrounds that offer the potential to uncover defects in damage tolerance by enabling lesions to persist (Bjedov ct al. 2007).

4 DNA Metabolism in Bacterial Pathogenesis

In general, DNA repair pathways in different organisms have been cataloged with reference to the *E. coli* model. For the most part, this has provided a useful starting point; however, the identification of repair pathways and systems that are present in other organisms but do not possess homologs in *E. coli* constitutes a major limitation of this approach. For example, although NHEJ was originally thought to be limited to eukaryotes owing to its apparent absence in *E. coli*, homologs of the bacterial NHEJ system were described in multiple organisms following its first identification in *M. tuberculosis* (Shuman and Glickman 2007) (Table 1). As an aside, it is interesting to note that an alternative end-joining pathway has subsequently been characterized in *E. coli* that differs from conventional NHEJ in its dependence on RecBCD (Chayot et al. 2010), again highlighting the potential existence of as-yet unrecognized repair pathways and/or alternative functions for known repair proteins.

Even where specific pathways or repair components are present in multiple organisms, there is a need to establish whether they fulfil identical functions. Bacterial genomes are characterized by significant differences in size, structure (including the presence of episomal genetic content), organization (e.g., the distribution of essential versus non-essential genes, the relative proportion of operons and intergenic regions), and nucleotide composition (e.g., G + C content, repeat regions, and homopolymeric runs) and, in the case of pathogens, are located in organisms with very specific host and tissue tropisms. The selective pressures determining genome architecture remain poorly understood (Koonin 2009); however, it seems likely that the composition and function of DNA replication and repair components have co-evolved with intrinsic features to ensure the propagation of specific properties (Zhao et al. 2007).

Notwithstanding this complexity, comparative genomics of the 3R complements of selected bacterial pathogens revealed some common themes (Ambur et al. 2009): for example, only a limited number of pathways are conserved across all species and, for many organisms, a reduced number of genes characterizes each repair class relative to the *E. coli* model. Notably, these analyses identified DNA replication, NER, and recombinational repair pathways as common to the pathogen genomes analyzed (Ambur et al. 2009). While this seems to contradict the very limited DNA metabolic complement of the inferred "minimal bacterial gene set" (Gil et al. 2004), it perhaps attests to the fact that the ability to sustain genotoxic stress represents a major component of the pathogen arsenal. Although untested, this interpretation is consistent with recent evidence associating DNA repair with the evolution of *M. tuberculosis* pathogenesis (McGuire et al. 2012), as well as previous work implicating selection on 3R genes as a major force in the differentiation of modern *M. tuberculosis* lineages (Dos Vultos et al. 2008).

As noted above, there are relatively few examples of the contribution of a single DNA repair pathway to mycobacterial pathogenesis. This is not unique to *M. tuberculosis*: direct evidence of a dominant role for specific DNA repair pathways has proven difficult to obtain for many pathogens, probably owing in part to the operation in most organisms of multiple mechanisms for the repair and/or tolerance of a specific DNA lesion (Ambur et al. 2009). That said, the need for active DNA repair has been cogently demonstrated in a number of infection models, providing insight into the functions that enable continued survival within—and adaptation to—disparate host environments, as well as the nature of the genotoxic stresses encountered. For example, *Salmonella enterica* requires active recombination repair (Buchmeier et al. 1993, 1995) and BER pathways for infection of mice (Richardson et al. 2009). Functional recombination repair (Loughlin et al. 2003; Amundsen et al. 2008, 2009) and BER (O'Rourke et al. 2003) enzymes are also essential for *Helicobacter pylori* to colonize mouse stomachs, while loss of BER limits *Neisseria meningitidis* bacteraemia in rats (Carpenter et al. 2007). Similarly, disruption of key BER and MMR components impairs the ability of

Vibrio cholerae to survive passage through the mouse gastrointestinal tract (Davies et al. 2011). While these examples highlight the deleterious consequences of impaired DNA repair function, there is also evidence that, in some cases, loss of specific repair components can result in improved pathogen outcomes. For example, targeted deletion of selected BER genes increases the virulence of *Streptococcus mutans* in a larval model of infection, probably, as a result of a hypermutator phenotype (Gonzalez et al. 2012). In most cases, however, this represents a short-term benefit: even where virulence is not negatively affected, susceptibility to extinction as a result of genetic bottlenecks, and a reduced capacity to adapt to secondary environments, can limit the ability of mutator populations to persist through successive infectious cycles (Merino et al. 2002).

In general, detectable phenotypes are limited to gross changes in bacterial cell number (CFU counts) or effects on host survival—the standard measures of virulence. Genetic complexity and a shortage of suitable assays to measure repair function (or the consequences of its loss) have even hindered in vitro approaches to resolving questions of redundancy for a number of repair pathways. The application of a sensitive genetic reporter assay to elucidate the roles of different mycobacterial recombination and end-joining pathways to DSB repair (Gupta et al. 2011) might provide an instructive example in this regard: although the DSBs induced in the experimental system were lethal (fewer than $\sim$0.03 % cells survived), this provided a pool of surviving mutants which was sufficiently large to enable the detection and quantification of discrete repair events. In *M. tuberculosis*, the increasing appreciation that bacilli can occupy discrete lesions (Russell et al. 2010)—and, perhaps, distinct environments (Barrios-Payan et al. 2012)—within a single host, raises the additional complication that different genotoxic stresses might be encountered depending on the site and stage of infection. In turn, this suggests that loss of a specific DNA repair function might be associated with subtler—and, perhaps, longer term—effects that could demand a more sensitive measure of genotoxicity. It is possible, for example, that consequences for adaptability and fitness, albeit significant for global strain prevalence and disease epidemiology (Gagneux 2012), cannot be assessed within the lifespan of a single, experimental infection.

5 Targeting DNA Metabolism for New TB Drug Discovery?

The predicted role of DNA metabolism in pathogenesis suggests the possibility of targeting selected mycobacterial pathways with novel chemotherapeutic agents. Antibiotic drug development prioritizes pathways or functions that are essential for bacterial growth and survival; for this reason, components of the DNA replication machinery represent the preferred 3R targets. To date, however, the targeting of bacterial "replisome" proteins has failed to delivered candidate drugs, a shortfall that has prompted a call for the more vigorous application of new technologies—including structure-based drug design and fragment-based lead generation—to assess the tractability of replication components for chemical inhibition (Sanyal and Doig 2012). In contrast to the essential DNA metabolic enzymes, repair components which are dispensable for normal growth in vitro but critical for pathogenesis in vivo offer an alternative class of drug target. This strategy is attractive for several reasons: repair components are required only in response to DNA damage and, although untested, there is an idea that for conditionally essential functions, the selective pressure to mutate to antibiotic resistance might not be as great given that the pathway is essential for pathogenesis but not survival.

For *M. tuberculosis*, candidate DNA repair targets in this category would include UvrB, the Fpg/Nei-family formamidopyrimidine-DNA glycosylase (*fpg*), and endonuclease VIII (*nei*), based on the phenotypes of knockout mutants in animal models of infection (Sassetti and Rubin 2003; Darwin and Nathan 2005; Dutta et al. 2010). In support of the attractiveness of these components and the pathways in which they function, a micromolar inhibitor of the mycobacterial NER pathway was recently described which selectively targets the UvrABC complex in damaged cells, preventing the recovery of *M. smegmatis* following UV-induced DNA damage (Mazloum et al. 2011). Although the efficacy of this compound remains to be validated in vivo, the demonstration that a high-throughput phenotypic screen can be successfully applied to identify a candidate inhibitor of DNA repair holds promise for the targeting of other mycobacterial 3R functions. There is also some evidence to suggest that DNA repair components might be profitably targeted by "co-drugs" designed to potentiate existing compounds by inhibiting proteins/pathways whose operation undermines the efficacy of frontline therapies. In a clever screen of the *E. coli* Keio collection, knockout mutants were identified which exhibited increased susceptibility to specific compounds relative to the wild-type strain (Liu et al. 2010). Notably, many of the genes that were implicated in the intrinsic "resistome" for many antibiotics would not have been predicted based on current knowledge of that drug's mode of action. Moreover, a number of those fell within the category of DNA repair, which suggests the utility of this approach in identifying unexpected target combinations and, simultaneously, its potential to reveal unknown functions of classic 3R components. Very recent evidence implicating the lethal incorporation of oxidized guanine into DNA as a major cause of antibiotic-induced bacterial cell death (Foti et al. 2012) provides further support for the idea that DNA replication and repair pathways contribute significantly to intrinsic drug resistance and, for that reason, should be considered as viable targets for novel therapies.

The role of damage tolerance in bacterial survival and, possibly, adaptation was discussed in a preceding section. Consistent with their inferred importance for mycobacterial pathogenesis, tolerance pathways might offer an additional option for novel antibacterial therapies. As discussed previously (Warner 2010), the possibility of inhibiting tolerance mechanisms—particularly inducible mutagenesis pathways—represents a subtle deviation from the concept of designing co-drugs to potentiate existing antibiotics, in this case by developing compounds that can be added to existing therapies to protect current drugs by targeting the mechanisms that underlie the evolution of resistance (Smith and Romesberg 2007). In some respects, this approach can be considered analogous to inhibiting efflux pathways (Adams et al. 2011): on its own, a specific efflux pump(s) represents a useless target but, in combination with the appropriate frontline drug, its inhibition might be critical to efficacy by ensuring that the active compound is maintained at an elevated intracellular concentration. While there is some evidence to suggest the feasibility of this approach (Georgescu et al. 2008; Wigle et al. 2009), the utility of "anti-evolution" compounds remains to be demonstrated in the context of an infectious disease such as tuberculosis.

6 Genome Dynamics in *M. tuberculosis*

The increasing availability of relatively cheap platforms for the generation of whole-genome sequence (WGS) data has enabled significant advances in the understanding of *M. tuberculosis* evolution, epidemiology, population structure, and strain diversity (Ford et al. 2012). It seems likely, therefore, that the application of WGS to in vivo infection models (Ford et al. 2011) and clinical isolates from strain collections (Hershberg et al. 2008; Ioerger et al. 2010; Casali et al. 2012; Namouchi et al. 2012) as well as individuals presenting with active disease (Saunders et al. 2011) might similarly enable an unprecedented glimpse into mycobacterial genome dynamics and the role of DNA repair during host infection.

The motivation for genome-wide characterization of clinical isolates is multifaceted: as noted above, *M. tuberculosis* is non-transformable and non-conjugative yet, despite (or because of) its relative genetic isolation, is a very successful pathogen (Russell 2001). At the levels of both species and strains (lineages), the application of comparative genomics has revealed unexpected genetic diversity in *M. tuberculosis* and, therefore, might be applied to correlate particular strain lineages (phylogenies) with disease outcomes (virulence) (Parwati et al. 2010) and disease type (eg., pulmonary versus extrapulmonary; Click et al. 2012). So, what can WGS reveal about the microevolution of *M. tuberculosis* strains within the human host? In the most compelling example to date of the application of these techniques to investigate in vivo genome dynamics, Sarah Fortune and colleagues analyzed the mutational events that occurred during both active and latent infection in a non-human primate model (Ford et al. 2011). In accordance with the idea that oxidative stress constitutes a major source of DNA damage in *M. tuberculosis* during host infection, their analysis revealed that the few mutations detected were indeed consistent with lesions caused by oxidative damage. Moreover, by applying a flexible range of predicted in vivo replication rates, they were able to estimate the rate at which the mutations arose. Unexpectedly, the calculated mutation rate was very similar to rates inferred from in vitro analyses. Furthermore, the mutation rate did not seem to differ between active and latently infected animals. Although this may indicate that bacilli replicate throughout latent infection, these results might be contrasted with evidence of apparent genomic stasis in sub-clinical human tuberculosis infections (Yang et al. 2011b).

Is there any evidence to support a role for mutators in mycobacterial evolution? For many bacterial pathogens, the absence of key DNA repair functions might facilitate adaptation during specific stages of the life cycle; however, the loss of repair machinery can impact long-term colonization and transmission. To some extent, this balance can be offset by the ability to re-acquire genes encoding DNA repair functions through mechanisms that enable horizontal gene transfer. However, evidence from comparative genomic studies seems to eliminate horizontal gene transfer as a contributor to the modern evolution of *M. tuberculosis*, at least in the case of drug-resistant isolates (Casali et al. 2012). The potential contribution of inducible mutagenesis mechanisms to bacterial pathogenesis is less clear, and might be difficult to infer (or measure) given the predicted role of mutations in providing the genetic diversity that enables the adaptive evolution of the invading organism to its host through multiple infection stages. Moreover, recent evidence suggesting the ability of *M. tuberculosis* to reside in multiple extrapulmonary sites and cell types during subclinical infection (Barrios-Payan et al. 2012) implies that the operation of maintenance versus adaptive functions might be separated spatially and temporally. That is, genome diversification might be favored during active disease (and transmission), whereas genome maintenance might be paramount during subclinical persistence.

7 Conclusions

There is increasing evidence that the DNA metabolic pathways in *M. tuberculosis* have evolved to ensure *M. tuberculosis* survival in the face of the metabolic and immune-mediated stresses encountered through multiple cycles of infection, replication, transmission, and persistence. Understanding the emergence and propagation of *M. tuberculosis* strains better adapted to host pathogenesis, including isolates resistant to one or more frontline anti-tubercular drugs, will require the elucidation of the specific molecular mechanisms governing mycobacterial DNA replication fidelity and maintenance. For that reason, additional work will be needed to give detail to the existing models of mycobacterial DNA replication and repair. Future research questions might therefore include the following: In the absence of MMR, what are the mechanisms that ensure replication fidelity in *M. tuberculosis*? Does *M. tuberculosis* modulate its mutation rate in response to specific environmental cues? What is the role of the DinB homologs in damage tolerance or repair? Does the presence of multiple homologs and/or analogs of specific repair components indicate redundant function or is it indicative of the specialist adaptation of specific repair proteins?

References

Achtman M (2012) Insights from genomic comparisons of genetically monomorphic bacterial pathogens. Philos Trans R Soc Lond B Biol Sci 367:860–867

Adams KN, Takaki K, Connolly LE et al (2011) Drug tolerance in replicating mycobacteria mediated by a macrophage-induced efflux mechanism. Cell 145:39–53

Almeida Da Silva PE, Palomino JC (2011) Molecular basis and mechanisms of drug resistance in *Mycobacterium tuberculosis*: classical and new drugs. J Antimicrob Chemother 66: 1417–1430

Ambur OH, Davidsen T, Frye SA, Balasingham SV, Lagesen K, Rognes T, Tonjum T (2009) Genome dynamics in major bacterial pathogens. FEMS Microbiol Rev 33:453–470

Amundsen SK, Fero J, Hansen LM, Cromie GA, Solnick JV, Smith GR, Salama NR (2008) *Helicobacter pylori* addAB Helicase-nuclease and RecA promote recombination-related DNA repair and survival during stomach Colonization. Mol Microbiol 69:994–1007

Amundsen SK, Fero J, Salama NR, Smith GR (2009) Dual nuclease and helicase activities of *Helicobacter pylori* AddAB are required for DNA repair, recombination, and mouse infectivity. J Biol Chem 284:16759–16766

Andersson DI, Koskiniemi S, Hughes D (2010) Biological roles of translesion synthesis DNA polymerases in eubacteria. Mol Microbiol 77:540–548

Aniukwu J, Glickman MS, Shuman S (2008) The pathways and outcomes of mycobacterial NHEJ depend on the structure of the broken DNA ends. Genes Dev 22:512–527

Balasingham SV, Zegeye ED, Homberset H, Rossi ML, Laerdahl JK, Bohr VA, Tonjum T (2012) Enzymatic activities and DNA substrate specificity of *Mycobacterium tuberculosis* DNA helicase XPB. PLoS One 7:e36960

Barrios-Payan J, Saqui-Salces M, Jeyanathan M, Vazquez AA, Arreola MC, Rook G, Hernandez-Pando R (2012) Extrapulmonary location of *Mycobacterium tuberculosis* DNA during latent infection. J Infect Dis. 206:1194

Biswas T, Pero JM, Joseph CG, Tsodikov OV (2009) DNA-dependent ATPase activity of bacterial XPB helicases. Biochemistry 48:2839–2848

Bjedov I, Dasgupta CN, Slade D, Le Blastier S, Selva M, Matic I (2007) Involvement of *Escherichia coli* DNA polymerase IV in tolerance of cytotoxic alkylating DNA lesions in vivo. Genetics 176:1431–1440

Boshoff HI, Reed MB, Barry CE 3rd, Mizrahi V (2003) DnaE2 polymerase contributes to in vivo survival and the emergence of drug resistance in *Mycobacterium tuberculosis*. Cell 113:183–193

Buchmeier NA, Lipps CJ, So MY, Heffron F (1993) Recombination-deficient mutants of Salmonella typhimurium are avirulent and sensitive to the oxidative burst of macrophages. Mol Microbiol 7:933–936

Buchmeier NA, Libby SJ, Xu Y, Loewen PC, Switala J, Guiney DG, Fang FC (1995) DNA repair is more important than catalase for Salmonella virulence in mice. J Clin Invest 95:1047–1053

Carpenter EP, Corbett A, Thomson H et al (2007) AP endonuclease paralogues with distinct activities in DNA repair and bacterial pathogenesis. EMBO J 26:1363–1372

Casali N, Nikolayevskyy V, Balabanova Y et al (2012) Microevolution of extensively drug-resistant tuberculosis in Russia. Genome Res 22:735–745

Chayot R, Montagne B, Mazel D, Ricchetti M (2010) An end-joining repair mechanism in Escherichia coli. Proc Natl Acad Sci U S A 107:2141–2146

Click ES, Moonan PK, Winston CA, Cowan LS, Oeltmann JE (2012) Relationship between *Mycobacterium tuberculosis* phylogenetic lineage and clinical site of tuberculosis. Clin Infect Dis 54:211–219

Cohen SE, Godoy VG, Walker GC (2009) Transcriptional modulator NusA interacts with translesion DNA polymerases in *Escherichia coli*. J Bacteriol 191:665–672

Cohen SE, Lewis CA, Mooney RA, Kohanski MA, Collins JJ, Landick R, Walker GC (2010) Roles for the transcription elongation factor NusA in both DNA repair and damage tolerance pathways in *Escherichia coli*. Proc Natl Acad Sci U S A 107:15517–15522

Coin F, Oksenych V, Egly JM (2007) Distinct roles for the XPB/p52 and XPD/p44 subcomplexes of TFIIH in damaged DNA opening during nucleotide excision repair. Mol Cell 26:245–256

Cole ST, Brosch R, Parkhill J et al (1998) Deciphering the biology of *Mycobacterium tuberculosis* from the complete genome sequence. Nature 393:537–544

Comas I, Homolka S, Niemann S, Gagneux S (2009) Genotyping of genetically monomorphic bacteria: DNA sequencing in *Mycobacterium tuberculosis* highlights the limitations of current methodologies. PLoS One 4:e7815

Cordone A, Audrain B, Calabrese I, Euphrasie D, Reyrat JM (2011) Characterization of a *Mycobacterium smegmatis uvrA* mutant impaired in dormancy induced by hypoxia and low carbon concentration. BMC Microbiol 11:231

Darwin KH, Nathan CF (2005) Role for nucleotide excision repair in virulence of *Mycobacterium tuberculosis*. Infect Immun 73:4581–4587

Davies BW, Bogard RW, Dupes NM, Gerstenfeld TA, Simmons LA, Mekalanos JJ (2011) DNA damage and reactive nitrogen species are barriers to *Vibrio cholerae* colonization of the infant mouse intestine. PLoS pathogens 7:e1001295

Davis EO, Forse LN (2009) DNA repair: key to survival. In: Parish T, Brown A (eds) Mycobacterium Genomics and Molecular Biology. Caister Academic Press, Norfolk, UK, pp 79–117

Davis EO, Jenner PJ, Brooks PC, Colston MJ, Sedgwick SG (1992) Protein splicing in the maturation of *M. tuberculosis* recA protein: a mechanism for tolerating a novel class of intervening sequence. Cell 71:201–210

Della M, Palmbos PL, Tseng HM et al (2004) Mycobacterial Ku and ligase proteins constitute a two-component NHEJ repair machine. Science 306:683–685

deRojas-Walker T, Tamir S, Ji H, Wishnok JS, Tannenbaum SR (1995) Nitric oxide induces oxidative damage in addition to deamination in macrophage DNA. Chem Res Toxicol 8:473–477

Dos Vultos T, Mestre O, Rauzier J et al (2008) Evolution and diversity of clonal bacteria: the paradigm of *Mycobacterium tuberculosis*. PLoS One 3:e1538

Dos Vultos T, Mestre O, Tonjum T, Gicquel B (2009) DNA repair in *Mycobacterium tuberculosis* revisited. FEMS Microbiol Rev 33:471–487

Durbach SI, Springer B, Machowski EE et al (2003) DNA alkylation damage as a sensor of nitrosative stress in *Mycobacterium tuberculosis*. Infect Immun 71:997–1000

Dutta NK, Mehra S, Didier PJ et al (2010) Genetic requirements for the survival of tubercle bacilli in primates. J Infect Dis 201:1743–1752

Ebrahimi-Rad M, Bifani P, Martin C et al (2003) Mutations in putative mutator genes of *Mycobacterium tuberculosis* strains of the W-Beijing family. Emerg Infect Dis 9:838–845

Ford CB, Lin PL, Chase MR et al (2011) Use of whole genome sequencing to estimate the mutation rate of *Mycobacterium tuberculosis* during latent infection. Nat Genet 43:482–486

Ford C, Yusim K, Ioerger T et al (2012) *Mycobacterium tuberculosis*–heterogeneity revealed through whole genome sequencing. Tuberculosis (Edinb) 92:194–201

Foti JJ, Devadoss B, Winkler JA, Collins JJ, Walker GC (2012) Oxidation of the guanine nucleotide pool underlies cell death by bactericidal antibiotics. Science 336:315–319

Gagneux S (2012) Host-pathogen coevolution in human tuberculosis. Philos Trans R Soc Lond B Biol Sci 367:850–859

Gamulin V, Cetkovic H, Ahel I (2004) Identification of a promoter motif regulating the major DNA damage response mechanism of *Mycobacterium tuberculosis*. FEMS Microbiol Lett 238:57–63

Georgescu RE, Yurieva O, Kim SS, Kuriyan J, Kong XP, O'Donnell M (2008) Structure of a small-molecule inhibitor of a DNA polymerase sliding clamp. Proc Natl Acad Sci U S A 105:11116–11121

Gil R, Silva FJ, Pereto J, Moya A (2004) Determination of the core of a minimal bacterial gene set. Microbiol Mol Biol Rev 68:518–537

Gong C, Martins A, Bongiorno P, Glickman M, Shuman S (2004) Biochemical and genetic analysis of the four DNA ligases of mycobacteria. J Biol Chem 279:20594–20606

Gong C, Bongiorno P, Martins A, Stephanou NC, Zhu H, Shuman S, Glickman MS (2005) Mechanism of nonhomologous end-joining in mycobacteria: a low-fidelity repair system driven by Ku, ligase D and ligase C. Nat Struct Mol Biol 12:304–312

Gonzalez K, Faustoferri RC, Quivey RG Jr (2012) Role of DNA base excision repair in the mutability and virulence of *Streptococcus mutans*. Mol Microbiol 85:361–377

Goodman MF (2002) Error-prone repair DNA polymerases in prokaryotes and eukaryotes. Annu Rev Biochem 71:17–50

Gorna AE, Bowater RP, Dziadek J (2010) DNA repair systems and the pathogenesis of *Mycobacterium tuberculosis*: varying activities at different stages of infection. Clin Sci (Lond) 119:187–202

Guo Y, Bandaru V, Jaruga P et al (2010) The oxidative DNA glycosylases of *Mycobacterium tuberculosis* exhibit different substrate preferences from their *Escherichia coli* counterparts. DNA Repair (Amst) 9:177–190

Gupta R, Barkan D, Redelman-Sidi G, Shuman S, Glickman MS (2011) Mycobacteria exploit three genetically distinct DNA double-strand break repair pathways. Mol Microbiol 79:316–330

Guthlein C, Wanner RM, Sander P et al (2009) Characterization of the mycobacterial NER system reveals novel functions of the *uvrD1* helicase. J Bacteriol 191:555–562

Hershberg R, Lipatov M, Small PM et al (2008) High functional diversity in *Mycobacterium tuberculosis* driven by genetic drift and human demography. PLoS Biol 6:e311

Huang F, He ZG (2012) Characterization of a Conserved Interaction between DNA Glycosylase and ParA in *Mycobacterium smegmatis* and *M. tuberculosis*. PLoS One 7:e38276

Ioerger TR, Feng Y, Chen X et al (2010) The non-clonality of drug resistance in Beijing-genotype isolates of *Mycobacterium tuberculosis* from the Western Cape of South Africa. BMC Genomics 11:670

Jarosz DF, Godoy VG, Delaney JC, Essigmann JM, Walker GC (2006) A single amino acid governs enhanced activity of DinB DNA polymerases on damaged templates. Nature 439:225–228

Jiang Q, Karata K, Woodgate R, Cox MM, Goodman MF (2009) The active form of DNA polymerase V is UmuD'(2)C-RecA-ATP. Nature 460:359–363

Kana BD, Abrahams GL, Sung N et al (2010) Role of the DinB homologs Rv1537 and Rv3056 in *Mycobacterium tuberculosis*. J Bacteriol 192:2220–2227

Koonin EV (2009) Evolution of genome architecture. Int J Biochem Cell Biol 41:298–306

Kurthkoti K, Varshney U (2011) Base excision and nucleotide excision repair pathways in mycobacteria. Tuberculosis (Edinb) 91:533–543

Kurthkoti K, Varshney U (2012) Distinct mechanisms of DNA repair in mycobacteria and their implications in attenuation of the pathogen growth. Mech Ageing Dev 133:138–146

Li Z, Wen J, Lin Y et al (2011) A Sir2-like protein participates in mycobacterial NHEJ. PLoS One 6:e20045

Lillebaek T, Dirksen A, Baess I, Strunge B, Thomsen VO, Andersen AB (2002) Molecular evidence of endogenous reactivation of *Mycobacterium tuberculosis* after 33 years of latent infection. J Infect Dis 185:401–404

Liu A, Tran L, Becket E et al (2010) Antibiotic sensitivity profiles determined with an *Escherichia coli* gene knockout collection: generating an antibiotic bar code. Antimicrob Agents Chemother 54:1393–1403

Loughlin MF, Barnard FM, Jenkins D, Sharples GJ, Jenks PJ (2003) *Helicobacter pylori* mutants defective in RuvC Holliday junction resolvase display reduced macrophage survival and spontaneous clearance from the murine gastric mucosa. Infect Immun 71:2022–2031

Machowski EE, Barichievy S, Springer B, Durbach SI, Mizrahi V (2007) In vitro analysis of rates and spectra of mutations in a polymorphic region of the Rv0746 PE_PGRS gene of *Mycobacterium tuberculosis*. J Bacteriol 189:2190–2195
Matson SW, Bean DW, George JW (1994) DNA helicases: enzymes with essential roles in all aspects of DNA metabolism. Bioessays 16:13–22
Mazloum N, Stegman MA, Croteau DL et al (2011) Identification of a chemical that inhibits the mycobacterial UvrABC complex in nucleotide excision repair. Biochemistry 50:1329–1335
McGuire AM, Weiner B, Park ST et al (2012) Comparative analysis of Mycobacterium and related Actinomycetes yields insight into the evolution of *Mycobacterium tuberculosis* pathogenesis. BMC Genomics 13:120
McHenry CS (2011a) Bacterial replicases and related polymerases. Curr Opin Chem Biol 15:587–594
McHenry CS (2011b) Breaking the rules: bacteria that use several DNA polymerase IIIs. EMBO Rep 12:408–414
Merino D, Reglier-Poupet H, Berche P, Charbit A (2002) A hypermutator phenotype attenuates the virulence of *Listeria monocytogenes* in a mouse model. Mol Microbiol 44:877–887
Mizrahi V, Andersen SJ (1998) DNA repair in *Mycobacterium tuberculosis*. What have we learnt from the genome sequence? Mol Microbiol 29:1331–1339
Mizrahi V, Dawes SS, Rubin H (2000) DNA Replication. In: Hatfull G, Jacobs WR Jr (eds) Molecular Genetics of Mycobacteria. ASM Press, Washington, D.C, pp 159–172
Namouchi A, Didelot X, Schock U, Gicquel B, Rocha EP (2012) After the bottleneck: Genome-wide diversification of the *Mycobacterium tuberculosis* complex by mutation, recombination, and natural selection. Genome Res 22:721–734
Nathan C, Shiloh MU (2000) Reactive oxygen and nitrogen intermediates in the relationship between mammalian hosts and microbial pathogens. Proc Natl Acad Sci U S A 97:8841–8848
Nouvel LX, Dos Vultos T, Kassa-Kelembho E, Rauzier J, Gicquel B (2007) A non-sense mutation in the putative anti-mutator gene *ada/alkA* of *Mycobacterium tuberculosis* and *M. bovis* isolates suggests convergent evolution. BMC Microbiol 7:39
Ohmori H, Friedberg EC, Fuchs RP et al (2001) The Y-family of DNA polymerases. Mol Cell 8:7–8
Olsen I, Balasingham SV, Davidsen T et al (2009) Characterization of the major formamido-pyrimidine-DNA glycosylase homolog in *Mycobacterium tuberculosis* and its linkage to variable tandem repeats. FEMS Immunol Med Microbiol 56:151–161
O'Rourke EJ, Chevalier C, Pinto AV, Thiberge JM, Ielpi L, Labigne A, Radicella JP (2003) Pathogen DNA as target for host-generated oxidative stress: role for repair of bacterial DNA damage in *Helicobacter pylori* colonization. Proc Natl Acad Sci U S A 100:2789–2794
O'Sullivan DM, McHugh TD, Gillespie SH (2005) Analysis of *rpoB* and *pncA* mutations in the published literature: an insight into the role of oxidative stress in *Mycobacterium tuberculosis* evolution? J Antimicrob Chemother 55:674–679
Parwati I, Alisjahbana B, Apriani L et al (2010) *Mycobacterium tuberculosis* Beijing genotype is an independent risk factor for tuberculosis treatment failure in Indonesia. J Infect Dis 201:553–557
Pitcher RS, Green AJ, Brzostek A, Korycka-Machala M, Dziadek J, Doherty AJ (2007) NHEJ protects mycobacteria in stationary phase against the harmful effects of desiccation. DNA Repair (Amst) 6:1271–1276
Prabha S, Rao DN, Nagaraja V (2011) Distinct properties of hexameric but functionally conserved *Mycobacterium tuberculosis* transcription-repair coupling factor. PLoS One 6:e19131
Rachman H, Strong M, Ulrichs T et al (2006) Unique transcriptome signature of *Mycobacterium tuberculosis* in pulmonary tuberculosis. Infect Immun 74:1233–1242
Rand L, Hinds J, Springer B, Sander P, Buxton RS, Davis EO (2003) The majority of inducible DNA repair genes in *Mycobacterium tuberculosis* are induced independently of RecA. Mol Microbiol 50:1031–1042

Richardson AR, Soliven KC, Castor ME, Barnes PD, Libby SJ, Fang FC (2009) The Base Excision Repair system of *Salmonella enterica* serovar typhimurium counteracts DNA damage by host nitric oxide. PLoS Pathog 5:e1000451

Rohde KH, Veiga DF, Caldwell S, Balazsi G, Russell DG (2012) Linking the Transcriptional Profiles and the Physiological States of Mycobacterium tuberculosis during an Extended Intracellular Infection. PLoS Pathog 8:e1002769

Russell DG (2001) Mycobacterium tuberculosis: here today, and here tomorrow. Nat Rev Mol Cell Biol 2:569–577

Russell DG (2007) Who puts the tubercle in tuberculosis? Nat Rev Microbiol 5:39–47

Russell DG, Barry CE 3rd, Flynn JL (2010) Tuberculosis: what we don't know can, and does, hurt us. Science 328:852–856

Sandgren A, Strong M, Muthukrishnan P, Weiner BK, Church GM, Murray MB (2009) Tuberculosis drug resistance mutation database. PLoS Med 6:e2

Sanyal G, Doig P (2012) Bacterial DNA replication enzymes as targets for antibacterial drug discovery. Expert Opin Drug Discov 7:327–339

Sassetti CM, Rubin EJ (2003) Genetic requirements for mycobacterial survival during infection. Proc Natl Acad Sci U S A 100:12989–12994

Sassetti CM, Boyd DH, Rubin EJ (2003) Genes required for mycobacterial growth defined by high density mutagenesis. Mol Microbiol 48:77–84

Saunders NJ, Trivedi UH, Thomson ML, Doig C, Laurenson IF, Blaxter ML (2011) Deep resequencing of serial sputum isolates of *Mycobacterium tuberculosis* during therapeutic failure due to poor compliance reveals stepwise mutation of key resistance genes on an otherwise stable genetic background. J Infect 62:212–217

Schmid SR, Linder P (1992) D-E-A-D protein family of putative RNA helicases. Mol Microbiol 6:283–291

Shuman S, Glickman MS (2007) Bacterial DNA repair by non-homologous end joining. Nat Rev Microbiol 5:852–861

Singh P, Patil KN, Khanduja JS et al (2010) *Mycobacterium tuberculosis* UvrD1 and UvrA proteins suppress DNA strand exchange promoted by cognate and noncognate RecA proteins. Biochemistry 49:4872–4883

Sinha KM, Stephanou NC, Gao F, Glickman MS, Shuman S (2007) Mycobacterial UvrD1 is a Ku-dependent DNA helicase that plays a role in multiple DNA repair events, including double-strand break repair. J Biol Chem 282:15114–15125

Sinha KM, Stephanou NC, Unciuleac MC, Glickman MS, Shuman S (2008) Domain requirements for DNA unwinding by mycobacterial UvrD2, an essential DNA helicase. Biochemistry 47:9355–9364

Sinha KM, Glickman MS, Shuman S (2009a) Mutational analysis of Mycobacterium UvrD1 identifies functional groups required for ATP hydrolysis, DNA unwinding, and chemomechanical coupling. Biochemistry 48:4019–4030

Sinha KM, Unciuleac MC, Glickman MS, Shuman S (2009b) AdnAB: a new DSB-resecting motor-nuclease from mycobacteria. Genes Dev 23:1423–1437

Smith PA, Romesberg FE (2007) Combating bacteria and drug resistance by inhibiting mechanisms of persistence and adaptation. Nat Chem Biol 3:549–556

Smollett KL, Smith KM, Kahramanoglou C, Arnvig KB, Buxton RS, Davis EO (2012) Global Analysis of the Regulon of the Transcriptional Repressor LexA, a Key Component of SOS Response in *Mycobacterium tuberculosis*. J Biol Chem 287:22004–22014

Springer B, Sander P, Sedlacek L, Hardt WD, Mizrahi V, Schar P, Bottger EC (2004) Lack of mismatch correction facilitates genome evolution in mycobacteria. Mol Microbiol 53:1601–1609

Srinath T, Bharti SK, Varshney U (2007) Substrate specificities and functional characterization of a thermo-tolerant uracil DNA glycosylase (UdgB) from *Mycobacterium tuberculosis*. DNA Repair (Amst) 6:1517–1528

Talaat AM, Lyons R, Howard ST, Johnston SA (2004) The temporal expression profile of *Mycobacterium tuberculosis* infection in mice. Proc Natl Acad Sci U S A 101:4602–4607

Wang Y, Huang Y, Xue C, He Y, He ZG (2011) ClpR protein-like regulator specifically recognizes RecA protein-independent promoter motif and broadly regulates expression of DNA damage-inducible genes in mycobacteria. J Biol Chem 286:31159–31167

Wanner RM, Guthlein C, Springer B, Bottger EC, Ackermann M (2008) Stabilization of the genome of the mismatch repair deficient *Mycobacterium tuberculosis* by context-dependent codon choice. BMC Genomics 9:249

Wanner RM, Castor D, Guthlein C, Bottger EC, Springer B, Jiricny J (2009) The uracil DNA glycosylase UdgB of *Mycobacterium smegmatis* protects the organism from the mutagenic effects of cytosine and adenine deamination. J Bacteriol 191:6312–6319

Warner DF (2010) The role of DNA repair in *M. tuberculosis* pathogenesis. Drug Discov Today Dis Mech 7:e5–e11

Warner DF, Mizrahi V (2006) Tuberculosis chemotherapy: the influence of bacillary stress and damage response pathways on drug efficacy. Clin Microbiol Rev 19:558–570

Warner DF, Mizrahi V (2011) Making ends meet in mycobacteria. Mol Microbiol 79:283–287

Warner DF, Ndwandwe DE, Abrahams GL, Kana BD, Machowski EE, Venclovas C, Mizrahi V (2010) Essential roles for imuA'- and imuB-encoded accessory factors in DnaE2-dependent mutagenesis in *Mycobacterium tuberculosis*. Proc Natl Acad Sci U S A 107:13093–13098

Werngren J, Hoffner SE (2003) Drug-susceptible Mycobacterium tuberculosis Beijing genotype does not develop mutation-conferred resistance to rifampin at an elevated rate. J Clin Microbiol 41:1520–1524

Westblade LF, Campbell EA, Pukhrambam C, Padovan JC, Nickels BE, Lamour V, Darst SA (2010) Structural basis for the bacterial transcription-repair coupling factor/RNA polymerase interaction. Nucleic Acids Res 38:8357–8369

Wigle TJ, Sexton JZ, Gromova AV et al (2009) Inhibitors of RecA activity discovered by high-throughput screening: cell-permeable small molecules attenuate the SOS response in *Escherichia coli*. J Biomol Screen 14:1092–1101

Williams A, Guthlein C, Beresford N, Bottger EC, Springer B, Davis EO (2011) UvrD2 is essential in *Mycobacterium tuberculosis*, but its helicase activity is not required. J Bacteriol 193:4487–4494

Yakovleva L, Shuman S (2012) Mycobacterium smegmatis SftH exemplifies a distinctive clade of superfamily II DNA-dependent ATPases with 3' to 5' translocase and helicase activities. Nucleic Acids Res. 40(15):7465–7475

Yang W, Woodgate R (2007) What a difference a decade makes: insights into translesion DNA synthesis. Proc Natl Acad Sci U S A 104:15591–15598

Yang M, Aamodt RM, Dalhus B et al (2011a) The ada operon of *Mycobacterium tuberculosis* encodes two DNA methyltransferases for inducible repair of DNA alkylation damage. DNA Repair (Amst) 10:595–602

Yang Z, Rosenthal M, Rosenberg NA et al (2011b) How dormant is *Mycobacterium tuberculosis* during latency? A study integrating genomics and molecular epidemiology. Infect Genet Evol 11:1164–1167

Zahrt TC, Deretic V (2002) Reactive nitrogen and oxygen intermediates and bacterial defenses: unusual adaptations in *Mycobacterium tuberculosis*. Antioxid Redox Signal 4:141–159

Zhao X, Zhang Z, Yan J, Yu J (2007) GC content variability of eubacteria is governed by the pol III α subunit. Biochem Biophys Res Commun 356:20–25

Molecular Biology of Drug Resistance in *Mycobacterium tuberculosis*

Tasha Smith, Kerstin A. Wolff and Liem Nguyen

Abstract Tuberculosis (TB) has become a curable disease, thanks to the discovery of antibiotics. However, it has remained one of the most difficult infections to treat. Most current TB regimens consist of 6–9 months of daily doses of four drugs that are highly toxic to patients. The purpose of these lengthy treatments is to completely eradicate *Mycobacterium tuberculosis*, notorious for its ability to resist most antibacterial agents, thereby preventing the formation of drug resistant mutants. On the contrary, the prolonged therapies have led to poor patient adherence. This, together with a severe limit of drug choices, has resulted in the emergence of strains that are increasingly resistant to the few available antibiotics. Here, we review our current understanding of molecular mechanisms underlying the profound drug resistance of *M. tuberculosis*. This knowledge is essential for the development of more effective antibiotics, which are not only potent against drug resistant *M. tuberculosis* strains but also help shorten the current treatment courses required for drug susceptible TB.

T. Smith · K. A. Wolff
Case Western Reserve University School of Medicine,
Cleveland, OH 44106, USA

L. Nguyen (✉)
Department of Molecular Biology and Microbiology, School of Medicine,
Case Western Reserve University, 10900 Euclid Avenue,
Cleveland, OH 44106-4960, USA
e-mail: liem.nguyen@case.edu

Current Topics in Microbiology and Immunology (2012) 374: 53–80
DOI: 10.1007/82_2012_279

Published Online: 21 November 2012

Contents

1 Introduction

Tuberculosis (TB) has plagued humans since antiquity. During the pre-antibiotic era, patients diagnosed with this so-called "consumption" disease would have felt hopeless, much like how people feel about many cancers today. TB was treated with gold, arsenic, cod liver oil, herbs, bed rest, sunshine and fresh air, etc. (Birnbaum et al. 1891), but none of these therapies were really effective. It was therefore one of the most deadly diseases during this long period of time. The recent discovery of antibiotics, dubbed "magic bullets" because of their powerful potency against pathogenic bacteria, has brought about a real revolution in TB chemotherapy. Starting with streptomycin in 1943, series of potent TB drugs were introduced to clinical practice during this "golden age of antibiotics" (1940s–1960s). The implementation of these drugs to TB treatment immediately resulted in a sharp decline of TB incidence throughout the world. In fact, the euphoria of the great success brought a sense of total control over this disease. It was commonly thought that TB was no longer a public health concern in many developed countries. Some even went further to predict that TB would be soon eradicated from the world (Myers 1963).

Yet, the disease came back in the 1980s with outbreaks of multidrug resistant (MDR) strains, often associated with the spreading epidemic of the acquired immune deficiency syndrome (AIDS). It is currently estimated that *M. tuberculosis* is infecting around 2 billion people worldwide, equivalent to one-third of the world population (Corbett et al. 2003). Besides an ability to persist in the face of host defense mechanisms, allowing its wide spread, *M. tuberculosis* is capable of resistance to most antimicrobial agents available (Nguyen and Pieters 2009). As a result, the existing options for TB chemotherapy are severely restricted. The repeated use of the same drugs, together with prolonged regimens that often lead to poor patient compliance, has resulted in the emergence of strains that are increasingly resistant to the available drugs. From the first isolated strains that were resistant to single drugs, sequential accumulation of resistance mutations has led to the emergence of MDR, extensively drug resistant (XDR), and most recently totally drug resistant (TDR) *M. tuberculosis* strains (Ormerod 2005; Dorman and Chaisson 2007; Udwadia 2012). Infections with some of the latter strains are essentially incurable by the current TB drugs. Therefore, these resistant strains of *M. tuberculosis* pose a serious threat to worldwide TB control programs. To tackle the current epidemic of drug resistant TB, novel therapeutic interventions are urgently needed. Besides the efforts to develop completely new antibiotics that are not affected by the existing resistance mechanisms, other nontraditional approaches such as targeting resistance mechanisms or repurposing old drugs need to be further investigated. For these approaches to be successful, drug resistance mechanisms in *M. tuberculosis* should be thoroughly studied and well understood.

2 Acquired Antibiotic Resistance Mechanisms

Pathogenic bacteria including *M. tuberculosis* are able to acquire resistance to a particular antibiotic to which they were previously susceptible. The concept here referred to as "acquired antibiotic resistance" is in contrast with the intrinsic resistance discussed later in this chapter. Acquired resistances might occur through either mutation or horizontal gene transfer. In *M. tuberculosis*, horizontal transfer of resistance genes via plasmids or transposon elements has not been reported. By contrast, all currently known acquired resistances are mediated through chromosomal mutations that arise under selective pressure of antibiotic use. A summary of *M. tuberculosis* genes to which mutations confer TB drug resistance is presented in Table 1.

2.1 Genetic Mutation

Darwin's theory of evolution is perfectly epitomized by the progression of drug resistance in *M. tuberculosis*. Like any new traits arising during selective evolution, antibiotic resistant strains only become predominant in *M. tuberculosis* populations if the resistance phenotypes provide the mutants with survival advantages over their susceptible counterparts. The prolonged drug exposure due to lengthy regimens might have greatly contributed to the continued progression of the selective evolution of resistant strains that otherwise would hardly predominate the population because of their reduced fitness compared to susceptible strains. The ever-increasing drug use in response to growing TB incidences has resulted in a steady evolution of *M. tuberculosis* strains that are progressively resistant to the available drugs. Besides the matter of selective survival, recent studies show that exposure of bacterial cells to sublethal levels of bactericidal antibiotics promotes

Table 1 Genes involved in acquired antibiotic resistance in *M. tuberculosis*

	Drug	Mode of action	Gene	Gene function	Role
First-line	Isoniazid	Inhibition of mycolic acid biosynthesis and other metabolic processes	*katG*	Catalase-peroxidase	Prodrug activation
			inhA	Enoyl ACP reductase	Drug target
			ndh	NADH dehydrogenase II	Activity modulation
			ahpC	Alkyl hydroperoxidase	Resistance marker
	Rifampicin	Inhibition of transcription	*rpoB*	β-subunit of RNA polymerase	Drug target
	Pyrazinamide	Inhibition of trans-translation	*pncA*	Pyrazinamidase	Prodrug activation
			rpsA	S1 ribosomal protein	Drug target
	Ethambutol	Inhibition of arabinogalactan synthesis	*embCAB*	Arabinosyltransferases	Drug target
			embR	*embCAB* transcription regulator	Drug target expression
	Streptomycin	Inhibition of translation	*rpsL*	S12 ribosomal protein	Drug target
			rrs	16S rRNA	Drug target
			gidB	16S rRNA methyltransferase	Target modification
Second-line	Amikacin/ Kanamycin	Inhibition of translation	*rrs*	16S rRNA	Drug target
			eis	Acetyltransferase	Drug modification
	Ethionamide	Inhibition of mycolic acid biosynthesis	*ethA*	Flavin monooxygenase	Prodrug activation
			inhA	Enoyl ACP reductase	Drug target
			ethR	*ethA* transcription repressor	Prodrug activator expression
			ndh	NADH dehydrogenase II	Activity modulation
			mshA	Glycosyltransferase	Prodrug activation
	Fluoroquinolones	Inhibition of DNA gyrase	*gyrA*	DNA gyrase subunit A	Drug target
			gyrB	DNA gyrase subunit B	Drug binding

cellular mutagenesis leading to increased mutation rates in other drug resistance genes (Kohanski et al. 2010a). This phenomenon, possibly mediated through the increased production of reactive oxygen species triggered by bactericidal antibiotics (Kohanski et al. 2007), might play a key role in the rapid emergence of multidrug resistance phenotypes in pathogenic bacteria such as *M. tuberculosis*. In light of this knowledge, many current TB drugs might well represent this double-edged sword. Both isoniazid and ethionamide require activation by redox enzymes in the mycobacterial cytoplasm to become inhibitory. This process produces reactive oxygen and/or radicals that exert the mycobactericidal activity (Fig. 1) (Ito et al. 1992; Wang et al. 1998). But once a mutant survives the killing action of reactive oxygen and radicals, these same chemical matters would enhance its mutability leading to the acquisition of additional drug resistance mutations. Whether or not, and to what level, reactive oxygen and radical species contribute to the rise of resistance mutations remains to be understood in *M. tuberculosis*.

2.1.1 New Insights into Pyrazinamide Action and Resistance

Pyrazinamide is a first-line TB antibiotic that is commonly used in combination with other drugs to shorten treatment regimens. Despite its widespread applications in TB chemotherapy, the precise mechanism underlying its action, as well as the mycobacterial mechanisms conferring pyrazinamide resistance had been elusive until recently. Similar to isoniazid and ethionamide, pyrazinamide is a prodrug and conversion to its active form pyrazinoic acid is catalyzed by the mycobacterial enzyme pyrazinamidase, encoded by the *pncA* gene (Scorpio and Zhang 1996). Loss of pyrazinamidase activity leads to pyrazinamide resistance, whereas overexpression confers increased susceptibility (Boshoff and Mizrahi 2000; Bamaga et al. 2002). Mapping of pyrazinamide resistance in *M. tuberculosis* clinical isolates found most mutations associated with *pncA* (Scorpio et al. 1997). However, there is a small subset of low-level pyrazinamide resistant strains that have no mutation in *pncA*, suggesting that these strains might carry mutations in genes encoding targets of pyrazinamide. In a search for *M. tuberculosis* proteins that bind the active molecule pyrazinoic acid, the 30S ribosomal protein S1 (RpsA) was identified (Shi et al. 2011). Overexpression of *rpsA* results in increased pyrazinamide resistance (Shi et al. 2011). Sequencing of a non-*pncA* pyrazinamide resistant strain revealed a 3-base pair in-frame deletion that leads to loss of Alanine 438 at the C terminus of RpsA (Shi et al. 2011). These observations suggested that RpsA is involved in *M. tuberculosis* pyrazinamide resistance.

Besides its essential ribosomal function required for protein translation, RpsA was previously found to be involved in trans-translation through its specific binding to transfer-messenger RNA (tmRNA) (Wower et al. 2000; Saguy et al. 2007). Trans-translation is a cellular process involved in rescuing ribosomes that are stalled during translation as well as in the degradation of the incomplete polypeptide chain and its messenger RNA (Keiler 2008). In this process, stalled

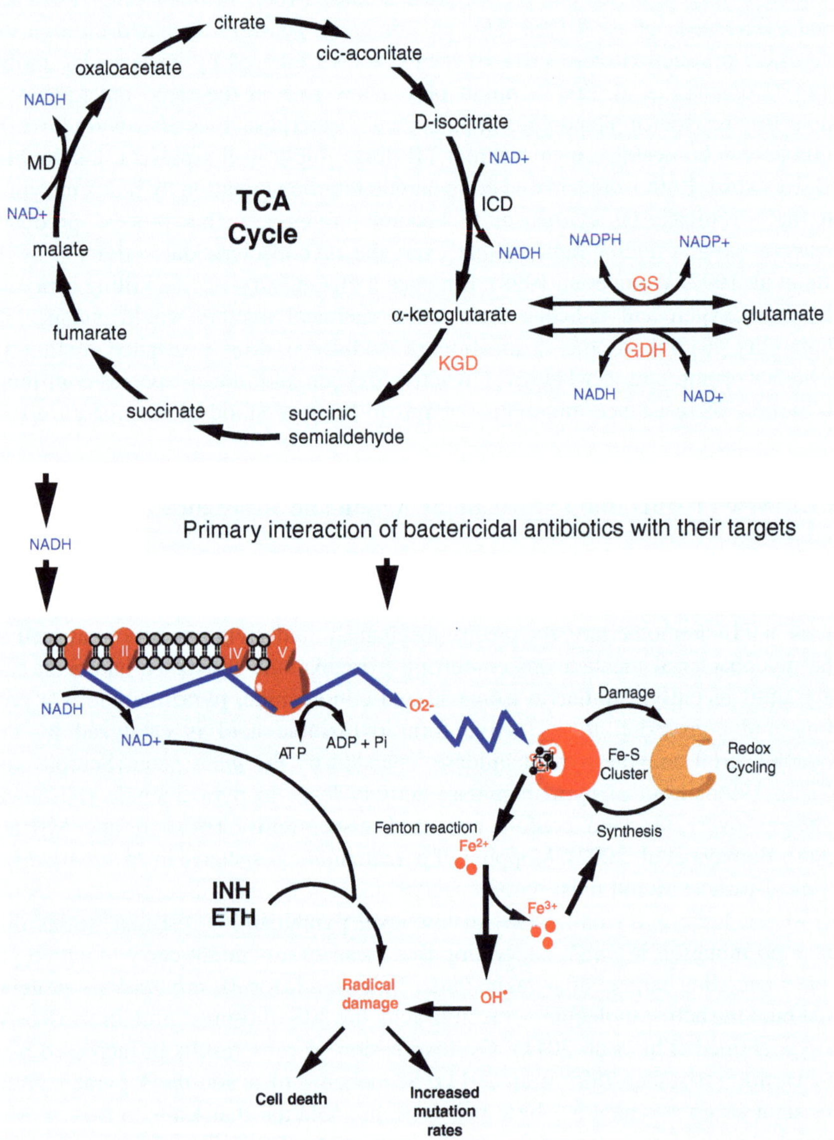
citrate
cic-aconitate
D-isocitrate
NAD+
ICD
NADH
oxaloacetate
NADH
MD
NAD+
malate
TCA
Cycle
fumarate
succinate
succinic
semialdehyde
α-ketoglutarate
KGD
NADPH
NADP+
GS
glutamate
GDH
NADH
NAD+
NADH
Primary interaction of bactericidal antibiotics with their targets
I
II
IV
V
NADH
NAD+
ATP
ADP + Pi
O2-
Damage
Fe-S
Cluster
Redox
Cycling
Synthesis
Fenton reaction
Fe2+
Fe3+
INH
ETH
Radical
damage
OH•
Cell death
Increased
mutation
rates

◀ **Fig. 1** Possible interconnections of the TCA cycle, cellular redox homeostasis, and action of bactericidal antibiotics in *M. tuberculosis*. The primary interactions between bactericidal antibiotics and their targets trigger oxidation of NADH, produced in the TCA cycle, through the electron transport chain. This leads to increased production of superoxide that destroys iron–sulfur clusters yielding iron for oxidation of the Fenton reaction. The Fenton reaction results in the formation of hydroxyl radicals that damage nucleic acids, proteins, and lipids, eventually leading to cell death. However, if a cell survives, these hydroxyl radicals increase its mutagenesis thus promoting the emergence of MDR strains. The current TB drugs isoniazid (INH) and ethionamide (ETH) kill mycobacteria via direct conversion to free radicals that may contribute to the formation of MDR *M. tuberculosis* strains. The recently discovered MDR determinant PknG might regulate activity of many enzymes of the TCA cycle via its phosphorylation of GarA. These enzymes (KGD, GS, and GDH, in *red*) might affect the cellular NADH pool required for the downstream electron transport chain triggered by bactericidal antibiotics thus leading to cell death. MD: malate dehydrogenase, ICD: isocitrate dehydrogenase. Redrawn with modifications from (Kohanski et al. 2007)

mRNA is displaced by tmRNA that encodes a short peptide tagging the stalled protein for subsequent degradation.

In vitro binding experiments confirmed that pyrazinoic acid, but not pyrazinamide prevents interaction between RpsA and tmRNA (Shi et al. 2011). In addition, the 3-base pair deletion found in *rpsA* of the non-*pncA* pyrazinamide resistant strain mentioned above abolishes pyrazinoic acid binding (Shi et al. 2011). Also, pyrazinoic acid was shown to inhibit trans-translation of the reporter protein dihydrofolate reductase (Shi et al. 2011). These results suggested that the molecular target of pyrazinamide is the trans-translation process that might be essential for stress survival and recovery from nutrient starvation (Keiler 2008). Large-scale sequencing needs to be done to evaluate the significance of *rpsA* mutations in clinical pyrazinamide resistant *M. tuberculosis* strains.

2.2 Fitness Compensation and the Inevitable Evolution to the Totally Drug Resistant Phenotype

Acquired antibiotic resistance mechanisms are often associated with a reduced fitness of the resistant mutants (Andersson and Levin 1999; Andersson 2006; Andersson and Hughes 2010). This "fitness cost" affects the development, as well as the stability and domination of the resistance phenotypes. It has been generally observed that epidemics of drug resistant *M. tuberculosis* strains mainly circulate among HIV positive patients, raising the hope that these mutants might never spread widely among healthy people. However, first studies using mathematical modeling, as well as those carried out in other organisms, predict that the fitness cost of *M. tuberculosis* drug resistant strains could be reduced by the subsequent appearance of "compensatory mutations", i.e., mutations that correct the fitness cost due to earlier mutations (Reynolds 2000; Andersson and Hughes 2010; Sergeev et al. 2012). This compensatory evolution, restoring fitness of resistant strains, might stabilize the epidemic of resistance phenotypes after they emerge. Indeed, in vitro studies suggest that resistant *M. tuberculosis* isolates are able to restore their fitness after prolonged exposure to antibiotics (Gillespie et al. 2002; Gagneux et al. 2006), although the identity of those mutations was not known. A recent study using whole-genome sequencing identified a set of compensatory mutations in the RNA polymerase genes of rifampicin resistant *M. tuberculosis* strains (Comas et al. 2012). In vitro growth competition assays confirmed that these mutations restore fitness of the rifampicin resistant *M. tuberculosis* strains compared to their susceptible counterparts. More convincingly, these mutations were mapped in 30 % of MDR *M. tuberculosis* strains isolated from regions of MDR TB prevalence (Comas et al. 2012).

TDR *M. tuberculosis* strains have been confirmed in India this year (Ormerod 2005; Dorman and Chaisson 2007; Udwadia 2012). Like MDR and XDR *M. tuberculosis* strains, the appearance of these TDR isolates is possibly inevitable with the increasing use of TB antibiotics. It is probably just a matter of time before these deadly bacteria appear more frequently in other places. The crucial question is whether compensatory evolution might allow stabilization of these drug resistance phenotypes, thus enhancing transmission rates of the resistant strains. The coevolution of drug resistance and fitness traits in *M. tuberculosis* therefore represents a deadly menace to humans.

3 Intrinsic Antibiotic Resistance Mechanisms

Besides the ability to acquire new resistance through the acquisition of chromosomal mutations, *M. tuberculosis* is endowed with an array of intrinsic resistance mechanisms that allow active neutralization of antibiotic actions. These resistance mechanisms provide a high resistance background that not only limits the application of available antibiotics to TB treatment, but also hampers the development of new drugs. *M. tuberculosis* intrinsic drug resistance can be divided into two categories: passive resistance and specialized resistance mechanisms.

3.1 Passive Resistance Mechanisms

Similar to the problems encountered in drug development and chemotherapy of Gram-negative bacteria, the impermeable cell wall of mycobacteria functions as an effective barrier for the penetration of antibiotics. Although mycobacteria are classified as Gram-positive bacteria, their cell wall is extremely thick and multilayered with varied hydrophobicity. These layers create an interlayer space similar to the periplasm of the Gram-negative bacterial cell wall (Hoffmann et al. 2008; Zuber et al. 2008). The peptidoglycan sacculus is covered by an arabinogalactan layer, both of which are hydrophilic, thus preventing the transport of hydrophobic molecules (Brennan and Nikaido 1995). These two layers are covalently linked to an outward layer of mycolic acids, long chain fatty acids that form a waxy, nonfluid barrier restricting penetration of both hydrophobic and hydrophilic molecules (Liu et al. 1995). For example, diffusion of β-lactams through the mycobacterial cell wall is hundreds of fold slower than the penetration through that of *Escherichia coli* (Kasik and Peacham 1968; Chambers et al. 1995).

The role of the mycobacterial cell wall in intrinsic antibiotic resistance is well demonstrated by the studies of mutants defective in cell wall biosynthesis. Pioneering work by the Nikaido laboratory identified a *M. smegmatis* mutant defective in mycolate biosynthesis. This mutant exhibited increased uptake of and sensitivity to erythromycin, chloramphenicol, novobiocin, and rifampicin (Liu and Nikaido 1999). Also, transposon mutagenesis studies confirmed the role of cell wall integrity in mycobacterial intrinsic drug resistance (Gao et al. 2003; Philalay et al. 2004). For example, transposon insertions in *kasB* or the *virS-mymA* operon (*rv3082 to rv3089*), genes that were defined to be involved in mycolic acid biosynthesis, lead to increased chemical penetration and sensitivity to various antibiotics (rifampicin, ciprofloxacin, isoniazid and pyrazinamide) (Gao et al. 2003; Singh et al. 2003, 2005). The ligation of mycolic acids to sugar moieties (arabinogalactan or trehalose) in the cell wall is catalyzed by a family of redundant mycolyltransferase enzymes initially known as "the antigen 85 complex" (Belisle et al. 1997). Deletion of the *fbpA* gene encoding one of the mycolyltransferases results in reduced levels of trehalose dimycolates and increased sensitivity to antibiotics widely used for antibacterial chemotherapy (Nguyen et al. 2005). These observations confirm that the mycobacterial cell wall plays an important role in mycobacterial intrinsic resistance against antibiotics. However, because the doubling time of *M. tuberculosis* is extremely long, the slow penetration rate might in some cases be high enough to allow antibiotics to accumulate to inhibitory levels well before cell division occurs, thus making cell wall permeability an important but not a decisive determinant of drug resistance (Brennan and Nikaido 1995; Chambers et al. 1995; Quinting et al. 1997).

Similar to the cell wall of Gram-negative bacteria, porins are mounted to outer layers of the mycobacterial cell wall, thus allowing import of nutrients and small molecules required for growth (Niederweis 2003). These porins might also play a role in the import of antibiotics through the outer layer of the mycobacterial cell wall (Danilchanka et al. 2008). *In trans* expression of the major porin MspA from *M. smegmatis* increases susceptibility of *M. tuberculosis* and *M. bovis* to β-lactams, isoniazid, ethambutol, and streptomycin (Stephan et al. 2004). By contrast, deletion of *mspA* or *mspC* in *M. smegmatis* leads to increased resistance to not only hydrophilic, but also hydrophobic and large antibiotics such as vancomycin, erythromycin, and rifampicin (Stephan et al. 2004; Danilchanka et al. 2008). Although *M. tuberculosis* encodes at least two porin-like proteins, OmpA and Rv1698 (Senaratne et al. 1998; Siroy et al. 2008), and *in trans* expression of *rv1698* restores antibiotic susceptibility to a *M. smegmatis mspA* mutant (Siroy et al. 2008), the role of porins in antibiotic uptake and susceptibility has not been demonstrated directly in *M. tuberculosis*.

3.2 Specialized Resistance Mechanisms

Besides the cell wall barrier that helps slow down the penetration of antibiotics (Nikaido 1994), *M. tuberculosis* and other mycobacterial species operate multiple specialized resistance machineries that allow active detoxification of drugs once they reach the cytoplasmic space.

3.2.1 Modification of Drug Targets

Pathogenic bacteria are able to avoid antibacterial activity of antibiotics through structural modifications of their targets, thereby reducing antibiotic binding affinity. The mechanism that confers the intrinsic resistance of *M. tuberculosis* to macrolide and lincosamide antibiotics represents a typical example for this type of resistance. *M. tuberculosis* and other mycobacterial species are naturally resistant to macrolides and lincosamides. These antibiotics stop the growth of bacterial cells through inhibitory action on the protein synthetic machinery. They bind reversibly to a specific site of the ribosomal RNA in the 50S subunit of bacterial ribosomes, thus inhibiting translocation of peptidyl-tRNA (Buriankova et al. 2004). It has been observed that the Pasteur vaccine strain Bacillus of Calmette and Guérin (BCG) is uniquely susceptible to many macrolides and lincosamides, whereas its parental strain *M. bovis* and other vaccine strains remain resistant to the antibiotics. Through comparative genomics, it was realized that the sensitivity of BCG to macrolides and lincosamides is due to its chromosomal deletion of the *erm37* gene encoding a ribosomal RNA methyltransferase (Buriankova et al. 2004). *erm37* is located within a larger chromosomal locus known as Region of Difference 2 (RD2) which was deleted in BCG during its culture passage. Genetic experiments showed that the macrolide–lincosamide resistance of BCG could be restored to the level of its parental *M. bovis* strain by *in trans* expression of the *erm37* gene from *M. tuberculosis* (Buriankova et al. 2004). Chemical analyses showed that this enzyme alters structures of the *M. tuberculosis* ribosomes via methylation of 23S ribosomal RNA (Madsen et al. 2005). In vitro macrolide binding assays also confirmed that Erm37 activity reduces affinity of macrolides to the ribosomes, thus lowering the inhibitory activity of macrolides on protein synthesis (Buriankova et al. 2004). Other *erm* genes conferring macrolide and lincosamide resistance were also identified in *M. smegmatis* and *M. fortuitum* (Nash 2003; Nash et al. 2005). Interestingly, expression of *erm* genes in mycobacteria is inducible by exposure to macrolides and lincosamide antibiotics (Nash 2003; Nash et al. 2005; Andini and Nash 2006), suggesting that their function is specialized for macrolide and lincosamide resistance. This inducible expression of Erm37 is most likely controlled by the MDR transcription regulator WhiB7 (Morris et al. 2005; Burian et al. 2012), which is discussed elsewhere in this chapter.

Another example demonstrating the ability of *M. tuberculosis* to neutralize drug action through enzymatic modification of its target is the methylation of ribosomal RNA that mediates resistance to cyclic peptide antibiotics. Capreomycin and viomycin are commonly used to treat MDR TB, but clinical strains resistant to these drugs have been isolated. Genetic studies carried out with *M. smegmatis* and *M. tuberculosis* found that mutations associated with capreomycin resistance are mapped to the *tlyA* gene that encodes a 2'-O-methyltransferase (Maus et al. 2005). TlyA methylates both 16S and 23S ribosomal RNA at nucleotide C1409 and C1920, respectively (Johansen et al. 2006). These methylations render mycobacterial ribosomes susceptible to the binding of capreomycin and viomycin (Maus et al. 2005; Johansen et al. 2006). Inactivation of *tlyA* led to increased resistance to these cyclic peptides, whereas *in trans* complementation restored capreomycin susceptibility (Maus et al. 2005).

3.2.2 Chemical Modification of Drugs

Mycobacteria are also able to inactivate antibiotics via direct chemical modifications. Recent studies have revealed the importance of acetylation on mycobacterial resistance to aminoglycosides. Aminoglycosides are broad-spectrum antibiotics that can act either as bactericidal or bacteriostatic drugs depending on concentration. This group of antibiotics constitutes an important position in the history of TB chemotherapy. Whereas streptomycin was the first effective remedy for TB, kanamycin and amikacin are currently used as last resorts to treat MDR TB cases. In addition, resistance to these drugs in MDR strains is the hallmark to define XDR TB. Aminoglycosides may have multiple potential mechanisms of action. Although some of them have been shown to act as inhibitors of protein biosynthesis, the precise mechanism of action of aminoglycosides remains to be understood. Early studies in *M. smegmatis* and *M. fortuitum* identified homologs of aminoglycoside 2'-N-acetyltransferase (*aac*) that confer resistance to gentamicin, dibekacin, tobramycin, and netilmicin (Ainsa et al. 1996). Although a homolog of AAC is apparently present in *M. tuberculosis* (Vetting et al. 2003), its function in aminoglycoside resistance has not been demonstrated. Surprisingly, the intrinsic resistance of *M. tuberculosis* to aminoglycosides has been recently attributed to a different acetyltransferase (Zaunbrecher et al. 2009). The protein termed Enhanced Intracellular Survival (EIS) was first discovered as a determinant of mycobacterial survival in host macrophages (Wei et al. 2000).

High-level kanamycin resistance in *M. tuberculosis* isolates has been mapped to the 16S ribosomal RNA gene *rrs*. However, the majority of isolates that display lower levels of kanamycin resistance have no mutation in *rrs*. To study mechanisms underlying kanamycin resistance in these isolates, a cosmid library was constructed using genomic DNA from one of these kanamycin resistant *M. tuberculosis* strains (Zaunbrecher et al. 2009). The cosmid library was transformed into a kanamycin susceptible *M. tuberculosis* strain and kanamycin resistant transformants were obtained. Further mapping of the cosmid, which conferred kanamycin resistance, identified mutations within the *eis* promoter region, which increased transcription of *eis* by 180-fold (Zaunbrecher et al. 2009). Importantly, these mutations were found in 80 % of low-level kanamycin resistant clinical isolates (Zaunbrecher et al. 2009; Campbell et al. 2011; Engstrom et al. 2011), as well as in MDR *M. tuberculosis* isolates (Huang et al. 2011). In addition, in vitro studies showed that Eis acetylates multiple amine groups of aminoglycosides using acetyl-coenzyme A as an acetyl donor (Chen et al. 2011), thereby inactivating the antibiotics. The dual function of EIS in the protection of *M. tuberculosis* against mycobactericidal mechanisms of both host immunity and antibiotics indicates a sinister coevolution of these two traits under the pressure of the recent antibiotic use.

3.2.3 Enzymatic Degradation of Drugs

Another pathway that pathogenic bacteria commonly use to subvert antibiotics is to directly degrade them using hydrolases. This mechanism is best studied in the case of β-lactams. These antibiotics bind and inhibit the activity of penicillin binding proteins (PBPs) involved in the assembly of the peptidoglycan network, thereby disrupting cell wall biosynthesis and leading to cell death. There are at least four major PBPs encoded in the genome of *M. tuberculosis*. These proteins bind β-lactams at clinically achievable concentrations (Chambers et al. 1995), indicating that target affinity is not an important determinant of β-lactam resistance in mycobacteria. The cell wall of mycobacterial species clearly contributes to the reduced accessibility of β-lactams to their targets. In this regard, the slow growth rate of *M. tuberculosis* serves both as a negative and positive factor in contributing to β-lactam resistance. For example, carbapenems are relatively unstable, and therefore lose activity much faster than the mycobacterial growth rate. However, daily uptake could lead to lethal concentrations that are sufficient to inhibit the slow cell division machinery of mycobacteria (Watt et al. 1992). This makes the cell wall barrier an important but not the principal determinant of β-lactam resistance.

β-lactamases, which hydrolyze the β-lactam ring of β-lactams, have been proven to provide the paramount β-lactam resistance in mycobacteria (Chambers et al. 1995). This conclusion is most supported by studies in *M. fallax*, the only *Mycobacterium* species highly susceptible to β-lactams (Kasik 1979; Quinting et al. 1997). Permeability assays first showed that β-lactam penetration through the cell wall of *M. fallax* is similar to those observed in other mycobacteria. This barrier apparently slows down accessibility of β-lactams to their targets, but half equilibration times should allow accumulation of the drugs at concentrations lethal to the bacterium (Quinting et al. 1997). However, when the β-lactamase from *M. fortuitum* was *in trans* expressed in *M. fallax*, its resistance was elevated to levels comparable to other mycobacterial species (Quinting et al. 1997). It is important to note that mycobacterial β-lactamases are generally considered less active than those of other pathogenic bacteria. However, the slow penetration of β-lactams across the thick cell wall of mycobacteria renders this low β-lactamase activity effective enough to protect mycobacteria from *β*-lactam action (Jarlier et al. 1991).

The most important *β*-lactamase in *M. tuberculosis* is BlaC, which belongs to the Ambler class-A *β*-lactamases (Voladri et al. 1998; Wang et al. 2006), whose enzymology and structures have been thoroughly characterized (Voladri et al. 1998; Wang et al. 2006). BlaC exhibits broad substrate specificity, possibly due to its large and flexible substrate-binding site (Wang et al. 2006). In addition, *M. tuberculosis* BlaC displays measurable activity to carbapenems, which are generally resistant to *β*-lactamases of other pathogenic bacteria (Hugonnet and Blanchard 2007; Tremblay et al. 2010). Also, *β*-lactamase inhibitors such as clavulanic acid are less effective against BlaC compared to other class A enzymes. Besides BlaC, *M. tuberculosis* encodes at least three more *β*-lactamase genes: *blaS*, *rv0406c,* and *rv3677c*, which were shown to provide *M. tuberculosis* with lower *β*-lactamase activities (Flores et al. 2005; Nampoothiri et al. 2008). Expression of BlaC in *M. tuberculosis* is inducible by *β*-lactams (Sala et al. 2009), indicating that this system has been specialized for *β*-lactam resistance. The induction of BlaC was recently shown to be mediated through BlaI, a winged helix regulator that functions as a transcriptional repressor of *blaC* expression. In the absence of *β*-lactams, BlaI forms homodimers that bind the promoter of the *blaC* gene, thus inhibiting *blaC* transcription (Sala et al. 2009). When *M. tuberculosis* is exposed to *β*-lactams, BlaI is dissociated from its DNA binding site, thereby derepressing the transcription of *blaC* that confers increased *β*-lactam resistance (Sala et al. 2009).

3.2.4 Molecular Mimicry of Drug Targets

Molecular mimicry represents a fascinating mechanism that *M. tuberculosis* might use to neutralize action of fluoroquinolones. These synthetic antibiotics have recently become important drugs for the treatment of drug resistant TB cases (Duncan and Barry 2004). Fluoroquinolones are bactericidal drugs that kill bacteria through inhibition of DNA replication, transcription, and repair. These antibiotics interact with DNA gyrase or topoisomerase in complexes with DNA, thereby stabilizing DNA breaks while inhibiting resealing of DNA strands. These events eventually result in DNA degradation and cell death (Andriole 2005).

Similar to other bacteria, acquired fluoroquinolone resistance in *M. tuberculosis* is commonly mapped to mutations in the genes that encode DNA gyrases, *gyrA,* and *gyrB* (Table 1). However, the molecular mechanisms responsible for the intrinsic fluoroquinolone resistance in mycobacteria have not been well understood. A *M. smegmatis* protein (MfpA) was first identified to confer low-level resistance to fluoroquinolones. Overexpression of *mfpA* from a multicopy plasmid results in increased resistance to ciprofloxacin and sparfloxacin in *M. smegmatis* and *M. bovis* (Montero et al. 2001). By contrast, deletion of *mfpA* leads to reduced fluoroquinolone resistance, indicating that the resistance level is dependent on *mfpA* expression. The sequence of MfpA has highest homology to pentapeptide repeat proteins in which every fifth amino acid is either leucine or phenylalanine. Interestingly, structure of the *M. tuberculosis* MfpA resembles the 3D structure of the DNA double helix (Ferber 2005; Hegde et al. 2005), with the tandem of pentapeptide repeats coiling around in a right-handed helix of the same width as DNA (Morais Cabral et al. 1997; Hegde et al. 2005). It was suggested that MfpA mimics DNA structure to sequester fluoroquinolones in the cytoplasm, thus setting DNA free from the drug attack (Ferber 2005). However, the physiological function of MfpA and how MfpA contributes to fluoroquinolone resistance in the clinic remain to be established. Nevertheless, this finding reveals a fascinating capability of pathogenic bacteria in evolving novel resistance mechanisms against antibiotics.

3.2.5 Drug Deportation By Efflux Pumps

An active mechanism that commonly provides protection against antibiotics in pathogenic bacteria is to expel the reagents using efflux pumps. Most of these membrane-spanning proteins play roles in the physiology or metabolism of bacteria; for example, transporting of nutrients, toxins, wastes, or signaling molecules through the cell wall. Therefore, the functions in antibiotic resistance of many transporters might be secondary and due to nonspecific transportation. Indeed, 20 out of the total of 36 genes encoding membrane transport proteins in the *E. coli* genome confer some levels of resistance to one or more antibiotics (Nishino and Yamaguchi 2001). However, these transporters are commonly controlled by regulatory systems that might have evolved to respond to antibiotics, thus making transport activities specialized for antibiotic resistance. For example, a major MDR determinant in *E. coli* is AcrB, a transporter of a broad substrate specificity. However, activity of AcrB has evolved as its expression is controlled by three antibiotic-responsive regulatory systems: Mar, Sox, and Rob (Alekshun and Levy 1997).

There are at least 18 transporters thus far characterized to be involved in antibiotic susceptibility of mycobacterial species (Viveirosa et al. 2012). For more specific information related to these transporters and their role in antibiotic resistance, readers are referred to many recent excellent reviews (da Silva et al. 2011; Viveirosa et al. 2012). Most of these transporters confer low-level antibiotic resistance to mycobacteria. Similar to *E. coli*, activities of some of these transporters are coupled to regulatory systems that are responsive to antibiotics. For example, two mycobacterial transporters, encoded by the *iniBAC* and *efpA* genes, are negatively controlled by the transcription regulator Lsr2 (Colangeli et al. 2007). Whereas IniBAC confers resistance to isoniazid and ethambutol, EfpA is rather a nonspecific MDR transporter (Colangeli et al. 2005, 2007). Importantly, the transcriptional control of IniBAC and EfpA by Lsr2 is inducible by isoniazid or ethambutol (Colangeli et al. 2007), indicating a functional evolution toward antibiotic resistance. Another example is the positive regulation of Tap, a transporter of aminoglycosides and tetracycline, by the MDR regulatory protein WhiB7 (Ainsa et al. 1998; Morris et al. 2005). Expression of WhiB7 and its regulon including Tap is induced when *M. tuberculosis* is exposed to tetracycline, streptomycin, or erythromycin (Morris et al. 2005; Burian et al. 2012). Interestingly, a recent study showed that expression of multiple efflux pumps including Tap is induced in mycobacteria residing within host granulomas, indicating that these molecular pumps may contribute to the drug tolerance of *M. tuberculosis* in latent TB (Adams et al. 2011).

3.3 Redox Homeostasis and the Mycobactericidal Activity of Drugs

Pioneering work in *E. coli* and *Salmonella* established the relationship of antibiotic resistance and oxidative stress responses (Demple 2005). These two phenotypes are controlled by common regulatory proteins such as SoxRS, MarRAB, or Rob. Although a clear reason for this regulatory overlap has not been established, it is conserved in both *E. coli* and *Salmonella* and believed to provide evolutionary advantages to these bacteria in the defense against general toxic environments (Demple 2005). Growing evidence recently revealed a tight relationship of killing mechanisms of antibiotics and cellular redox homeostasis in bacteria (Kohanski et al. 2007, 2010a, b), which could help explain the co-evolution of regulatory systems to control these two phenotypes. In *E. coli*, exposure to bactericidal antibiotics stimulates the production of hydroxyl radicals that ultimately lead to cell death (Kohanski et al. 2007). The terminal production of hydroxyl radicals triggered by bactericidal antibiotics is mediated through complex sequential events starting with the tricarboxylic acid (TCA) cycle. First, NADH produced in the TCA cycle is oxidized via complex electron transport chains, which results in the production of superoxide. Superoxide then damages iron–sulfur clusters, thereby donating ferrous iron to the oxidation of the Fenton reaction that produces hydroxyl radicals. Finally, hydroxyl radicals damage DNA, proteins, and lipids, resulting in cell death (Fig. 1) (Kohanski et al. 2007).

In mycobacteria, the relationship between oxidative stress responses and antibiotic resistance has also been implicated through the studies of the prodrugs isoniazid and ethionamide. Oxidative stress response proteins such as KatG and AhpC were found to be important for the oxidative activation of these antibiotics (Zhang et al. 1992, 1996; Sherman et al. 1996). In addition, expression of the stress responsive sigma factor F (SigF) is induced by antibiotics (Michele et al. 1999). Also, mycothiol, the actinobacterial specific thiol molecule involved in mycobacterial protection against oxygen toxicity, is also required for antibiotic resistance (Rawat et al. 2002; Buchmeier et al. 2003; Vilcheze et al. 2008). More recently, the relationship between redox homeostasis and antibiotic resistance has become clearer from the studies of the MDR systems Lsr2 and WhiB7 (Morris et al. 2005; Colangeli et al. 2009; Burian et al. 2012). Lsr2 was previously reported to control expression of the drug efflux pump IniBAC discussed above (Colangeli et al. 2005, 2007). Interestingly, a recent paper suggested that Lsr2 protects *M. tuberculosis* from reactive oxygen intermediates, possibly through its histone-like activity that protects DNA from the attacks of hydroxyl radicals (Colangeli et al. 2009). WhiB7 is a MDR determinant found only in actinobacteria including streptomycetes and mycobacteria. Deletion of *whiB7* in mycobacteria leads to increased susceptibility to multiple antibiotics, whereas overexpression results in elevated resistance levels, indicating that WhiB7 activity is directly linked to the multidrug resistance phenotype (Morris et al. 2005). The *M. tuberculosis* WhiB7 protein is a 122-amio acid iron–sulfur cluster carrying protein. Transcriptomic analysis showed that WhiB7 affects transcription of multiple structural antibiotic resistance genes including *eis*, *erm37*, and *tap* that are discussed elsewhere in this chapter. Interestingly, expression of *whiB7* could be induced by exposure to not only antibiotics such as erythromycin and tetracycline but also the reducing reagent dithiothreitol (Burian et al. 2012), indicating its function in antibiotic resistance is responsive to cellular redox fluctuations. Combination of antibiotics and dithiothreitol further elevates *whiB7* expression. By contrast, an oxidative reagent diamide inhibits *whiB7* induction (Burian et al. 2012). These experiments indicate the importance of a reductive environment for WhiB7 activity. Besides WhiB7 and Lsr2, another MDR protein whose function might relate to oxidative stress and the TCA cycle was recently reported in mycobacteria. The eukaryotic-type protein kinase G (PknG) is required for the intrinsic multidrug resistance in *M. smegmatis* and *M. tuberculosis* (Wolff et al. 2009). Either genetic deletion or chemical inhibition of PknG kinase activity results in increased susceptibility of mycobacteria to multiple antibiotics (Wolff et al. 2009). Interestingly, PknG was previously reported to mediate survival of pathogenic mycobacteria in host macrophages (Walburger et al. 2004). These findings indicate that traits related to antibiotic resistance and virulence might be interconnected in *M. tuberculosis*. Although the precise molecular mechanism underlying the function of PknG in mycobacterial antibiotic resistance has remained unknown, the possible involvement of this kinase in the control of the TCA cycle might provide an explanation (Fig. 1). Together with some other kinases, PknG was found to phosphorylate OdhI, an inhibitor of the 2-oxoglutarate dehydrogenase (ODH) complex of the TCA cycle, in *Corynebacterium glutamicum* (Niebisch et al. 2006). ODH catalyzes the NAD^+-dependent conversion of 2-oxoglutarate (or α-ketoglutarate) to succinyl CoA. Phosphorylated OdhI is no longer able to inhibit ODH (Niebisch et al. 2006). In *M. tuberculosis*, the TCA cycle lacks ODH activity. Instead, α-ketoglutarate is first converted into succinic semialdehyde by α-ketoglutarate decarboxylase (KGD), before further conversion to succinate (Fig. 1) (Tian et al. 2005). The *M. tuberculosis* OdhI homolog, termed GarA, is also phosphorylated by multiple mycobacterial kinases including PknG. Unphosphorylated GarA was suggested to act as an inhibitor of KGD and NAD^+-dependent glutamate dehydrogenase (GDH), yet as an enhancer of glutamate synthase (GS) activity (Nott et al. 2009; Chao et al. 2010). These enzymes are required for the interconversion of NADH and NAD^+ (Fig. 1). Therefore, PknG kinase activity may affect the cellular NADH pool that is required for the superoxide production triggered by bactericidal antibiotics leading to eventual cell death (Kohanski et al. 2007,

2010b). It remains to be established whether this model is correct and how PknG affects this NADH-mediated mycobactericidal action of TB antibiotics.

4 Epigenetic Drug Tolerance

Besides the acquired and intrinsic resistance mechanisms that are defined by genetic determinants, epigenetic drug tolerance has been observed in *M. tuberculosis*, often associated with latent or relapsed TB cases. This phenotypic drug tolerance is associated with the phenomenon of persisters, a small subpopulation of cells with distinct but as-yet-unknown metabolic or physiological states. Persisters are genetically identical with susceptible counterparts and able to convert into susceptible cells upon restoration of normal environments (Lewis 2008; Zhang 2012).

Whereas drug resistance mechanisms result in reduced access of an antibiotic to its target, drug tolerance displayed by persisters might be associated with low metabolic activities or cell division during dormancy. This leads to a reduction of the cellular requirement for the proteins or machineries that are targeted by the antibiotic (Lewis 2008). Therefore, binding and inhibition of antibiotics to targets might occur normally, but the inhibition is no longer lethal to the dormant bacterium (Lewis 2008). In light of the recent knowledge of how bactericidal antibiotics kill bacterial cells (Kohanski et al. 2007, 2010b), inactive metabolic pathways such as a stalled TCA cycle might prevent cell death caused by those antibiotics. Once a persister re-enters active growth, cellular requirement for the targeted proteins is resumed, thus resensitizing the metabolically active cell to antibiotics. *M. tuberculosis* enters a dormant state during latent infection that is characterized by a shutdown of most of its metabolism, leading to increased antibiotic tolerance (Gomez and McKinney 2004; Gengenbacher and Kaufmann 2012). Indeed, latent infections possibly contain persister cells that are more difficult to treat than active TB normally caused by actively dividing bacilli (Gomez and McKinney 2004; Gengenbacher and Kaufmann 2012). *M. tuberculosis* isolates obtained from TB patients that had shown prolonged persistence or treatment relapse showed significantly increased tolerance to isoniazid, ethambutol, and rifampicin compared to isolates from regular patients, but this tolerance was strictly phenotypic (Wallis et al. 1999). Also, persister-like bacilli that replicate very slowly or completely stop growth have been isolated from TB sputum (Garton et al. 2008). Transcriptome analysis of these cells indicated that drug tolerance is due to low metabolic rates rather than resistance mutations (Garton et al. 2008).

A dormant-like state could be induced in vitro by mimicking environmental aspects of *M. tuberculosis* persisters surviving during latent TB. Probably, the best known of such methods is the Wayne model in which gradual oxygen depletion results in a stepwise transition to a dormant cell population that exhibits very low or absent metabolic activities and increased drug tolerance (Wayne and Hayes 1996). Another system uses nutrient starvation to obtain nonreplicating *M. tuberculosis* cells, which display down-regulated respiration and global metabolism but high levels of multidrug tolerance (Betts et al. 2002; Xie et al. 2005). In addition, exposure of *M. tuberculosis* to antibiotics such as D-cycloserine allows isolation of drug tolerant persisters whose transcriptome displays a downregulation of metabolic pathways characteristic of bacterial persisters (Keren et al. 2011). While all of these models share certain features, such as reduced intracellular ATP level and increased lipid metabolism, it is not surprising that other markers vary, possibly because of different inducing conditions (Keren et al. 2011; Gengenbacher and Kaufmann 2012). Nevertheless, these studies confirmed the relationship of phenotypic drug resistance and low metabolic activity during *M. tuberculosis* dormancy in the host.

When bacteria are growing in vitro in the absence of antibiotics, persisters can be steadily isolated with frequencies increasing in stationary phase (Keren et al. 2004, 2011; Hansen et al. 2008; Lewis 2008). Thus, persister formation is likely an intrinsic characteristic of a bacterial population that is related to community heterogeneity. It is thought that persisters function as cells that sacrifice active growth in order to ensure survival of the population in disastrous events such as antibiotic exposure (Keren et al. 2004; Lewis 2008). A recent study demonstrated culture heterogeneity in *M. smegmatis* arisen from asymmetric cell division and cell aging (Fig. 2) (Aldridge et al. 2012). After a cell division event, only one of the daughter cells contains the pre-existing division pole. and thus quickly progresses through another cycle of division, whereas the second daughter cell must first form a new growing pole of its own (Aldridge et al. 2012). The latter cell also elongates more slowly than the former, but it is more tolerant to antibiotics until the next division event (Fig. 2) (Aldridge et al. 2012). Distinct subpopulations of the diversified cultures thereby display varied levels of antibiotic susceptibility (Aldridge et al. 2012) that might be related to the formation of persisters triggered by antibiotic exposures.

The precise molecular mechanisms underlying the formation of drug tolerant persisters in vivo remain largely unknown. While it is extremely difficult to study the *M. tuberculosis* persisters residing in granulomas of human TB infections, animal studies and in vitro dormancy systems might help to advance our understanding. In the Cornell mouse model originally described in the 1950s, infected mice are treated with isoniazid and pyrazinamide until mice show no sign of active disease and no detectable bacilli by organ culture (McCune et al. 1956; McCune and Tompsett 1956). Reactivation of the disease occurs spontaneously following cessation of the treatment or triggered by immunosuppression, indicating the existence of drug tolerant persisters that are not cleared by the regimens (McCune et al. 1956; McCune and Tompsett 1956; Scanga et al. 1999). A similar mouse model infected with transposon insertion *M. tuberculosis* mutants was recently used to screen for mycobacterial genes involved with isoniazid persistence (Dhar and McKinney 2010). Using this animal infection model, genes required for isoniazid persistence have been identified. For example, *cydC*, encoding a putative ATP-binding subunit of an ABC transporter, which is upregulated during chronic infections in mice (Shi et al. 2005), was found to be required for survival in isoniazid treatment (Dhar and McKinney 2010). Importantly, survival of *cydC*

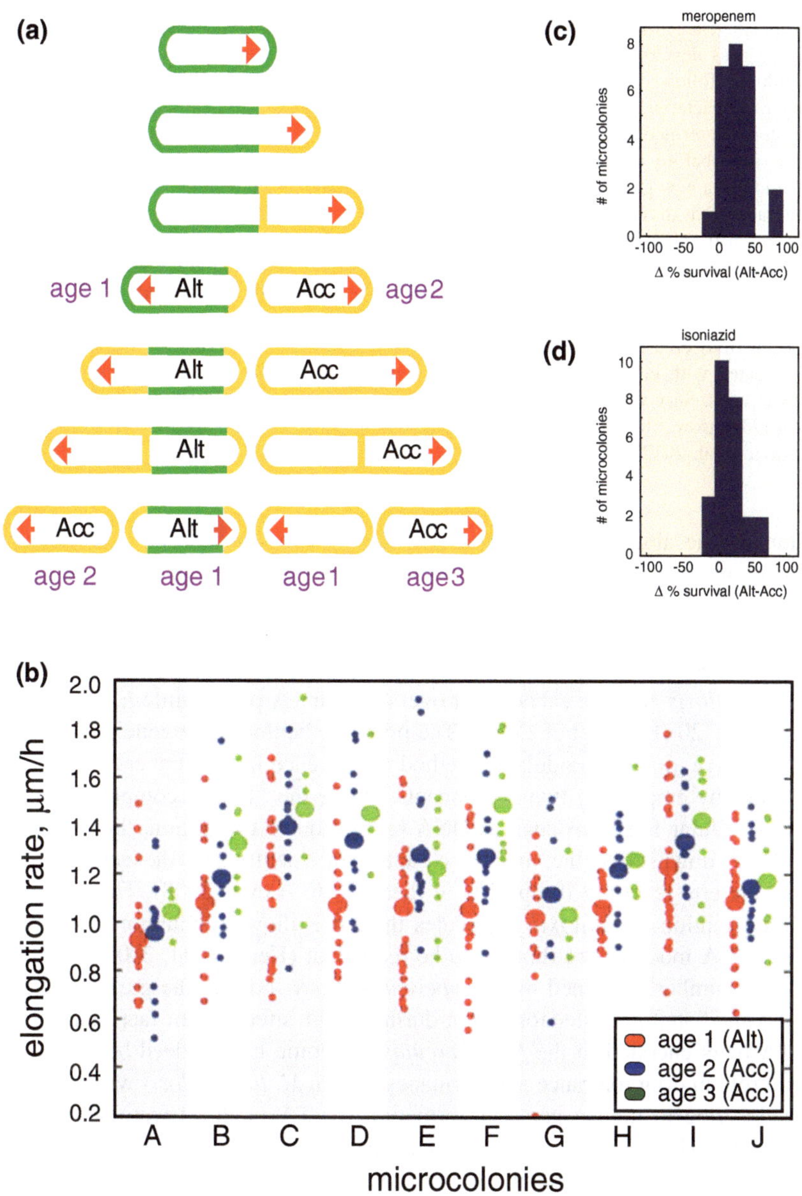
(a)
age 1
Alt
Acc
age2
Alt
Acc
Alt
Acc
Acc
Alt
Acc
age 2
age 1
age1
age3
(c)
meropenem
# of microcolonies
Δ % survival (Alt-Acc)
(d)
isoniazid
# of microcolonies
Δ % survival (Alt-Acc)
(b)
elongation rate, μm/h
A
B
C
D
E
F
G
H
I
J
microcolonies
age 1 (Alt)
age 2 (Acc)
age 3 (Acc)

◀ **Fig. 2** Asymmetric cell division introduces population heterogeneity in growth rate and drug tolerance levels. A microfluidic chamber, which allows culturing mycobacterial cells for live-cell imaging, was developed. Growth from single cells (microcolonies) up to five generations were individually followed and analyzed. **a** Schematic model of growth and division starting from a single mycobacterial cell. Elongation occurs at one pole (*red arrow*) that is inherited by only one daughter cell, termed accelerator (Acc). The second daughter cell, or alternator (Alt), has to form a new pole before elongation and cell division can occur again. This requirement for the formation of a new growing pole results in a downshift of the cell's elongation rate compared to accelerators that divide at unchanged speed due to the inherited pole. **b** The elongation rate of cells within 10 microcolonies is grouped dependent on cell pole ages (see **a**). Means of elongation rates are indicated by the larger ovals, distributions are indicated by smaller ovals. **c** and **d** Microcolonies in which alternators outnumber accelerators (*green area, right*) exhibit increased drug persistence compared to those in which accelerators dominate (*orange area, left*). This is likely due to decreased elongation rates and metabolic activity of alternator cells. Microcolonies were treated with either meropenem (**c**) or isoniazid (**d**) at concentrations lethal for actively growing cells. Survival was determined by measuring the number of cells in each microcolony that could regrow after the termination of antibiotic treatment. Reproduced with permission from (Aldridge et al. 2012)

mutant in the absence of isoniazid was identical to that of wild-type *M. tuberculosis*, indicating the function of the CydC is specialized for antibiotic persistence.

In other bacteria, toxin–antitoxin (TA) systems have been identified as the most obvious mechanism likely involved in persister formation. Interestingly, the *M. tuberculosis* genome encodes for over 65 such TA pairs, while *E. coli* encodes for roughly 20 (Keren et al. 2011). The principle behind these gene pairs consists of a sensitive, quickly modulated method to influence large numbers of genes and pathways by employing two components. While the "toxin" component inhibits critical cellular functions such as DNA replication or translation, thus leading to metabolic downshifts, the matching "antitoxin" neutralizes the activity of the toxin, thereby resulting in upshifts of metabolism (Lewis 2008). The production and degradation of antitoxin molecules therefore allows regulation of dormancy triggers. TA modules are thought to be redundant (Keren et al. 2004, 2011), and the high number contained by *M. tuberculosis* may indicate the great importance for TA-mediated persister formation during latent infection. In fact, many of the 65 TA pairs encoded in the *M. tuberculosis* genome have indeed been shown to play a role in drug tolerance and dormancy. When *M. tuberculosis* was exposed to D-cycloserine to induce persister formation, 10 TA modules were shown to be upregulated (Keren et al. 2004, 2011). Among those induced was *rv2866* that encodes the toxin component of the TA pair Rv2865–Rv2866. Rv2866 is a homolog of the mRNA endonuclease RelE, which was shown to induce dormancy by shutting down translation in *E. coli* (Keren et al. 2004, 2011). Another study showed that VapC, the toxin component of the TA module VapC–VapBC, acts through specific binding and cleavage of RNA, thus downregulating metabolism (McKenzie et al. 2012; Sharp et al. 2012). Similarly, the toxin Rv1102c of the TA module Rv1102c–Rv1103c acts through its ribonuclease activity, which can be neutralized by Rv1103c that forms a complex with the toxin (Han et al. 2010).

When Rv1102c was expressed without its antitoxin in *M. smegmatis*, growth was arrested and the formation of persisters tolerant to kanamycin and gentamycin was increased (Han et al. 2010).

Phosphates were recently implicated in *M. tuberculosis* drug persistence. Specifically, PhoY2, the *M. tuberculosis* homolog of *E. coli* PhoU that regulates the *pst* operon, acts as a negative regulator of phosphate uptake (Li and Zhang 2007; Shi and Zhang 2010). Deletion of *phoY2* led to decreased tolerance of pyrazinamide and rifampicin in *M. tuberculosis* (Shi and Zhang 2010). How phosphate uptake directly impacts persister formation and drug tolerance remains largely unknown. However, recent studies suggested that inorganic polyphosphate might play a role in the transition from vegetative growth to dormancy and drug tolerance in *M. tuberculosis* (Thayil et al. 2011). Polyphosphate might act by inducing expression of the sigma factor RpoS that regulates expression of some 50 genes responsible for downregulation of metabolism and cell division (Shiba et al. 1997; Hengge-Aronis 2002).

In an interesting departure from the common theme of drug tolerance-metabolism correlation, a recent study suggested the involvement of efflux pumps in drug tolerance of *M. tuberculosis* (Adams et al. 2011). This study using an in vivo model of zebrafish infected with *M. marinum* followed by treatment with TB drugs reported the emergence and enrichment of a multidrug tolerant *M. marinum* population that was later disseminated into granulomas (Adams et al. 2011). Interestingly, drug tolerance in these bacilli could be reduced by co-treatment with efflux pump inhibitors, indicating that the observed drug tolerance was at least partially mediated by activity of efflux pumps (Adams et al. 2011).

5 Conclusions and Perspectives

It has only been less than 100 years since antibiotics were first used to treat bacterial infections. This time period is very short considering the pre-antibiotic era dated back thousands of years during which infectious diseases might have acted as selective forces of human evolution (Wang et al. 2012). For a long time, humans and bacteria had co-evolved in duels. However, everything has changed since antibiotics were discovered and applied in mass amounts to modern medicine: bacteria now must evolve under the additional selective pressure of these killing molecules. Evidence thus far indicates that pathogenic bacteria such as *M. tuberculosis* are well able to cope with this pressure and that they have evolved to become progressively resistant to antibiotics. Acquired resistances due to mutations in genes encoding target proteins or genes required for drug activities have allowed rapid evolution of mutants that become newly resistant to the antibiotics used. Accumulation of these resistance mutations has led to the emergence of *M. tuberculosis* strains that are more and more resistant to the available antibiotics. More alarmingly, recent studies indicate that these drug resistant mutants are able to evolve to regain fitness via compensatory mutations, thus enhancing their transmissibility and/or virulence. In addition to the acquired resistances caused by chromosomal mutations, *M. tuberculosis* is naturally resistant to most antibiotics. The profound intrinsic drug resistance in *M. tuberculosis* includes both passive and specialized mechanisms, the latter of which are able to respond to the presence of antibiotics. In fact, structural proteins capable of antibiotic resistance might have existed long before the clinical applications of these molecules (D'Costa et al. 2006, 2011). While these "drug resistance" proteins might still play roles in the physiology or metabolism of *M. tuberculosis* and other bacteria, inducible expression upon antibiotic exposure allows activities of these proteins important for antibiotic resistance. In this regard, the evolution of regulatory systems toward antibiotic responses might play a key role in specializing the function of those drug resistance systems. In addition to these two types of drug resistance, transition of *M. tuberculosis* from active growth into a dormant state confers increased phenotypic drug tolerance, which reflects the therapeutic recalcitrance of latent TB. These resistance and tolerance mechanisms hamper not only the clinical application of available antibiotics but also the development of new drugs. Nevertheless, understanding the resistance mechanisms could aid in the discovery of novel chemotherapeutic methods. For example, a recently emerging strategy termed "targeting resistance" uses inhibitors of resistance mechanisms to (re)sensitize resistant bacteria to the inactivated antibiotics. This strategy might help not only to recharge the inactivated drugs but also to synergize drugs that are less active because of intrinsic resistance mechanisms (Nguyen 2012).

Acknowledgments Work in the Nguyen laboratory is supported by NIH Grant R01 AI087903.

References

Adams KN, Takaki K, Connolly LE et al (2011) Drug tolerance in replicating mycobacteria mediated by a macrophage-induced efflux mechanism. Cell 145:39–53

Ainsa JA, Blokpoel MC, Otal I, Young DB, De Smet KA, Martin C (1998) Molecular cloning and characterization of Tap, a putative multidrug efflux pump present in *Mycobacterium fortuitum* and *Mycobacterium tuberculosis*. J Bacteriol 180:5836–5843

Ainsa JA, Martin C, Gicquel B, Gomez-Lus R (1996) Characterization of the chromosomal aminoglycoside 2'-N-acetyltransferase gene from *Mycobacterium fortuitum*. Antimicrob Agents Chemother 40:2350–2355

Aldridge BB, Fernandez-Suarez M, Heller D, Ambravaneswaran V, Irimia D, Toner M, Fortune SM (2012) Asymmetry and aging of mycobacterial cells lead to variable growth and antibiotic susceptibility. Science 335:100–104

Alekshun MN, Levy SB (1997) Regulation of chromosomally mediated multiple antibiotic resistance: the mar regulon. Antimicrob Agents Chemother 41:2067–2075

Andersson DI (2006) The biological cost of mutational antibiotic resistance: any practical conclusions? Curr Opin Microbiol 9:461–465

Andersson DI, Hughes D (2010) Antibiotic resistance and its cost: is it possible to reverse resistance? Nature reviews. Microbiology 8:260–271

Andersson DI, Levin BR (1999) The biological cost of antibiotic resistance. Curr Opin Microbiol 2:489–493

Andini N, Nash KA (2006) Intrinsic macrolide resistance of the *Mycobacterium tuberculosis* complex is inducible. Antimicrob Agents Chemother 50:2560–2562
Andriole VT (2005) The quinolones: past, present, and future. Clin Infect Dis 41(Suppl 2):S113–S119
Bamaga M, Wright DJ, Zhang H (2002) Selection of *in vitro* mutants of pyrazinamide-resistant *Mycobacterium tuberculosis*. Int J Antimicrob Agents 20:275–281
Belisle JT, Vissa VD, Sievert T, Takayama K, Brennan PJ, Besra GS (1997) Role of the major antigen of *Mycobacterium tuberculosis* in cell wall biogenesis. Science 276:1420–1422
Betts JC, Lukey PT, Robb LC, McAdam RA, Duncan K (2002) Evaluation of a nutrient starvation model of *Mycobacterium tuberculosis* persistence by gene and protein expression profiling. Mol Microbiol 43:717–731
Birnbaum M, Koch R, Brendecke F (1891) Prof. Koch's method to cure tuberculosis popularly treated. H.E. Haferkorn, Milwaukee
Boshoff HI, Mizrahi V (2000) Expression of *Mycobacterium smegmatis* pyrazinamidase in *Mycobacterium tuberculosis* confers hypersensitivity to pyrazinamide and related amides. J Bacteriol 182:5479–5485
Brennan PJ, Nikaido H (1995) The envelope of mycobacteria. Annu Rev Biochem 64:29–63
Buchmeier NA, Newton GL, Koledin T, Fahey RC (2003) Association of mycothiol with protection of *Mycobacterium tuberculosis* from toxic oxidants and antibiotics. Mol Microbiol 47:1723–1732
Burian J, Ramon-Garcia S, Sweet G, Gomez-Velasco A, Av-Gay Y, Thompson CJ (2012) The mycobacterial transcriptional regulator *whiB7* gene links redox homeostasis and intrinsic antibiotic resistance. J Biol Chem 287:299–310
Buriankova K, Doucet-Populaire F, Dorson O, Gondran A, Ghnassia JC, Weiser J, Pernodet JL (2004) Molecular basis of intrinsic macrolide resistance in the *Mycobacterium tuberculosis* complex. Antimicrob Agents Chemother 48:143–150
Campbell PJ, Morlock GP, Sikes RD et al (2011) Molecular detection of mutations associated with first- and second-line drug resistance compared with conventional drug susceptibility testing of *Mycobacterium tuberculosis*. Antimicrob Agents Chemother 55:2032–2041
Chambers HF, Moreau D, Yajko D et al (1995) Can penicillins and other beta-lactam antibiotics be used to treat tuberculosis? Antimicrob Agents Chemother 39:2620–2624
Chao J, Wong D, Zheng X, Poirier V, Bach H, Hmama Z, Av-Gay Y (2010) Protein kinase and phosphatase signaling in *Mycobacterium tuberculosis* physiology and pathogenesis. Biochim Biophys Acta 1804:620–627
Chen W, Biswas T, Porter VR, Tsodikov OV, Garneau-Tsodikova S (2011) Unusual regioversatility of acetyltransferase Eis, a cause of drug resistance in XDR-TB. Proc Natl Acad Sci USA 108:9804–9808
Colangeli R, Haq A, Arcus VL et al (2009) The multifunctional histone-like protein Lsr2 protects mycobacteria against reactive oxygen intermediates. Proc Natl Acad Sci USA 106:4414–4418
Colangeli R, Helb D, Sridharan S et al (2005) The *Mycobacterium tuberculosis iniA* gene is essential for activity of an efflux pump that confers drug tolerance to both isoniazid and ethambutol. Mol Microbiol 55:1829–1840
Colangeli R, Helb D, Vilcheze C et al (2007) Transcriptional regulation of multi-drug tolerance and antibiotic-induced responses by the histone-like protein Lsr2 in *M. tuberculosis*. PLoS Pathog 3:e87
Comas I, Borrell S, Roetzer A et al (2012) Whole-genome sequencing of rifampicin-resistant *Mycobacterium tuberculosis* strains identifies compensatory mutations in RNA polymerase genes. Nat Genet 44:106–110
Corbett EL, Watt CJ, Walker N, Maher D, Williams BG, Raviglione MC, Dye C (2003) The growing burden of tuberculosis: global trends and interactions with the HIV epidemic. Arch Intern Med 163:1009–1021
D'Costa VM, King CE, Kalan L et al (2011) Antibiotic resistance is ancient. Nature 477:457–461
D'Costa VM, McGrann KM, Hughes DW, Wright GD (2006) Sampling the antibiotic resistome. Science 311:374–377

da Silva PE, Von Groll A, Martin A, Palomino JC (2011) Efflux as a mechanism for drug resistance in *Mycobacterium tuberculosis*. FEMS Immunol Med Microbiol 63:1–9
Danilchanka O, Pavlenok M, Niederweis M (2008) Role of porins for uptake of antibiotics by *Mycobacterium smegmatis*. Antimicrobial Agents Chemother 52:3127–3134
Demple B (2005) The Nexus of oxidative stress responses and antibiotic resistance mechanisms in *Escherichia coli* and *Salmonella*. In: White DG, Alekshun MN, McDermott PF, Levy SB (eds) Frontiers in antimicrobial resistance : a tribute to Stuart B. Levy. American Society for Microbiology, Washington, DC, pp 191–197
Dhar N, McKinney JD (2010) *Mycobacterium tuberculosis* persistence mutants identified by screening in isoniazid-treated mice. Proc National Acad Sci USA 107:12275–12280
Dorman SE, Chaisson RE (2007) From magic bullets back to the magic mountain: the rise of extensively drug-resistant tuberculosis. Nat Med 13:295–298
Duncan K, Barry CE 3rd (2004) Prospects for new antitubercular drugs. Curr Opin Microbiol 7:460–465
Engstrom A, Perskvist N, Werngren J, Hoffner SE, Jureen P (2011) Comparison of clinical isolates and *in vitro* selected mutants reveals that *tlyA* is not a sensitive genetic marker for capreomycin resistance in *Mycobacterium tuberculosis*. J Antimicrob Chemother 66:1247–1254
Ferber D (2005) Biochemistry. Protein that mimics DNA helps tuberculosis bacteria resist antibiotics. Science 308:1393
Flores AR, Parsons LM, Pavelka MS Jr (2005) Genetic analysis of the beta-lactamases of *Mycobacterium tuberculosis* and *Mycobacterium smegmatis* and susceptibility to beta-lactam antibiotics. Microbiology 151:521–532
Gagneux S, Long CD, Small PM, Van T, Schoolnik GK, Bohannan BJ (2006) The competitive cost of antibiotic resistance in *Mycobacterium tuberculosis*. Science 312:1944–1946
Gao LY, Laval F, Lawson EH et al (2003) Requirement for *kasB* in *Mycobacterium* mycolic acid biosynthesis, cell wall impermeability and intracellular survival: implications for therapy. Mol Microbiol 49:1547–1563
Garton NJ, Waddell SJ, Sherratt AL et al (2008) Cytological and transcript analyses reveal fat and lazy persister-like bacilli in tuberculous sputum. PLoS medicine 5:e75
Gengenbacher M, Kaufmann SH (2012) *Mycobacterium tuberculosis*: success through dormancy. FEMS Microbiol Rev 36:514–532
Gillespie SH, Billington OJ, Breathnach A, McHugh TD (2002) Multiple drug-resistant *Mycobacterium tuberculosis*: evidence for changing fitness following passage through human hosts. Microbial Drug Resist 8:273–279
Gomez JE, McKinney JD (2004) *M. tuberculosis* persistence, latency, and drug tolerance. Tuberculosis 84:29–44
Han JS, Lee JJ, Anandan T et al (2010) Characterization of a chromosomal toxin-antitoxin, Rv1102c-Rv1103c system in *Mycobacterium tuberculosis*. Biochemical Biophys Res Commun 400:293–298
Hansen S, Lewis K, Vulic M (2008) Role of global regulators and nucleotide metabolism in antibiotic tolerance in *Escherichia coli*. Antimicrobial Agents Chemother 52:2718–2726
Hegde SS, Vetting MW, Roderick SL, Mitchenall LA, Maxwell A, Takiff HE, Blanchard JS (2005) A fluoroquinolone resistance protein from *Mycobacterium tuberculosis* that mimics DNA. Science 308:1480–1483
Hengge-Aronis R (2002) Signal transduction and regulatory mechanisms involved in control of the sigma(S) (RpoS) subunit of RNA polymerase. Microbiol Mol Biol Rev: MMBR 66:373–395
Hoffmann C, Leis A, Niederweis M, Plitzko JM, Engelhardt H (2008) Disclosure of the mycobacterial outer membrane: cryo-electron tomography and vitreous sections reveal the lipid bilayer structure. Proc Natl Acad Sci USA 105:3963–3967
Huang WL, Chi TL, Wu MH, Jou R (2011) Performance assessment of the GenoType MTBDRsl test and DNA sequencing for detection of second-line and ethambutol drug resistance among

patients infected with multidrug-resistant *Mycobacterium tuberculosis*. J Clin Microbiol 49:2502–2508

Hugonnet JE, Blanchard JS (2007) Irreversible inhibition of the *Mycobacterium tuberculosis* beta-lactamase by clavulanate. Biochemistry 46:11998–12004

Ito K, Yamamoto K, Kawanishi S (1992) Manganese-mediated oxidative damage of cellular and isolated DNA by isoniazid and related hydrazines: non-Fenton-type hydroxyl radical formation. Biochemistry 31:11606–11613

Jarlier V, Gutmann L, Nikaido H (1991) Interplay of cell wall barrier and beta-lactamase activity determines high resistance to beta-lactam antibiotics in *Mycobacterium chelonae*. Antimicrob Agents Chemother 35:1937–1939

Johansen SK, Maus CE, Plikaytis BB, Douthwaite S (2006) Capreomycin binds across the ribosomal subunit interface using tlyA-encoded 2'-O-methylations in 16S and 23S rRNAs. Mol Cell 23:173–182

Kasik JE (1979) Mycobacterial Beta-Lactamases. In: Hamilton-Miller JMT, Smith JT (eds) Beta-Lactamases. Academic Press, London, p 500

Kasik JE, Peacham L (1968) Properties of beta-lactamases produced by three species of mycobacteria. Biochem J 107:675–682

Keiler KC (2008) Biology of trans-translation. Annual Rev Microbiol 62:133–151

Keren I, Minami S, Rubin E, Lewis K (2011) Characterization and transcriptome analysis of *Mycobacterium tuberculosis* persisters. mBio 2:e00100-00111

Keren I, Shah D, Spoering A, Kaldalu N, Lewis K (2004) Specialized persister cells and the mechanism of multidrug tolerance in *Escherichia coli*. J Bacteriol 186:8172–8180

Kohanski MA, DePristo MA, Collins JJ (2010a) Sublethal antibiotic treatment leads to multidrug resistance via radical-induced mutagenesis. Mol Cell 37:311–320

Kohanski MA, Dwyer DJ, Collins JJ (2010b) How antibiotics kill bacteria: from targets to networks. Nat Rev Microbiol 8:423–435

Kohanski MA, Dwyer DJ, Hayete B, Lawrence CA, Collins JJ (2007) A common mechanism of cellular death induced by bactericidal antibiotics. Cell 130:797–810

Lewis K (2008) Multidrug tolerance of biofilms and persister cells. Curr Top Microbiol Immunol 322:107–131

Li Y, Zhang Y (2007) PhoU is a persistence switch involved in persister formation and tolerance to multiple antibiotics and stresses in *Escherichia coli*. Antimicrobial Agents Chemother 51:2092–2099

Liu J, Nikaido H (1999) A mutant of *Mycobacterium smegmatis* defective in the biosynthesis of mycolic acids accumulates meromycolates. Proc Natl Acad Sci U S A 96:4011–4016

Liu J, Rosenberg EY, Nikaido H (1995) Fluidity of the lipid domain of cell wall from *Mycobacterium chelonae*. Proc Natl Acad Sci USA 92:11254–11258

Madsen CT, Jakobsen L, Buriankova K, Doucet-Populaire F, Pernodet JL, Douthwaite S (2005) Methyltransferase Erm(37) slips on rRNA to confer atypical resistance in *Mycobacterium tuberculosis*. J Biol Chem 280:38942–38947

Maus CE, Plikaytis BB, Shinnick TM (2005) Mutation of *tlyA* confers capreomycin resistance in *Mycobacterium tuberculosis*. Antimicrobial Agents Chemother 49:571–577

McCune RM Jr, McDermott W, Tompsett R (1956) The fate of *Mycobacterium tuberculosis* in mouse tissues as determined by the microbial enumeration technique. II. The conversion of tuberculous infection to the latent state by the administration of pyrazinamide and a companion drug. J Exp Med 104:763–802

McCune RM Jr, Tompsett R (1956) Fate of *Mycobacterium tuberculosis* in mouse tissues as determined by the microbial enumeration technique. I. The persistence of drug-susceptible tubercle bacilli in the tissues despite prolonged antimicrobial therapy. J Exp Med 104:737–762

McKenzie JL, Robson J, Berney M et al (2012) A VapBC toxin-antitoxin module is a posttranscriptional regulator of metabolic flux in mycobacteria. J Bacteriol 194:2189–2204

Michele TM, Ko C, Bishai WR (1999) Exposure to antibiotics induces expression of the *Mycobacterium tuberculosis sigF* gene: implications for chemotherapy against mycobacterial persistors. Antimicrobial Agents Chemother 43:218–225

Montero C, Mateu G, Rodriguez R, Takiff H (2001) Intrinsic resistance of *Mycobacterium smegmatis* to fluoroquinolones may be influenced by new pentapeptide protein MfpA. Antimicrob Agents Chemother 45:3387–3392

Morais Cabral JH, Jackson AP, Smith CV, Shikotra N, Maxwell A, Liddington RC (1997) Crystal structure of the breakage-reunion domain of DNA gyrase. Nature 388:903–906

Morris RP, Nguyen L, Gatfield J et al (2005) Ancestral antibiotic resistance in *Mycobacterium tuberculosis*. Proc Natl Acad Sci USA 102:12200–12205

Myers A (1963) Can tuberculosis be eradicated? Chest 43:327–329

Nampoothiri KM, Rubex R, Patel AK, Narayanan SS, Krishna S, Das SM, Pandey A (2008) Molecular cloning, overexpression and biochemical characterization of hypothetical beta-lactamases of *Mycobacterium tuberculosis* H37Rv. J Appl Microbiol 105:59–67

Nash KA (2003) Intrinsic macrolide resistance in *Mycobacterium smegmatis* is conferred by a novel *erm* gene, *erm(38)*. Antimicrob Agents Chemother 47:3053–3060

Nash KA, Zhang Y, Brown-Elliott BA, Wallace RJ Jr (2005) Molecular basis of intrinsic macrolide resistance in clinical isolates of *Mycobacterium fortuitum*. J Antimicrob Chemother 55:170–177

Nguyen L (2012) Targeting antibiotic resistance mechanisms in *Mycobacterium tuberculosis*: recharging the old magic bullets. Expert Rev Anti Infect Ther 10:963–965

Nguyen L, Chinnapapagari S, Thompson CJ (2005) FbpA-Dependent biosynthesis of trehalose dimycolate is required for the intrinsic multidrug resistance, cell wall structure, and colonial morphology of *Mycobacterium smegmatis*. J Bacteriol 187:6603–6611

Nguyen L, Pieters J (2009) Mycobacterial subversion of chemotherapeutic reagents and host defense tactics: challenges in tuberculosis drug development. Ann Rev Pharmacol Toxicol 49:427–453

Niebisch A, Kabus A, Schultz C, Weil B, Bott M (2006) Corynebacterial protein kinase G controls 2-oxoglutarate dehydrogenase activity via the phosphorylation status of the OdhI protein. J Biol Chem 281:12300–12307

Niederweis M (2003) Mycobacterial porins-new channel proteins in unique outer membranes. Mol Microbiol 49:1167–1177

Nikaido H (1994) Prevention of drug access to bacterial targets: permeability barriers and active efflux. Science 264:382–388

Nishino K, Yamaguchi A (2001) Analysis of a complete library of putative drug transporter genes in *Escherichia coli*. J Bacteriol 183:5803–5812

Nott TJ, Kelly G, Stach L, et al (2009) An intramolecular switch regulates phosphoindependent FHA domain interactions in *Mycobacterium tuberculosis*. Science Signal 2:ra12

Ormerod LP (2005) Multidrug-resistant tuberculosis (MDR-TB): epidemiology, prevention and treatment. Br Med Bull 73–74:17–24

Philalay JS, Palermo CO, Hauge KA, Rustad TR, Cangelosi GA (2004) Genes required for intrinsic multidrug resistance in *Mycobacterium avium*. Antimicrob Agents Chemother 48:3412–3418

Quinting B, Reyrat JM, Monnaie D et al (1997) Contribution of beta-lactamase production to the resistance of mycobacteria to beta-lactam antibiotics. FEBS Lett 406:275–278

Rawat M, Newton GL, Ko M, Martinez GJ, Fahey RC, Av-Gay Y (2002) Mycothiol-deficient *Mycobacterium smegmatis* mutants are hypersensitive to alkylating agents, free radicals, and antibiotics. Antimicrob Agents Chemother 46:3348–3355

Reynolds MG (2000) Compensatory evolution in rifampin-resistant *Escherichia coli*. Genetics 156:1471–1481

Saguy M, Gillet R, Skorski P, Hermann-Le Denmat S, Felden B (2007) Ribosomal protein S1 influences trans-translation *in vitro* and *in vivo*. Nucleic Acids Res 35:2368–2376

Sala C, Haouz A, Saul FA, Miras I, Rosenkrands I, Alzari PM, Cole ST (2009) Genome-wide regulon and crystal structure of BlaI (Rv1846c) from *Mycobacterium tuberculosis*. Mol Microbiol 71:1102–1116

Scanga CA, Mohan VP, Joseph H, Yu K, Chan J, Flynn JL (1999) Reactivation of latent tuberculosis: variations on the Cornell murine model. Infect Immun 67:4531–4538

Scorpio A, Lindholm-Levy P, Heifets L, Gilman R, Siddiqi S, Cynamon M, Zhang Y (1997) Characterization of *pncA* mutations in pyrazinamide-resistant *Mycobacterium tuberculosis*. Antimicrobial Agents Chemother 41:540–543

Scorpio A, Zhang Y (1996) Mutations in *pncA*, a gene encoding pyrazinamidase/nicotinamidase, cause resistance to the antituberculous drug pyrazinamide in tubercle bacillus. Nat Med 2:662–667

Senaratne RH, Mobasheri H, Papavinasasundaram KG, Jenner P, Lea EJ, Draper P (1998) Expression of a gene for a porin-like protein of the OmpA family from *Mycobacterium tuberculosis* H37Rv. J Bacteriol 180:3541–3547

Sergeev R, Colijn C, Murray M, Cohen T (2012) Modeling the dynamic relationship between HIV and the risk of drug-resistant tuberculosis. Sci Transl Med 4:135ra167

Sharp JD, Cruz JW, Raman S, Inouye M, Husson RN, Woychik NA (2012) Growth and translation inhibition through sequence-specific RNA binding by *Mycobacterium tuberculosis* VapC toxin. J Biol Chem 287:12835–12847

Sherman DR, Mdluli K, Hickey MJ, Arain TM, Morris SL, Barry CE 3rd, Stover CK (1996) Compensatory *ahpC* gene expression in isoniazid-resistant *Mycobacterium tuberculosis*. Science 272:1641–1643

Shi L, Sohaskey CD, Kana BD, Dawes S, North RJ, Mizrahi V, Gennaro ML (2005) Changes in energy metabolism of *Mycobacterium tuberculosis* in mouse lung and under *in vitro* conditions affecting aerobic respiration. Proc Natl Acad Sci USA 102:15629–15634

Shi W, Zhang X, Jiang X et al (2011) Pyrazinamide inhibits trans-translation in *Mycobacterium tuberculosis*. Science 333:1630–1632

Shi W, Zhang Y (2010) PhoY2 but not PhoY1 is the PhoU homologue involved in persisters in *Mycobacterium tuberculosis*. J Antimicrob Chemother 65:1237–1242

Shiba T, Tsutsumi K, Yano H et al (1997) Inorganic polyphosphate and the induction of *rpoS* expression. Proc Natl Acad Sci USA 94:11210–11215

Singh A, Gupta R, Vishwakarma RA, Narayanan PR, Paramasivan CN, Ramanathan VD, Tyagi AK (2005) Requirement of the *mymA* operon for appropriate cell wall ultrastructure and persistence of *Mycobacterium tuberculosis* in the spleens of guinea pigs. J Bacteriol 187:4173–4186

Singh A, Jain S, Gupta S, Das T, Tyagi AK (2003) *mymA* operon of *Mycobacterium tuberculosis*: its regulation and importance in the cell envelope. FEMS Microbiol Lett 227:53–63

Siroy A, Mailaender C, Harder D et al (2008) Rv1698 of *Mycobacterium tuberculosis* represents a new class of channel-forming outer membrane proteins. J Biol Chem 283:17827–17837

Stephan J, Mailaender C, Etienne G, Daffe M, Niederweis M (2004) Multidrug resistance of a porin deletion mutant of *Mycobacterium smegmatis*. Antimicrob Agents Chemother 48:4163–4170

Thayil SM, Morrison N, Schechter N, Rubin H, Karakousis PC (2011) The role of the novel exopolyphosphatase MT0516 in *Mycobacterium tuberculosis* drug tolerance and persistence. PloS one 6:e28076

Tian J, Bryk R, Itoh M, Suematsu M, Nathan C (2005) Variant tricarboxylic acid cycle in *Mycobacterium tuberculosis*: identification of alpha-ketoglutarate decarboxylase. Proc Natl Acad Sci USA 102:10670–10675

Tremblay LW, Fan F, Blanchard JS (2010) Biochemical and structural characterization of *Mycobacterium tuberculosis* beta-lactamase with the carbapenems ertapenem and doripenem. Biochemistry 49:3766–3773

Udwadia ZF (2012) MDR, XDR, TDR tuberculosis: ominous progression. Thorax 67:286–288

Vetting M, Roderick SL, Hegde S, Magnet S, Blanchard JS (2003) What can structure tell us about *in vivo* function? The case of aminoglycoside-resistance genes. Biochem Soc Trans 31:520–522

Vilcheze C, Av-Gay Y, Attarian R et al (2008) Mycothiol biosynthesis is essential for ethionamide susceptibility in *Mycobacterium tuberculosis*. Molec Microbiol 69:1316–1329

Viveirosa M, Martins M, Rodrigues L, Machado D, Coutoa I, Ainsad J, Amaral L (2012) Inhibitors of mycobacterial efflux pumps as potential boosters for TB drugs. Expert Rev Anti Infect Ther 10:983–998

Voladri RK, Lakey DL, Hennigan SH, Menzies BE, Edwards KM, Kernodle DS (1998) Recombinant expression and characterization of the major beta-lactamase of *Mycobacterium tuberculosis*. Antimicrob Agents Chemother 42:1375–1381

Walburger A, Koul A, Ferrari G et al (2004) Protein kinase G from pathogenic mycobacteria promotes survival within macrophages. Science 304:1800–1804

Wallis RS, Patil S, Cheon SH et al (1999) Drug tolerance in *Mycobacterium tuberculosis*. Antimicrobial Agents Chemother 43:2600–2606

Wang F, Cassidy C, Sacchettini JC (2006) Crystal structure and activity studies of the *Mycobacterium tuberculosis* beta-lactamase reveal its critical role in resistance to beta-lactam antibiotics. Antimicrob Agents Chemother 50:2762–2771

Wang JY, Burger RM, Drlica K (1998) Role of superoxide in catalase-peroxidase-mediated isoniazid action against mycobacteria. Antimicrobial Agents Chemother 42:709–711

Wang X, Mitra N, Secundino I et al (2012) Specific inactivation of two immunomodulatory SIGLEC genes during human evolution. Proc Natl Acad Sci USA 109:9935–9940

Watt B, Edwards JR, Rayner A, Grindey AJ, Harris G (1992) *In vitro* activity of meropenem and imipenem against mycobacteria: development of a daily antibiotic dosing schedule. Tuber Lung Dis 73:134–136

Wayne LG, Hayes LG (1996) An *in vitro* model for sequential study of shiftdown of *Mycobacterium tuberculosis* through two stages of nonreplicating persistence. Infect Immun 64:2062–2069

Wei J, Dahl JL, Moulder JW, Roberts EA, O'Gaora P, Young DB, Friedman RL (2000) Identification of a *Mycobacterium tuberculosis* gene that enhances mycobacterial survival in macrophages. J Bacteriol 182:377–384

Wolff KA, Nguyen HT, Cartabuke RH, Singh A, Ogwang S, Nguyen L (2009) Protein kinase G is required for intrinsic antibiotic resistance in mycobacteria. Antimicrobial Agents Chemother 53:3515–3519

Wower IK, Zwieb CW, Guven SA, Wower J (2000) Binding and cross-linking of tmRNA to ribosomal protein S1, on and off the *Escherichia coli* ribosome. EMBO J 19:6612–6621

Xie Z, Siddiqi N, Rubin EJ (2005) Differential antibiotic susceptibilities of starved *Mycobacterium tuberculosis* isolates. Antimicrobial Agents Chemother 49:4778–4780

Zaunbrecher MA, Sikes RD Jr, Metchock B, Shinnick TM, Posey JE (2009) Overexpression of the chromosomally encoded aminoglycoside acetyltransferase eis confers kanamycin resistance in *Mycobacterium tuberculosis*. Proc Natl Acad Sci USA 106:20004–20009

Zhang Y (2012) Drug resistant and persistent tuberculosis: mechanisms and drug development. In: Dougherty TJ, Pucci MJ (ed) Antibiotic discovery and development. Springer Science + Business Media, pp 719–746

Zhang Y, Dhandayuthapani S, Deretic V (1996) Molecular basis for the exquisite sensitivity of *Mycobacterium tuberculosis* to isoniazid. Proc Natl Acad Sci USA 93:13212–13216

Zhang Y, Heym B, Allen B, Young D, Cole S (1992) The catalase-peroxidase gene and isoniazid resistance of *Mycobacterium tuberculosis*. Nature 358:591–593

Zuber B, Chami M, Houssin C, Dubochet J, Griffiths G, Daffe M (2008) Direct visualization of the outer membrane of mycobacteria and corynebacteria in their native state. J Bacteriol 190:5672–5680

Antimicrobial Efflux Pumps and *Mycobacterium tuberculosis* Drug Tolerance: Evolutionary Considerations

John D. Szumowski, Kristin N. Adams, Paul H. Edelstein and Lalita Ramakrishnan

Abstract The need for lengthy treatment to cure tuberculosis stems from phenotypic drug resistance, also known as drug tolerance, which has been previously attributed to slowed bacterial growth in vivo. We discuss recent findings that challenge this model and instead implicate macrophage-induced mycobacterial efflux pumps in antimicrobial tolerance. Although mycobacterial efflux pumps may have originally served to protect against environmental toxins, in the pathogenic mycobacteria, they appear to have been repurposed for intracellular growth. In this light, we discuss the potential of efflux pump inhibitors such as verapamil to shorten tuberculosis treatment by their dual inhibition of tolerance and growth.

Contents

J. D. Szumowski (✉) · L. Ramakrishnan
Department of Medicine (Division of Infectious Diseases),
University of Washington, Seattle, WA, USA
e-mail: jszumows@uw.edu

K. N. Adams · L. Ramakrishnan
Department of Microbiology, University of Washington, Seattle, WA, USA

P. H. Edelstein
Departments of Pathology, Laboratory Medicine,
and Medicine (Division of Infectious Diseases),
University of Pennsylvania, Philadelphia, PA, USA

L. Ramakrishnan
Department of Immunology, University of Washington, Seattle, WA, USA

Current Topics in Microbiology and Immunology (2012) 374: 81–108

DOI: 10.1007/82_2012_300

Published Online: 16 December 2012

1 Drug Tolerance is an Important Barrier to Shortening TB Treatment

The long duration of treatment required with current anti-tuberculous drugs presents a major challenge in tuberculosis (TB) management. At least 6 months of treatment are required to achieve acceptable cure and relapse rates for smear-positive TB (Connolly et al. 2007; Mitchison and Davies 2012). Although an important breakthrough when first introduced, such "short course" therapy is still too long. Adherence to months of TB therapy is difficult, with default rates of nearly 30 % reported in some series (Castelnuovo 2010). The consequences of poor adherence are serious both for the individual patient and for the community: drug resistance, treatment failure, and further TB transmission. Attempts to shorten treatment to 4 months have been thwarted by unacceptably high relapse rates (Johnson et al. 2009).

Why is lengthy treatment with current medications required to cure TB? The answer may be found in observations from landmark TB studies. For years, it has been recognized that when patients with drug-susceptible TB relapse, the bacilli typically remain genetically drug-susceptible and patients respond to their prior treatment regimens (British Medical Research Council 1972; Wallis et al. 1999). Complementary data from early bactericidal activity studies by Jindani and Mitchison demonstrated that during TB chemotherapy, sputum bacillary counts decrease in a characteristic biphasic manner (Jindani et al. 1980). For example with isoniazid, greater than 99 % of the initial sputum bacillary load is killed during the first 2 days of treatment, after which the rate of killing drops off markedly. The residual bacteria are a *phenotypically* resistant, "drug tolerant" population; TB drug minimum inhibitory concentrations are unchanged. Empirical studies have shown that it takes months of therapy to eradicate these bacteria and produce a stable cure (Mitchison and Davies 2012).

The phenomenon of antimicrobial tolerance was recognized in early experiments studying in vitro killing of streptococci and staphylococci by penicillin (Bigger 1944; Hobby et al. 1942) and was subsequently found to generalize to other bacteria, including *Mycobacterium tuberculosis* (Mtb) (McCune and Tompsett 1956; Wallis et al. 1999). Existing models of antimicrobial tolerance differ in their specifics but all invoke the presence of a metabolically quiescent, non-growing population. Older views focused on deterministic mechanisms such as hypoxia or nutrient starvation, conditions that are thought to occur in the tuberculous granuloma; more recent models implicate stochastic mechanisms whereby so-called "persister" cells arise independently of the growth environment (Dhar and McKinney 2007; Lewis 2010). Although tolerance models that emphasize a role for slow-growing or nongrowing bacteria are compatible with the observation that antimicrobials kill nongrowing TB poorly (Schaefer 1954), evidence from human treatment studies suggest that the drug-tolerant population may not in fact be quiescent. Serial radiological studies have demonstrated that existing lesions may enlarge and new lesions may develop despite an overall efficacious course of therapy, a phenomenon that may be explained by the presence of an enlarging, drug-tolerant Mtb population (Akira et al. 2000; Bobrowitz 1980).

2 Actively Growing Intracellular Mycobacteria Exhibit Multidrug Tolerance Mediated by Macrophage-Induced Bacterial Efflux Pumps

Recent insights from the zebrafish-*M. marinum* (Mm) model of TB offer potential explanations for these puzzling radiographic observations. Similar to the expansion of a subset of tuberculous lesions during human therapy, drug-tolerant Mm continue to expand and disseminate within macrophages during infection of zebrafish (Adams et al. 2011). Further investigation with macrophage-like cell lines revealed that subpopulations of both Mtb and Mm become tolerant to multiple classes of antimicrobials including isoniazid and rifampicin upon intracellular residence. The induction of drug tolerance in bacteria by the host macrophage environment has been previously described for *Legionella pneumophila* (Barker et al. 1995) and may be a more widespread phenomenon. Countering prior models, the mycobacterial work revealed that macrophage-induced tolerance is enriched in actively dividing bacteria (Fig. 1) (Adams et al. 2011). This surprising result was explained by the finding that in Mtb, macrophage-induced tolerance to rifampicin is mediated by a bacterial efflux pump, Rv1258c, that also promotes intracellular bacterial growth in the absence of antimicrobials (Table 1) (Adams et al. 2011). Rv1258c, a secondary transporter belonging to the major facilitator superfamily (MFS) of efflux pumps, is structurally related to MefA, a 12-membrane spanning MFS pump involved in macrolide resistance in *Streptococcus pneumoniae* (Ainsa et al. 1998; De Rossi et al. 2002; Li and Nikaido 2009; Saier et al. 2009). Rv1258c is transcriptionally induced following macrophage residence (Table 1) (Schnappinger et al. 2003) and appears to function as a virulence factor induced in the intracellular environment that pathogenic mycobacteria encounter (Chan et al. 2002; Clay et al. 2007; Dannenberg 1993; Ramakrishnan et al. 2000). Its association with rifampicin tolerance appears to be an epiphenomenon, and the identity of the "natural" substrate(s) of Rv1258c remains unknown. Indeed, despite decades of study there are still only a few clearly identified natural substrates of bacterial efflux pumps, such as spermidine in *B. subtilis*, bile

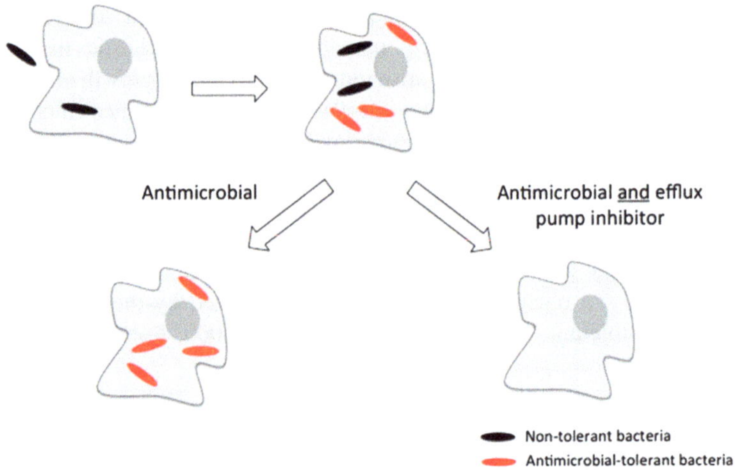

Fig. 1 *Model for efflux pump inhibitor action in Mycobacterium tuberculosis (Mtb)* Efflux pump expression is induced in Mtb following macrophage residence, perhaps stimulated by macrophage antimicrobial peptides. With antimicrobial treatment alone, nontolerant bacteria are killed, but tolerant bacteria survive and multiply within the macrophage. When antimicrobials are given in conjunction with an efflux pump inhibitor, the tolerant bacteria are killed along with nontolerant bacteria. Note that in this simplified diagram, there is no attempt to differentiate between mycobacterial residence in the cytoplasm versus the phagosome. Modified from (Adams et al. 2011)

salts in *Escherichia coli,* and cyclic-di-AMP in *Listeria monocytogenes* (Thanassi et al. 1997; Woodward et al. 2010; Woolridge et al. 1997).

The findings coupling intracellular bacterial growth and antimicrobial tolerance through induction of bacterial efflux pumps are compatible with the longstanding clinical observation that the duration of curative TB treatment is proportional to the organism burden. Indeed, the highest mycobacterial burden states, smear-positive and cavitary disease, require the longest therapy for a durable cure (Connolly et al. 2007; British Medical Research Council 1989; Zierski et al. 1980). Models of nonreplicating tolerance have attributed this association to high burden disease having increased numbers of nonreplicating as well as replicating bacteria (Connolly et al. 2007). However, the link between mycobacterial burden and treatment duration is equally well explained by the alternative model attributing tolerance to actively growing bacteria. Indeed, this view implicates increased efflux activity as the *driver* of both high mycobacterial burden disease and antimicrobial tolerance; high burden disease states should therefore be enriched in tolerant bacteria.

The role of efflux pumps in promoting drug tolerance opens up a potentially powerful approach for shortening TB treatment. The use of efflux pump inhibitors would target not only bacterial growth, but also drug tolerance. In the laboratory setting, macrophage-induced tolerance is inhibited by verapamil, a calcium channel antagonist in clinical use for years, which has been shown to also inhibit multiple bacterial efflux pumps in vitro (Adams et al. 2011; Marquez 2005; Rodrigues et al. 2011a). Consistent with the observation that macrophage-induced

Table 1 Macrophage-induced *Mycobacterium tuberculosis* efflux pumps

Drug efflux pump[a]	Transporter family	Macrophage growth attenuation[b]	Associated drug resistance	Homologs in other mycobacteria	References
Rv0194	ABC	Yes	STR	*M. marinum; M. ulcerans*	Braibant et al. (2000), Danilchanka et al. (2008)
Rv1218c	ABC	Yes		*M. smegmatis; M. marinum; M. avium; M. leprae; M. abscessus*	Balganesh et al. (2010, 2012), Braibant et al. (2000)
Rv1272c	ABC	Yes		*M. smegmatis; M.marinum; M. ulcerans, M. avium; M. leprae; M. abscessus*	Braibant et al. (2000)
Rv1273c	ABC	Yes		*M. smegmatis; M.marinum; M. ulcerans, M. avium; M. leprae; M. abscessus*	Braibant et al. (2000)
Rv1348	ABC	ND		*M. smegmatis; M.marinum; M. ulcerans; M. avium; M. abscessus*	Braibant et al. (2000), Farhana et al. (2008)
Rv1349	ABC	ND		*M. smegmatis; M.marinum; M. ulcerans; M. avium; M. abscessus*	Braibant et al. (2000), Farhana et al. (2008)
Rv1463	ABC	ND		*M. smegmatis; M.marinum; M. ulcerans, M. avium; M. leprae; M. abscessus*	Braibant et al. (2000)
Rv1687c	ABC	No		*M. smegmatis; M.marinum; M. ulcerans, M. avium; M. leprae*	Braibant et al. (2000)
Rv2686c	ABC	Yes	CIP	*M. smegmatis; M.marinum; M. ulcerans, M. leprae; M. abscessus*	Braibant et al. (2000), Louw et al. (2009), Pasca et al. (2004)
Rv2687c	ABC	ND	CIP	*M. smegmatis; M.marinum; M. ulcerans; M. abscessus*	Braibant et al. (2000), Pasca et al. (2004)
Rv2688c	ABC	No	CIP	*M. smegmatis; M.marinum; M. ulcerans; M. abscessus*	Braibant et al. (2000), Pasca et al. (2004)

(continued)

Table 1 (continued)

Drug efflux pump[a]	Transporter family	Macrophage growth attenuation[b]	Associated drug resistance	Homologs in other mycobacteria	References
Rv1258c	MFS	Yes	RIF, OFX, INH	*M. smegmatis; M.marinum; M. ulcerans, M. avium; M. leprae; M. abscessus*	Balganesh et al. (2012), De Rossi et al. (2002), Jiang et al. (2008), Rodrigues et al. (2011b), Siddiqi et al. (2004), Zhang et al. (2005)
Rv3239c	MFS	No		*M.marinum; M. ulcerans*	De Rossi et al. (2002), Louw et al. (2009)
Rv3728	MFS	No		*M.marinum; M. ulcerans; M. leprae*	De Rossi et al. (2002), Gupta et al. (2010), Louw et al. (2009)
Rv1183 (mmpL10)	RND	ND		*M. smegmatis; M.marinum, M. avium; M. leprae; M. abscessus*	Tekaia et al. (1999)
Rv1146 (mmpL13b)	RND	ND		*M. smegmatis; M.marinum; M. ulcerans, M. avium; M. abscessus*	Tekaia et al. (1999)
Rv3065 (mmr)	SMR	ND	ERM	*M.marinum; M. ulcerans, M. avium; M. leprae*	Balganesh et al. (2012), Gupta et al. (2010)
Rv0969 (ctpV)	Putative copper exporter	No		*M. smegmatis, M. marinum, M, ulcerans, M. avium; M. abscessus*	Ward et al. (2010)
Rv3578 (arsB2)	Probable arsenic pump	Yes		*M. smegmatis; M.marinum; M. ulcerans, M. avium; M. leprae*	Ordonez et al. (2005)

INH isoniazid; *RIF* rifampicin; *OFX* ofloxacin; *CIP* ciprofloxacin; *STR* streptomycin; *EMB* ethambutol; *ERM* erythromycin; *ABC* ATP-binding cassette; *MFS* major facilitator superfamily; *RND* resistance-nodulation cell division family; *SMR* small multidrug resistance family; *ND* not determined

[a] Efflux pumps that are significantly induced (1.3–2.6-fold) at 48 h in naive macrophages (Schnappinger et al. 2003). Those highlighted in bold have been tested experimentally for efflux activity, all others are predicted efflux pumps based on homology to efflux pumps in other organisms

[b] Macrophage growth attenuation from resting or IFNγ-activated macrophages (Rengarajan et al. 2005)

pumps mediate intracellular survival, verapamil also reduces intracellular mycobacterial growth in the absence of antibiotics (Adams et al. 2011; Martins et al. 2008). This chapter will discuss these findings in the context of a current understanding of antimicrobial efflux in mycobacteria and other bacteria, with an emphasis on teleological, functional, and therapeutic considerations.

3 Macrophage-Induced Mtb Efflux Pumps are Virulence Determinants

Although information about the natural substrates of efflux pumps is lacking, extensive in vitro studies have shown that efflux pumps in both prokaryotic and eukaryotic organisms can extrude a variety of toxic agents such as antimicrobials and chemotherapeutic agents (Ho and Kim 2005; Li and Nikaido 2009). Much effort has focused on the role of efflux in antimicrobial resistance (Li and Nikaido 2009). For example, the so-called multidrug resistance (MDR) pumps have been implicated in resistance to structurally diverse antimicrobial agents and contribute to the burden of bacterial drug resistance. Efflux-mediated antimicrobial resistance was initially reported in *E. coli* (Ball et al. 1980; McMurry et al. 1980), but has been subsequently recognized in a wide range of organisms, including the often recalcitrant *Pseudomonas aeruginosa* and *Acinetobacter baumannii* (Coyne et al. 2011; Li et al. 1995). With the identification of the LfrA pump in *M. smegmatis* as a mediator of fluoroquinolone resistance (Liu et al. 1996), there has been a growing interest in the contributions of drug efflux in mycobacteria (da Silva et al. 2011; Louw et al. 2009). Efflux has also been proposed to account for isoniazid-induced tolerance in Mtb and may be mediated by the isoniazid-induced protein IniA (Colangeli et al. 2005; Viveiros et al. 2002).

Our identification of Rv1258c as a mediator of intracellular growth led us to investigate if mycobacterial efflux pumps are widely used for this critical virulence trait. Mining the published literature reveals that 19 of the 55 annotated efflux pumps in the Mtb genome are transcriptionally induced in macrophages (Table 1) (Camus et al. 2002; Cole et al. 1998; Schnappinger et al. 2003). Of the 12 tested by mutational analysis, 7 are required for intracellular growth (Rengarajan et al. 2005). Thus several macrophage-induced efflux pumps serve nonredundant roles in promoting intracellular growth. Moreover, Mtb efflux pumps not found induced in the 48 hour macrophage infection assay have virulence phenotypes, in a 7-day macrophage infection assay and/or in mouse infection models (Bigi et al. 2004; Curry et al. 2005; Rengarajan et al. 2005; Sassetti and Rubin 2003; Schnappinger et al. 2003). These may represent virulence genes that are induced later in the course of macrophage residence or by specific environments in vivo such as the tuberculous granuloma (Chan et al. 2002; Ramakrishnan et al. 2000). Efflux pumps in Gram-negative bacteria have been linked to multiple virulence functions including gut colonization, and adherence and invasion of cultured cells (Table 2) (Piddock 2006b). It is likely that Mtb pumps participate in similar activities. That Mtb

Table 2 Bacterial efflux pumps associated with virulence

Pump family	Organism	Virulence phenotype	Proposed mechanism	Mediates antibiotic resistance	References
ABC					
DrrABC	Mtb	In vivo survival	Localization of phthiocerol dimycocerosate in cell wall	Yes	Camacho et al. (2001), Choudhuri et al. (2002), Sassetti and Rubin (2003)
MacAB	*Salmonella enterica serovar Typhimurium*	In vivo survival	May detoxify host-derived molecules	Yes	Nishino et al. (2006)
Rv1272c	Mtb	In vivo survival	Unknown	ND	Sassetti and Rubin (2003)
Rv1747	Mtb	Intracellular growth; in vivo survival	Substrate for PknF a serine threonine kinase involved in regulating glucose intake	ND	Molle et al. (2004), Sassetti and Rubin (2003), Spivey et al. (2011)
Rv3781	Mtb	In vivo survival	May be involved in arabinogalactan biosynthesis	ND	Dianiskova et al. (2011), Sassetti and Rubin (2003)
MFS					
MdrM	*Listeria monocytogenes*	In vivo growth	Secretion of c-di-AMP	Yes	Crimmins et al. (2008), Woodward et al. (2010)
MdrT	*Listeria monocytogenes*	In vivo growth	Secretion of c-di-AMP Cholic acid transporter	Yes	Crimmins et al. (2008), Quillin et al. (2011), Woodward et al. (2010)
NorA	*Staphylococcus aureus*	Host cell invasion	Unknown	Yes	Aeschlimann et al. (1999), DeMarco et al. (2007), Kalia et al. (2012)
NorB	*Staphylococcus aureus*	In vivo survival	Unknown	Yes	DeMarco et al. (2007), Ding et al. (2008)
P55 (Rv1410c)	Mtb, *Mycobacterium bovis*	Intracellular growth; in vivo survival	Preservation of cell wall	Yes	Bianco et al. (2011), Ramon-Garcia et al. (2009), Rengarajan et al. (2005), Sassetti and Rubin (2003)

(continued)

Table 2 (continued)

Pump family	Organism	Virulence phenotype	Proposed mechanism	Mediates antibiotic resistance	References
QacA	*Staphylococcus aureus*	In vivo persistence	Increased membrane fluidity	Yes	Bayer et al. (2006), Dhawan et al. (1997), Kupferwasser et al. (1999)
Rv0037c	Mtb	Intracellular growth	Unknown	ND	Rengarajan et al. (2005)
Rv0849	Mtb	Intracellular growth	Unknown	ND	Rengarajan et al. (2005)
Tap (Rv1258c)	Mtb	Intracellular growth	Unknown	Yes	Adams et al. (2011), Ainsa et al. (1998), Balganesh et al. (2012), Sharma et al. (2010), Siddiqi et al. (2004)
RND					
AcrAB	*Escherichia coli, Francisella tularensis, Klebsiella pneumoniae, Salmonella enterica serovar Typhimurium, Enterobacter cloacae*	In vivo survival	Efflux of bile acids	Yes	Bina et al. (2008a), Blair and Piddock (2009), Buckley et al. (2006), Helling et al. (2002), Ma et al. (1995), Padilla et al. (2010), Perez et al. (2012), Rosenberg et al. (2003), Thanassi et al. (1997)
BesC	*Borrelia burgdorferi*	In vivo survival	Possible component of type I secretion system	Yes	Bunikis et al. (2008)
BpeAB-OprB	*Burkholderia pseudomallei*	Host cell invasion	Quorum sensing	Yes	Chan and Chua (2005)
CmeABC	*Campylobacter jejuni*	In vivo colonization	Efflux of bile acids	Yes	Lin and Martinez (2006), Lin et al. (2003), Martinez and Lin (2006)
MexCD-OprJ	*Pseudomonas aeruginosa*	In vivo survival; hyperexpression compromises expression of type III secretion genes	Secretion of quorum sensing molecules	Yes	Join-Lambert et al. (2001), Linares et al. 2005)

(continued)

Table 2 (continued)

Pump family	Organism	Virulence phenotype	Proposed mechanism	Mediates antibiotic resistance	References
MexEF-OprN	*Pseudomonas aeruginosa*	In vivo survival; hyperexpression compromises expression of type III secretion genes	Secretion of quorum sensing molecules	Yes	Frisk et al. (2004), Join-Lambert et al. (2001), Kohler et al. (2001), Lamarche and Deziel (2011), Linares et al. (2005)
MmpL7	Mtb	Intracellular growth; in vivo survival	Translocation of phthiocerol dimycocerosate to cell wall	Yes	Camacho et al. (2001), Domenech et al. (2005), Lamichhane et al. (2005), Pasca et al. (2005), Rodrigues et al. (2011b); Sassetti and Rubin (2003)
MmpL10	Mtb	In vivo survival	Unknown	ND	Lamichhane et al. (2005), Sassetti and Rubin (2003)
MtrCDE	*Neisseria gonorrhoeae, Neisseria meningitidis*	In vivo survival	Resistance to host antimicrobial defenses	Yes	Hagman et al. (1995), Jerse et al. (2003), Shafer et al. (1995, 1998), Tzeng et al. (2005), Warner et al. (2008)
VexH	*Vibrio cholerae*	Intestinal colonization	Export of cholera toxin and toxin coregulated pilus	Yes	Bina et al. (2008b), Taylor et al. (2012)
TolC	*Brucella suis, Francisella tularensis, Legionella pneumophila, Salmonella enterica serovar Typhimurium, Salmonella enteritidis*	Intracellular growth; intestinal colonization	Might be involved in efflux of reactive oxygen species	Yes	Buckley et al. (2006), Ferhat et al. (2009), Nishino et al. (2006), Platz et al. (2010), Posadas et al. (2007), Stone and Miller (1995), Wu et al. (2012)
Other					
ArsB2	Mtb	Intracellular growth	Probable arsenic pump	ND	Rengarajan et al. (2005)
CopA	*Neisseria gonorrhoeae*	Invasion and survival	Export of copper ions	ND	Djoko et al. (2012)
CtpV	Mtb	Intracellular growth; in vivo survival	Putative copper exporter	ND	Rengarajan et al. (2005), Ward et al. (2010)

Mycobacterium tuberculosis (*Mtb*)

allocates a great number of efflux pumps to ensure intracellular survival is consistent with it being a central strategy for mycobacterial virulence (Cosma et al. 2003; Shepard 1957).

4 Mtb Macrophage-Induced Efflux Pumps: Signals and Substrates

The presence of distinct host-generated defenses within the macrophage may explain the observation that multiple macrophage-induced Mtb pumps are individually essential for intracellular growth. What are the stimuli that induce pump expression and what are the substrates? Do different pumps have shared stimuli but unique substrates? Although these answers are not clear, indirect clues suggest that tolerance-producing mycobacterial efflux pumps may be induced by antimicrobial peptides (AMPs) (Adams et al. 2011). Indeed, macrophage-induced *M. marinum* tolerance is not inhibited by dexamethasone, a glucocorticoid that reduces most macrophage defenses while sparing antimicrobial peptide expression (Adams et al. 2011; Duits et al. 2001; Ehrchen et al. 2007). This model has precedence: the macrophage-derived AMP LL-37 induces transcription of *mefE*, which encodes one component of a *Streptococcus pneumoniae* efflux pump related to Rv1258c (Zahner et al. 2010).

Could AMPs also be substrates of the macrophage-induced efflux pumps? Studies predominantly from Gram-negative bacteria suggest this could be the case (Bengoechea and Skurnik 2000; Brissette and Lukehart 2007; Padilla et al. 2010; Shafer et al. 1998; Tzeng et al. 2005; Warner and Levy 2010). For example, in *Neisseria gonorrhoeae,* mutation of the Mtr efflux pump and treatment with the chemical efflux pump inhibitor carbonyl cyanide-*m*-chlorophenylhydrazone (CCCP) both increase AMP sensitivity as well as intracellular AMP accumulation (Shafer et al. 1998). Similar to the Rv1258c pump, Mtr also confers antibiotic resistance (Hagman et al. 1995). While Rv1258c mediates tolerance to the hydrophobic antibiotic rifampicin but not the hydrophilic isoniazid, Mtr similarly mediates resistance to rifampicin and the hydrophobic erythromycin, but not the hydrophilic antibiotic streptomycin. Though many pumps have been noted to have broad substrate promiscuity (Neyfakh 2002), these examples suggest that hydrophobic compounds may be transported by a more limited subset of pumps.

Of course, other mechanisms likely contribute to AMP resistance aside from AMP efflux (Kraus and Peschel 2006). While a comprehensive screen of *Neisseria meningitidis* mutants with increased susceptibility to an AMP-like cyclic lipopeptide revealed a predominance of mutations in the *mtr* gene, mutations in other genes involved in lipid A and pilin synthesis were also identified (Tzeng et al. 2005). In addition, *S. aureus* strains overexpressing the QacA pump showed evidence of decreased membrane fluidity (Bayer et al. 2006). Thus, it would appear that resistance to AMPs can be mediated directly through efflux pumps as well as by compensatory mechanisms such as cell surface remodeling. In this context it is interesting that MmpL7, an Mtb efflux pump required for intracellular survival, is thought to exert its virulence effects by transporting phthiocerol dimycocerosates (PDIM) into the bacterial cell wall (Camacho et al. 2001; Cox et al. 1999). Similarly, a role in compensatory cell wall remodeling rather than direct drug transport may explain IniA's contribution to tolerance to isoniazid and ethambutol, drugs that act on the mycobacterial cell wall (Colangeli et al. 2005).

A consideration of the signals and substrates of the macrophage-induced Mtb pumps must account for two observations. First, only a subpopulation of intracellular bacteria exhibits antibiotic tolerance. The most likely explanation for this finding is that there is variation in efflux pump expression, with higher-expressing organisms attaining a drug-tolerant, macrophage growth-adapted phenotype. Why might pump expression vary? Variation could be stochastic or might occur if the pump inducing signal is accessible to only a subset of bacteria, such as the subpopulation of Mm and Mtb that exit the phagosome into the cytosol (Stamm et al. 2003; van der Wel et al. 2007). However, Mm lacking RD1/ESX-1, which is required for cytosolic translocation (Simeone et al. 2012), still become drug-tolerant after macrophage residence (Adams et al. 2011); pump inducing signals must therefore be present in the phagosome. Of note, AMPs are known to access phagosomal bacteria; the macrophage-derived cathelicidin LL-37 has been shown to effectively kill Mtb in cell culture (Liu et al. 2006, 2007).

Second, in advanced human TB, most bacteria reside in the granuloma's necrotic core, known as the caseum (Canetti 1955; Grosset 2003). However, the effects of macrophage-induced tolerance may still be relevant after Mtb has exited macrophages. Down-regulation of efflux pumps may occur relatively slowly and may be balanced by an influx of "freshly tolerant" Mtb brought in by phagocytes that traffic into and lyse within the necrotic caseum (Cosma et al. 2004; Dannenberg 2003). Alternatively, the original stimulus may persist after macrophage lysis; in support of this hypothesis is the finding that macrophage-induced tolerance lasts for at least 5 days in vitro following macrophage lysis (Adams et al. 2011). Finally, additional stimuli in the extracellular environment could also maintain tolerance. Again, AMPs remain viable candidates as they can be produced by a diversity of cell types, including the respiratory epithelium (Parker and Prince 2011).

5 Function and Regulation of Macrophage-Induced Efflux: A Teleological Perspective

It is remarkable that the majority of the Mtb macrophage-induced efflux pumps, including those demonstrated to mediate intracellular growth, are widely conserved among mycobacteria with divergent lifestyles, ranging from the environmental *Mycobacterium smegmatis* to the ultimately host-adapted *Mycobacterium leprae* that is incapable of axenic growth (Tables 1 and 3) (Cole et al. 2001;

Tsukamura 1976). Regulation of these pumps may also be conserved, as seen with Rv1258c. In Mtb, its expression is under the transcriptional control of WhiB7, which belongs to an ancient and highly-conserved family of transcriptional regulators found in multiple actinomycetes including the soil-dwelling *Streptomyces*, *Nocardia*, and both environmental and pathogenic mycobacteria (Morris et al. 2005). WhiB7 mediates the characteristic low-level intrinsic resistance of *Streptomyces* and mycobacteria to antimicrobials of multiple classes (Morris et al. 2005). It is induced in response to sub-inhibitory concentrations of antimicrobials and mediates Rv1258c transcription in these settings. Furthermore, WhiB7 is itself induced by macrophage residence (Larsson et al. 2012; Rohde et al. 2012), and would also be predicted to be required for Rv1258c transcriptional induction and bacterial survival in this context.

Despite the varied environments different mycobacterial species face, they may share signals and substrates for efflux pumps. Environmental mycobacteria like *M. smegmatis* may be using pumps to defend against small molecules such as lantibiotics and antibiotics produced by environmental competitors and perhaps enhance growth within free-living amoebae (Asaduzzaman and Sonomoto 2009; Lamrabet et al. 2012). The capacity to extrude AMP-like peptides may have allowed mycobacteria to expand further into intracellular niches, and thereby to a wide range of complex hosts. While the strictly host-adapted bacteria like Mtb and *M. leprae* have not been subjected to environmental antibiotic pressure for millennia, these skills again found use with introduction of antimicrobials into medical practice in the twentieth century.

6 Therapeutic Implications for Drug Tolerance

The conservation of these macrophage-induced pumps in a range of pathogenic mycobacteria suggests their inhibition may constitute a therapeutic strategy not only for TB, but for other difficult to treat mycobacterial diseases like leprosy, Buruli ulcer, and pulmonary infections with *M. avium* (Table 3). Indeed, Rv1258c has homologs in these species and rifampicin plays an important part in their treatment (Tables 1 and 3). Multiple drugs—verapamil, reserpine, phenothiazines such as thioridazine, and piperine—have been shown to inhibit bacterial efflux pumps in vitro (Kaatz 2005; Marquez 2005; Rodrigues et al. 2011a; Sharma et al. 2010). In general, the mechanisms by which these agents act are poorly understood. Several models have been proposed, such as: (1) direct binding and inhibition of pump assembly or function; (2) disruption of the transmembrane gradients utilized by secondary transporters; (3) inhibitor binding to the antimicrobial compound; (4) competition for efflux (Marquez 2005; Martins et al. 2008; Pages and Amaral 2009; Piddock 2006a). It is worth noting that some of these efflux pump inhibitors *may also* block macrophage antibiotic efflux, leading to increased intracellular drug levels (Cao et al. 1992), an effect that would potentiate their effect on the bacteria.

Table 3 Mycobacterial species with homologs of *Mycobacterium tuberculosis* macrophage-induced pumps

Species (Genome Size)[a]	Environmental Niche	Natural Vertebrate Host	Host Niche	Associated Human Disease(s)	Treatment[b]	References
M. smegmatis (7 MB)	Soil	None known	Not applicable	Extremely rare. Case reports primarily of localized disease, e.g. wound infections.	Optimal therapy unknown; resistant to multiple drugs including *rifampicin*.	Long et al. (2012), Pierre-Audigier et al. (1997), Tsukamura (1976), Wallace et al. (1988)
M. marinum (6.6 MB)	Water Amoebae	Fish and amphibians	Intra and extracellular	"Fish tank granuloma"	Clarithromycin or minocycline or *rifampicin* plus ethambutol	Linell and Norden (1954), Stinear et al. (2008), Yanong et al. (2010)
M. ulcerans (5.6 MB)	Aquatic insects	None known	Mainly extracellular after brief intracellular phase	Buruli ulcer	Surgery *rifampicin* and streptomycin	Doig et al. (2012), George et al. (1999), Wansbrough-Jones and Phillips (2006)
M. avium complex (5.5 MB)	Soil and water Amoebae insects, earthworms	Birds, domesticated and nondomesticated mammals	Intracellular	Pulmonary and systemic infections, especially in the immunocompromised.	Pulmonary disease: clarithromycin or azithromycin plus *rifampicin* and ethambutol, with or without an aminoglycoside	Beumer et al. (2010), Biet et al. (2005), Falkinham (2010), Falkinham et al. (2001), Yamazaki et al. (2006)

(continued)

Table 3 (continued)

Species (Genome Size)[a]	Environmental Niche	Natural Vertebrate Host	Host Niche	Associated Human Disease(s)	Treatment[b]	References
M. abscessus (5.1 MB)	Water Amoebae	Fish Amphibians	Intra and extracellular	Pneumonia, soft tissue infection, and disseminated infection in the immunocompromised	Multidrug resistant including to rifampicin Pulmonary and disseminated infection unlikely to be cured. Amikacin, imipenem, linezolid, tigecycline retain activity	Medjahed et al. (2010), Nessar et al. (2012), Ripoll et al. (2009)
M. tuberculosis (4.4 MB)	None known	Humans	Intra and extracellular	Tuberculosis	isoniazid, *rifampicin,* pyrazinamide, and ethambutol	Grosset (2003), Kumar and Rao (2011)
M. leprae (3.3 MB)	None known	Humans Recently introduced into armadillos	Intracellular	Leprosy	dapsone and *rifampicin*, with clofazimine added for multibacillary disease	Rodrigues and Lockwood (2011), Singh and Cole (2011)

[a] (Reddy et al. 2009) [b] (Chauty et al. 2007; Mandell et al. 2010)

Verapamil, a calcium channel antagonist long in clinical use, is perhaps the most promising inhibitor for further evaluation as an adjunctive TB agent given its ability to reverse macrophage-induced tolerance to rifampicin (Adams et al. 2011). Other candidates include piperine, a derivative of black pepper that has been proposed to inhibit Mtb efflux pumps including Rv1258c (Sharma et al. 2010; Srinivasan 2007); agents developed to counter Gram-negative efflux pumps such as the Phe-Arg-*β*-naphthylamine derivatives (Pages and Amaral 2009); and P-glycoprotein inhibitors originally studied in cancer such as tariquidar (Leitner et al. 2011). The greatest benefit may come from approaches that inhibit multiple pumps, either through broadly-acting inhibitors or a combination of more specific inhibitors.

Could clinically significant resistance to efflux pump inhibitors arise? Certainly these compounds could be vulnerable to many of the same mycobacterial defensive measures used against traditional antimicrobials such as decreased membrane permeability, chemical inactivation of the inhibitor, pump overexpression, pump mutation (Ahmed et al. 1993; Klyachko et al. 1997), or efflux of the inhibitor (Garvey and Piddock 2008). It is possible that the barrier to resistance for efflux pump inhibitors is higher than with traditional antimicrobials. For example, alteration of a binding site on one pump might not be sufficient to confer inhibitor resistance, if the inhibitor can target multiple bacterial pumps involved in tolerance. Moreover, inhibitors that additionally act on macrophage efflux may present a further barrier to evolution of resistance.

7 Efflux Pump Inhibition for Drug-Resistant TB

Appreciation for the potential of efflux pump inhibition strategies in drug-tolerant Mtb is recent, but joins a growing interest in developing this strategy for genetically drug-resistant Mtb (Amaral et al. 2010), where current treatment options are limited by even longer duration and increased toxicity (World Health Organization 2011). A substantial proportion of drug-resistant isolates have no identifiable mutations in known resistance-associated genes, and it appears that resistance in some of these isolates may result from increased efflux activity (Louw et al. 2009). In fact, multiple studies have reported increased efflux pump expression in Mtb clinical isolates (Gupta et al. 2006, 2010; Hao et al. 2011; Jiang et al. 2008; Siddiqi et al. 2004). Accordingly, efflux pump inhibitors have been shown to reduce isoniazid, ciprofloxacin, ofloxacin, streptomycin, and linezolid minimum inhibitory concentrations in resistant strains (Escribano et al. 2007; Machado et al. 2012; Richter et al. 2007; Rodrigues et al. 2012; Singh et al. 2011; Spies et al. 2008). Although in vivo data are limited, a promising recent study found that verapamil restored activity of isoniazid, rifampicin, and pyrazinamide against MDR-TB in mice (Louw et al. 2011).

Efflux has been generally associated with low-level intrinsic drug resistance, which may nevertheless exceed clinical breakpoints and can be further amplified by pumps of overlapping substrate specificities (Lee et al. 2000; Piddock 2006b).

Moreover, this resistance may confer a survival advantage during antibiotic treatment that allows other chromosomal mutations to accumulate, further increasing the degree of antimicrobial resistance (Srivastava et al. 2010). Efflux pump activity also appears able to induce cross resistance to structurally and mechanistically diverse compounds: rifampicin treatment of rifampicin-resistant Mtb induced resistance to ofloxacin, which could be reversed with efflux pump inhibitors (Louw et al. 2011). The clinical implications of this phenomenon are quite serious. In areas where access to mycobacterial cultures are limited, the standard TB regiments prescribed to patients with unrecognized drug-resistant TB may not only have minimal efficacy, they may serve to further limit treatment options. Thus, the addition of efflux pump inhibitors to overcome drug-tolerance may have the additional benefit of reducing emergence of genetically drug-resistant Mtb.

8 Conclusions

Long recognized as a common mechanism of genetic antimicrobial resistance, efflux pump activity may play a dual role in Mtb, contributing to both virulence and drug tolerance. These pumps may have originally served to defend against environmental toxins that included antibiotics, but came to be utilized by pathogenic mycobacteria for intracellular survival. It is intriguing that their ancient function has come "full circle"—these pumps provide Mtb with a survival advantage in the era of antituberculous chemotherapy. The demonstration of verapamil's activity against macrophage-induced tolerance in the laboratory warrants its assessment in TB patients to determine if it will permit treatment shortening. Further understanding of efflux-mediated drug tolerance may pave the way for new efflux pump inhibitors as well as complementary strategies to kill drug-tolerant mycobacteria.

Acknowledgments JDS is supported by NIH T32 5T32AI007044-36 and the Merle A. Sande-Pfizer Fellowship in International Infectious Diseases. KNA is supported by NIH T32 AI55396. PHE is supported by NIH 5R18HS020002-03. LR is supported by grants from the NIH and is the recipient of the NIH Director's Pioneer Award.

We thank Josh Woodward, Christine Cosma, and Mark Troll for helpful discussions and manuscript review.

References

Adams KN, Takaki K, Connolly LE, Wiedenhoft H, Winglee K, Humbert O, Edelstein PH, Cosma CL, Ramakrishnan L (2011) Drug tolerance in replicating mycobacteria mediated by a macrophage-induced efflux mechanism. Cell 145:39–53. doi:10.1016/j.cell.2011.02.022

Aeschlimann JR, Kaatz GW, Rybak MJ (1999) The effects of NorA inhibition on the activities of levofloxacin, ciprofloxacin and norfloxacin against two genetically related strains of *Staphylococcus aureus* in an in vitro infection model. J Antimicrob Chemother 44:343–349

Ahmed M, Borsch CM, Neyfakh AA, Schuldiner S (1993) Mutants of the *Bacillus subtilis* multidrug transporter Bmr with altered sensitivity to the antihypertensive alkaloid reserpine. J Biol Chem 268:11086–11089

Ainsa JA, Blokpoel MC, Otal I, Young DB, De Smet KA, Martin C (1998) Molecular cloning and characterization of tap, a putative multidrug efflux pump present in Mycobacterium fortuitum and Mycobacterium tuberculosis. J Bacteriol 180:5836–5843

Akira M, Sakatani M, Ishikawa H (2000) Transient radiographic progression during initial treatment of pulmonary tuberculosis: CT findings. J Comput Assist Tomogr 24:426–431

Amaral L, Boeree MJ, Gillespie SH, Udwadia ZF, van Soolingen D (2010) Thioridazine cures extensively drug-resistant tuberculosis (XDR-TB) and the need for global trials is now! Int J Antimicrob Agents 35:524–526. doi:10.1016/j.ijantimicag.2009.12.019

Asaduzzaman SM, Sonomoto K (2009) Lantibiotics: diverse activities and unique modes of action. J Biosci Bioeng 107:475–487. doi:10.1016/j.jbiosc.2009.01.003

Balganesh M, Dinesh N, Sharma S, Kuruppath S, Nair AV, Sharma U (2012) Efflux pumps of Mycobacterium tuberculosis Play a significant role in antituberculosis activity of potential drug candidates. Antimicrob Agents Chemother 56:2643–2651. doi:10.1128/AAC.06003-11

Balganesh M, Kuruppath S, Marcel N, Sharma S, Nair A, Sharma U (2010) Rv1218c, an ABC transporter of *Mycobacterium tuberculosis* with implications in drug discovery. Antimicrob Agents Chemother 54:5167–5172. doi:10.1128/AAC.00610-10

Ball PR, Shales SW, Chopra I (1980) Plasmid-mediated tetracycline resistance in Escherichia coli involves increased efflux of the antibiotic. Biochem Biophys Res Commun 93:74–81

Barker J, Scaife H, Brown MR (1995) Intraphagocytic growth induces an antibiotic-resistant phenotype of *Legionella pneumophila*. Antimicrob Agents Chemother 39:2684–2688

Bayer AS, Kupferwasser LI, Brown MH, Skurray RA, Grkovic S, Jones T, Mukhopadhay K, Yeaman MR (2006) Low-level resistance of *Staphylococcus aureus* to thrombin-induced platelet microbicidal protein 1 in vitro associated with qacA gene carriage is independent of multidrug efflux pump activity. Antimicrob Agents Chemother 50:2448–2454. doi:10.1128/AAC.00028-06

Bengoechea JA, Skurnik M (2000) Temperature-regulated efflux pump/potassium antiporter system mediates resistance to cationic antimicrobial peptides in yersinia. Mol Microbiol 37:67–80

Beumer A, King D, Donohue M, Mistry J, Covert T, Pfaller S (2010) Detection of Mycobacterium avium subsp. paratuberculosis in drinking water and biofilms by quantitative PCR. Appl Environ Microbiol 76:7367–7370. doi:10.1128/AEM.00730-10

Bianco MV, Blanco FC, Imperiale B, Forrellad MA, Rocha RV, Klepp LI, Cataldi AA, Morcillo N, Bigi F (2011) Role of P27–P55 operon from *Mycobacterium tuberculosis* in the resistance to toxic compounds. BMC Infect Dis 11:195. doi:10.1186/1471-2334-11-195

Biet F, Boschiroli ML, Thorel MF, Guilloteau LA (2005) Zoonotic aspects of *Mycobacterium bovis* and *Mycobacterium avium-intracellulare* complex (MAC). Vet Res 36:411–436. doi:10.1051/vetres:2005001

Bigger J (1944) Treatment of staphylococcal infections with penicillin. Lancet 244:497–500

Bigi F, Gioffre A, Klepp L, Santangelo MP, Alito A, Caimi K, Meikle V, Zumarraga M, Taboga O, Romano MI, Cataldi A (2004) The knockout of the lprG-Rv1410 operon produces strong attenuation of *Mycobacterium tuberculosis*. Microbes Infect Inst Pasteur 6:182–187. doi:10.1016/j.micinf.2003.10.010

Bina XR, Lavine CL, Miller MA, Bina JE (2008a) The AcrAB RND efflux system from the live vaccine strain of *Francisella tularensis* is a multiple drug efflux system that is required for virulence in mice. FEMS Microbiol Lett 279:226–233. doi:10.1111/j.1574-6968.2007.01033.x

Bina XR, Provenzano D, Nguyen N, Bina JE (2008b) Vibrio cholerae RND family efflux systems are required for antimicrobial resistance, optimal virulence factor production, and colonization of the infant mouse small intestine. Infect Immun 76:3595–3605. doi:10.1128/IAI.01620-07

Blair JM, Piddock LJ (2009) Structure, function and inhibition of RND efflux pumps in Gram-negative bacteria: an update. Curr Opin Microbiol 12:512–519. doi:10.1016/j.mib.2009.07.003

Bobrowitz ID (1980) Reversible roentgenographic progression in the initial treatment of pulmonary tuberculosis. Am Rev Respir Dis 121:735–742

Braibant M, Gilot P, Content J (2000) The ATP binding cassette (ABC) transport systems of *Mycobacterium tuberculosis*. FEMS Microbiol Rev 24:449–467

Brissette CA, Lukehart SA (2007) Mechanisms of decreased susceptibility to beta-defensins by *Treponema denticola*. Infect Immun 75:2307–2315. doi:10.1128/IAI.01718-06

Buckley AM, Webber MA, Cooles S, Randall LP, La Ragione RM, Woodward MJ, Piddock LJ (2006) The AcrAB-TolC efflux system of *Salmonella enterica* serovar Typhimurium plays a role in pathogenesis. Cell Microbiol 8:847–856. doi:10.1111/j.1462-5822.2005.00671.x

Bunikis I, Denker K, Ostberg Y, Andersen C, Benz R, Bergstrom S (2008) An RND-type efflux system in *Borrelia burgdorferi* is involved in virulence and resistance to antimicrobial compounds. PLoS Pathog 4:e1000009. doi:10.1371/journal.ppat.1000009

Camacho LR, Constant P, Raynaud C, Laneelle MA, Triccas JA, Gicquel B, Daffe M, Guilhot C (2001) Analysis of the phthiocerol dimycocerosate locus of *Mycobacterium tuberculosis*. Evidence that this lipid is involved in the cell wall permeability barrier. J Biol Chem 276:19845–19854. doi:10.1074/jbc.M100662200

Camus JC, Pryor MJ, Medigue C, Cole ST (2002) Re-annotation of the genome sequence of *Mycobacterium tuberculosis* H37Rv. Microbiology 148:2967–2973

Canetti G (1955) The tubercle bacillus in the pulmonary lesion of man, histobacteriology and its bearing on the therapy of pulmonary tuberculosis. Springer, New York

Cao CX, Silverstein SC, Neu HC, Steinberg TH (1992) J774 macrophages secrete antibiotics via organic anion transporters. J Infect Dis 165:322–328

Castelnuovo B (2010) A review of compliance to anti tuberculosis treatment and risk factors for defaulting treatment in Sub Saharan Africa. Afr Health Sci 10:320–324

Chan K, Knaak T, Satkamp L, Humbert O, Falkow S, Ramakrishnan L (2002) Complex pattern of *Mycobacterium marinum* gene expression during long-term granulomatous infection. Proc Natl Acad Sci U S A 99:3920–3925. doi:10.1073/pnas.002024599

Chan YY, Chua KL (2005) The *Burkholderia pseudomallei* BpeAB-OprB efflux pump: expression and impact on quorum sensing and virulence. J Bacteriol 187:4707–4719. doi:10.1128/JB.187.14.4707-4719.2005

Chauty A, Ardant MF, Adeye A, Euverte H, Guedenon A, Johnson C, Aubry J, Nuermberger E, Grosset J (2007) Promising clinical efficacy of streptomycin-rifampin combination for treatment of buruli ulcer (*Mycobacterium ulcerans* disease). Antimicrob Agents Chemother 51:4029–4035. doi:10.1128/AAC.00175-07

Choudhuri BS, Bhakta S, Barik R, Basu J, Kundu M, Chakrabarti P (2002) Overexpression and functional characterization of an ABC (ATP-binding cassette) transporter encoded by the genes drrA and drrB of *Mycobacterium tuberculosis*. Biochem J 367:279–285. doi:10.1042/BJ20020615

Clay H, Davis JM, Beery D, Huttenlocher A, Lyons SE, Ramakrishnan L (2007) Dichotomous role of the macrophage in early *Mycobacterium marinum* infection of the zebrafish. Cell Host Microbe 2:29–39. doi:10.1016/j.chom.2007.06.004

Colangeli R, Helb D, Sridharan S, Sun J, Varma-Basil M, Hazbon MH, Harbacheuski R, Megjugorac NJ, Jacobs WR Jr, Holzenburg A, Sacchettini JC, Alland D (2005) The *Mycobacterium tuberculosis* iniA gene is essential for activity of an efflux pump that confers drug tolerance to both isoniazid and ethambutol. Mol Microbiol 55:1829–1840. doi:10.1111/j.1365-2958.2005.04510.x

Cole ST, Brosch R, Parkhill J, Garnier T, Churcher C, Harris D, Gordon SV, Eiglmeier K, Gas S, Barry CE 3rd, Tekaia F, Badcock K, Basham D, Brown D, Chillingworth T, Connor R, Davies R, Devlin K, Feltwell T, Gentles S, Hamlin N, Holroyd S, Hornsby T, Jagels K, Krogh A, McLean J, Moule S, Murphy L, Oliver K, Osborne J, Quail MA, Rajandream MA, Rogers J, Rutter S, Seeger K, Skelton J, Squares R, Squares S, Sulston JE, Taylor K, Whitehead S, Barrell BG (1998) Deciphering the biology of *Mycobacterium tuberculosis* from the complete genome sequence. Nature 393:537–544. doi:10.1038/31159

Cole ST, Eiglmeier K, Parkhill J, James KD, Thomson NR, Wheeler PR, Honore N, Garnier T, Churcher C, Harris D, Mungall K, Basham D, Brown D, Chillingworth T, Connor R, Davies RM,

Devlin K, Duthoy S, Feltwell T, Fraser A, Hamlin N, Holroyd S, Hornsby T, Jagels K, Lacroix C, Maclean J, Moule S, Murphy L, Oliver K, Quail MA, Rajandream MA, Rutherford KM, Rutter S, Seeger K, Simon S, Simmonds M, Skelton J, Squares R, Squares S, Stevens K, Taylor K, Whitehead S, Woodward JR, Barrell BG (2001) Massive gene decay in the leprosy bacillus. Nature 409:1007–1011. doi:10.1038/35059006

Connolly LE, Edelstein PH, Ramakrishnan L (2007) Why is long-term therapy required to cure tuberculosis? PLoS Med 4:e120. doi:10.1371/journal.pmed.0040120

East African/British Medical Research Councils (1972) Controlled clinical trial of short-course (6-month) regimens of chemotherapy for treatment of pulmonary tuberculosis. Lancet 1:1079–1085

Hong Kong Chest Service/Tuberculosis Research Centre, Madras/British Medical Research Council (1989) A controlled trial of 3-month, 4-month, and 6-month regimens of chemotherapy for sputum-smear-negative pulmonary tuberculosis. Results at 5 years. Am Rev Respir Dis 139:871–876

Cosma CL, Humbert O, Ramakrishnan L (2004) Superinfecting mycobacteria home to established tuberculous granulomas. Nat Immunol 5:828–835. doi:10.1038/ni1091

Cosma CL, Sherman DR, Ramakrishnan L (2003) The secret lives of the pathogenic mycobacteria. Annu Rev Microbiol 57:641–676. doi:10.1146/annurev.micro.57.030502.091033

Cox JS, Chen B, McNeil M, Jacobs WR Jr (1999) Complex lipid determines tissue-specific replication of *Mycobacterium tuberculosis* in mice. Nature 402:79–83. doi:10.1038/47042

Coyne S, Courvalin P, Perichon B (2011) Efflux-mediated antibiotic resistance in Acinetobacter spp. Antimicrob Agents Chemother 55:947–953. doi:10.1128/AAC.01388-10

Crimmins GT, Herskovits AA, Rehder K, Sivick KE, Lauer P, Dubensky TW Jr, Portnoy DA (2008) *Listeria monocytogenes* multidrug resistance transporters activate a cytosolic surveillance pathway of innate immunity. Proc Natl Acad Sci U S A 105:10191–10196. doi:10.1073/pnas.0804170105

Curry JM, Whalan R, Hunt DM, Gohil K, Strom M, Rickman L, Colston MJ, Smerdon SJ, Buxton RS (2005) An ABC transporter containing a forkhead-associated domain interacts with a serine-threonine protein kinase and is required for growth of *Mycobacterium tuberculosis* in mice. Infect Immun 73:4471–4477. doi:10.1128/IAI.73.8.4471-4477.2005

da Silva PE, Von Groll A, Martin A, Palomino JC (2011) Efflux as a mechanism for drug resistance in *Mycobacterium tuberculosis*. FEMS Immunol Med Microbiol 63:1–9. doi:10.1111/j.1574-695X.2011.00831.x

Danilchanka O, Mailaender C, Niederweis M (2008) Identification of a novel multidrug efflux pump of *Mycobacterium tuberculosis*. Antimicrob Agents Chemother 52:2503–2511. doi:10.1128/AAC.00298-08

Dannenberg AM Jr (1993) Immunopathogenesis of pulmonary tuberculosis. Hospital Practice 28:51–58

Dannenberg AM Jr (2003) Macrophage turnover, division and activation within developing, peak and "healed" tuberculous lesions produced in rabbits by BCG. Tuberculosis 83:251–260

De Rossi E, Arrigo P, Bellinzoni M, Silva PA, Martin C, Ainsa JA, Guglierame P, Riccardi G (2002) The multidrug transporters belonging to major facilitator superfamily in *Mycobacterium tuberculosis*. Mol Med 8:714–724

DeMarco CE, Cushing LA, Frempong-Manso E, Seo SM, Jaravaza TA, Kaatz GW (2007) Efflux-related resistance to norfloxacin, dyes, and biocides in bloodstream isolates of Staphylococcus aureus. Antimicrob Agents Chemother 51:3235–3239. doi:10.1128/AAC.00430-07

Dhar N, McKinney JD (2007) Microbial phenotypic heterogeneity and antibiotic tolerance. Curr Opin Microbiol 10:30–38. doi:10.1016/j.mib.2006.12.007

Dhawan VK, Yeaman MR, Cheung AL, Kim E, Sullam PM, Bayer AS (1997) Phenotypic resistance to thrombin-induced platelet microbicidal protein in vitro is correlated with enhanced virulence in experimental endocarditis due to *Staphylococcus aureus*. Infect Immun 65:3293–3299

Dianiskova P, Kordulakova J, Skovierova H, Kaur D, Jackson M, Brennan PJ, Mikusova K (2011) Investigation of ABC transporter from mycobacterial arabinogalactan biosynthetic cluster. Gen Physiol Biophys 30:239–250. doi:10.4149/gpb_2011_03_239

Ding Y, Onodera Y, Lee JC, Hooper DC (2008) NorB, an efflux pump in *Staphylococcus aureus* strain MW2, contributes to bacterial fitness in abscesses. J Bacteriol 190:7123–7129. doi:10.1128/JB.00655-08

Djoko KY, Franiek JA, Edwards JL, Falsetta ML, Kidd SP, Potter AJ, Chen NH, Apicella MA, Jennings MP, McEwan AG (2012) Phenotypic characterization of a copA mutant of *Neisseria gonorrhoeae* identifies a link between copper and nitrosative stress. Infect Immun 80: 1065–1071. doi:10.1128/IAI.06163-11

Doig KD, Holt KE, Fyfe JA, Lavender CJ, Eddyani M, Portaels F, Yeboah-Manu D, Pluschke G, Seemann T, Stinear TP (2012) On the origin of *Mycobacterium ulcerans*, the causative agent of buruli ulcer. BMC Genomics 13:258. doi:10.1186/1471-2164-13-258

Domenech P, Reed MB, Barry CE 3rd (2005) Contribution of the *Mycobacterium tuberculosis* MmpL protein family to virulence and drug resistance. Infect Immun 73:3492–3501. doi:10.1128/IAI.73.6.3492-3501.2005

Duits LA, Rademaker M, Ravensbergen B, van Sterkenburg MA, van Strijen E, Hiemstra PS, Nibbering PH (2001) Inhibition of hBD-3, but not hBD-1 and hBD-2, mRNA expression by corticosteroids. Biochem Biophys Res Commun 280:522–525. doi:10.1006/bbrc.2000.4157

Ehrchen J, Steinmuller L, Barczyk K, Tenbrock K, Nacken W, Eisenacher M, Nordhues U, Sorg C, Sunderkotter C, Roth J (2007) Glucocorticoids induce differentiation of a specifically activated, anti-inflammatory subtype of human monocytes. Blood 109:1265–1274. doi:10.1182/blood-2006-02-001115

Escribano I, Rodriguez JC, Llorca B, Garcia-Pachon E, Ruiz M, Royo G (2007) Importance of the efflux pump systems in the resistance of *Mycobacterium tuberculosis* to fluoroquinolones and linezolid. Chemotherapy 53:397–401. doi:10.1159/000109769

Falkinham JO 3rd (2010) Hospital water filters as a source of *Mycobacterium avium* complex. J Med Microbiol 59:1198–1202. doi:10.1099/jmm.0.022376-0

Falkinham JO 3rd, Norton CD, LeChevallier MW (2001) Factors influencing numbers of *Mycobacterium avium*, *Mycobacterium intracellulare*, and other Mycobacteria in drinking water distribution systems. Appl Environ Microbiol 67:1225–1231

Farhana A, Kumar S, Rathore SS, Ghosh PC, Ehtesham NZ, Tyagi AK, Hasnain SE (2008) Mechanistic insights into a novel exporter-importer system of *Mycobacterium tuberculosis* unravel its role in trafficking of iron. PLoS One 3:e2087. doi:10.1371/journal.pone.0002087

Ferhat M, Atlan D, Vianney A, Lazzaroni JC, Doublet P, Gilbert C (2009) The TolC protein of *Legionella pneumophila* plays a major role in multi-drug resistance and the early steps of host invasion. PLoS One 4:e7732. doi:10.1371/journal.pone.0007732

Frisk A, Schurr JR, Wang G, Bertucci DC, Marrero L, Hwang SH, Hassett DJ, Schurr MJ (2004) Transcriptome analysis of *Pseudomonas aeruginosa* after interaction with human airway epithelial cells. Infect Immun 72:5433–5438. doi:10.1128/IAI.72.9.5433-5438.2004

Garvey MI, Piddock LJ (2008) The efflux pump inhibitor reserpine selects multidrug-resistant *Streptococcus pneumoniae* strains that overexpress the ABC transporters PatA and PatB. Antimicrob Agents Chemother 52:1677–1685. doi:10.1128/AAC.01644-07

George KM, Chatterjee D, Gunawardana G, Welty D, Hayman J, Lee R, Small PL (1999) Mycolactone: a polyketide toxin from *Mycobacterium ulcerans* required for virulence. Science 283:854–857

Grosset J (2003) *Mycobacterium tuberculosis* in the extracellular compartment: an underestimated adversary. Antimicrob Agents Chemother 47:833–836

World Health Organization (2011) Guidelines for the programmatic management of drug-resistant tuberculosis, 2011 update

Gupta AK, Chauhan DS, Srivastava K, Das R, Batra S, Mittal M, Goswami P, Singhal N, Sharma VD, Venkatesan K, Hasnain SE, Katoch VM (2006) Estimation of efflux mediated multi-drug resistance and its correlation with expression levels of two major efflux pumps in mycobacteria. J Commun Dis 38:246–254

Gupta AK, Katoch VM, Chauhan DS, Sharma R, Singh M, Venkatesan K, Sharma VD (2010) Microarray analysis of efflux pump genes in multidrug-resistant *Mycobacterium tuberculosis* during stress induced by common anti-tuberculous drugs. Microbial Drug Resist 16:21–28. doi:10.1089/mdr.2009.0054

Hagman KE, Pan W, Spratt BG, Balthazar JT, Judd RC, Shafer WM (1995) Resistance of *Neisseria gonorrhoeae* to antimicrobial hydrophobic agents is modulated by the mtrRCDE efflux system. Microbiology 141(Pt 3):611–622

Hao P, Shi-Liang Z, Ju L, Ya-Xin D, Biao H, Xu W, Min-Tao H, Shou-Gang K, Ke W (2011) The role of ABC efflux pump, Rv1456c-Rv1457c-Rv1458c, from *Mycobacterium tuberculosis* clinical isolates in China. Folia Microbiol 56:549–553. doi:10.1007/s12223-011-0080-7

Helling RB, Janes BK, Kimball H, Tran T, Bundesmann M, Check P, Phelan D, Miller C (2002) Toxic waste disposal in Escherichia coli. J Bacteriol 184:3699–3703

Ho RH, Kim RB (2005) Transporters and drug therapy: implications for drug disposition and disease. Clin Pharmacol Ther 78:260–277. doi:10.1016/j.clpt.2005.05.011

Hobby GL, Meyer K, Chaffee E (1942) Observations on the mechanism of action of penicillin. Proc Soc Exp Biol Med 50:281–288

Jerse AE, Sharma ND, Simms AN, Crow ET, Snyder LA, Shafer WM (2003) A gonococcal efflux pump system enhances bacterial survival in a female mouse model of genital tract infection. Infect Immun 71:5576–5582

Jiang X, Zhang W, Zhang Y, Gao F, Lu C, Zhang X, Wang H (2008) Assessment of efflux pump gene expression in a clinical isolate *Mycobacterium tuberculosis* by real-time reverse transcription PCR. Microbial Drug Resist 14:7–11. doi:10.1089/mdr.2008.0772

Jindani A, Aber VR, Edwards EA, Mitchison DA (1980) The early bactericidal activity of drugs in patients with pulmonary tuberculosis. Am Rev Respir Dis 121:939–949

Johnson JL, Hadad DJ, Dietze R, Maciel EL, Sewali B, Gitta P, Okwera A, Mugerwa RD, Alcaneses MR, Quelapio MI, Tupasi TE, Horter L, Debanne SM, Eisenach KD, Boom WH (2009) Shortening treatment in adults with noncavitary tuberculosis and 2-month culture conversion. Am J Respir Crit Care Med 180:558–563. doi:10.1164/rccm.200904-0536OC

Join-Lambert OF, Michea-Hamzehpour M, Kohler T, Chau F, Faurisson F, Dautrey S, Vissuzaine C, Carbon C, Pechere J (2001) Differential selection of multidrug efflux mutants by trovafloxacin and ciprofloxacin in an experimental model of *Pseudomonas aeruginosa* acute pneumonia in rats. Antimicrob Agents Chemother 45:571–576. doi:10.1128/AAC.45.2.571-576.2001

Kaatz GW (2005) Bacterial efflux pump inhibition. Curr Opin Invest Drugs 6:191–198

Kalia NP, Mahajan P, Mehra R, Nargotra A, Sharma JP, Koul S, Khan IA (2012) Capsaicin, a novel inhibitor of the NorA efflux pump, reduces the intracellular invasion of *Staphylococcus aureus*. J Antimicrob Chemother. doi:10.1093/jac/dks232

Klyachko KA, Schuldiner S, Neyfakh AA (1997) Mutations affecting substrate specificity of the *Bacillus subtilis* multidrug transporter Bmr. J Bacteriol 179:2189–2193

Kohler T, van Delden C, Curty LK, Hamzehpour MM, Pechere JC (2001) Overexpression of the MexEF-OprN multidrug efflux system affects cell-to-cell signaling in *Pseudomonas aeruginosa*. J Bacteriol 183:5213–5222

Kraus D, Peschel A (2006) Molecular mechanisms of bacterial resistance to antimicrobial peptides. Curr Top Microbiol Immunol 306:231–250

Kumar D, Rao KV (2011) Regulation between survival, persistence, and elimination of intracellular mycobacteria: a nested equilibrium of delicate balances. Microb Infect/Inst Pasteur 13:121–133. doi:10.1016/j.micinf.2010.10.009

Kupferwasser LI, Skurray RA, Brown MH, Firth N, Yeaman MR, Bayer AS (1999) Plasmid-mediated resistance to thrombin-induced platelet microbicidal protein in staphylococci: role of the qacA locus. Antimicrob Agents Chemother 43:2395–2399

Lamarche MG, Deziel E (2011) MexEF-OprN efflux pump exports the *Pseudomonas quinolone* signal (PQS) precursor HHQ (4-hydroxy-2-heptylquinoline). PLoS One 6:e24310. doi:10.1371/journal.pone.0024310

Lamichhane G, Tyagi S, Bishai WR (2005) Designer arrays for defined mutant analysis to detect genes essential for survival of *Mycobacterium tuberculosis* in mouse lungs. Infect Immun 73:2533–2540. doi:10.1128/IAI.73.4.2533-2540.2005

Lamrabet O, Mba Medie F, Drancourt M (2012) Acanthamoeba polyphaga-enhanced growth of *Mycobacterium smegmatis*. PLoS One 7:e29833. doi:10.1371/journal.pone.0029833

Larsson C, Luna B, Ammerman NC, Maiga M, Agarwal N, Bishai WR (2012) Gene expression of *Mycobacterium tuberculosis* putative transcription factors whiB1-7 in redox environments. PLoS One 7:e37516. doi:10.1371/journal.pone.0037516

Lee A, Mao W, Warren MS, Mistry A, Hoshino K, Okumura R, Ishida H, Lomovskaya O (2000) Interplay between efflux pumps may provide either additive or multiplicative effects on drug resistance. J Bacteriol 182:3142–3150

Leitner I, Nemeth J, Feurstein T, Abrahim A, Matzneller P, Lagler H, Erker T, Langer O, Zeitlinger M (2011) The third-generation P-glycoprotein inhibitor tariquidar may overcome bacterial multidrug resistance by increasing intracellular drug concentration. J Antimicrob Chemother 66:834–839. doi:10.1093/jac/dkq526

Lewis K (2010) Persister cells. Annu Rev Microbiol 64:357–372. doi:10.1146/annurev.micro.112408.134306

Li XZ, Nikaido H (2009) Efflux-mediated drug resistance in bacteria: an update. Drugs 69: 1555–1623. doi:10.2165/11317030-000000000-00000

Li XZ, Nikaido H, Poole K (1995) Role of mexA-mexB-oprM in antibiotic efflux in *Pseudomonas aeruginosa*. Antimicrob Agents Chemother 39:1948–1953

Lin J, Martinez A (2006) Effect of efflux pump inhibitors on bile resistance and in vivo colonization of *Campylobacter jejuni*. J Antimicrob Chemother 58:966–972. doi:10.1093/jac/dkl374

Lin J, Sahin O, Michel LO, Zhang Q (2003) Critical role of multidrug efflux pump CmeABC in bile resistance and in vivo colonization of *Campylobacter jejuni*. Infect Immun 71:4250–4259

Linares JF, Lopez JA, Camafeita E, Albar JP, Rojo F, Martinez JL (2005) Overexpression of the multidrug efflux pumps MexCD-OprJ and MexEF-OprN is associated with a reduction of type III secretion in *Pseudomonas aeruginosa*. J Bacteriol 187:1384–1391. doi:10.1128/JB.187.4.1384-1391.2005

Linell F, Norden A (1954) *Mycobacterium balnei*, a new acid fast bacillus occurring in swimming pools and capable of producing skin lesions in humans. Acta Tuberc Scand Suppl 33:1–84

Liu J, Takiff HE, Nikaido H (1996) Active efflux of fluoroquinolones in *Mycobacterium smegmatis* mediated by LfrA, a multidrug efflux pump. J Bacteriol 178:3791–3795

Liu PT, Stenger S, Li H, Wenzel L, Tan BH, Krutzik SR, Ochoa MT, Schauber J, Wu K, Meinken C, Kamen DL, Wagner M, Bals R, Steinmeyer A, Zugel U, Gallo RL, Eisenberg D, Hewison M, Hollis BW, Adams JS, Bloom BR, Modlin RL (2006) Toll-like receptor triggering of a vitamin D-mediated human antimicrobial response. Science 311:1770–1773. doi:10.1126/science.1123933

Liu PT, Stenger S, Tang DH, Modlin RL (2007) Cutting edge: vitamin D-mediated human antimicrobial activity against *Mycobacterium tuberculosis* is dependent on the induction of cathelicidin. J Immunol 179:2060–2063

Long Q, Zhou Q, Ji L, Wu J, Wang W, Xie J (2012) *Mycobacterium smegmatis* genomic characteristics associated with its saprophyte lifestyle. J Cell Biochem. doi:10.1002/jcb.24199

Louw GE, Warren RM, Gey van Pittius NC, Leon R, Jimenez A, Hernandez-Pando R, McEvoy CR, Grobbelaar M, Murray M, van Helden PD, Victor TC (2011) Rifampicin reduces susceptibility to ofloxacin in rifampicin-resistant *Mycobacterium tuberculosis* through efflux. Am J Respir Crit Care Med 184:269–276. doi:10.1164/rccm.201011-1924OC

Louw GE, Warren RM, Gey van Pittius NC, McEvoy CR, Van Helden PD, Victor TC (2009) A balancing act: efflux/influx in mycobacterial drug resistance. Antimicrob Agents Chemother 53:3181–3189. doi:10.1128/AAC.01577-08

Ma D, Cook DN, Alberti M, Pon NG, Nikaido H, Hearst JE (1995) Genes acrA and acrB encode a stress-induced efflux system of Escherichia coli. Mol Microbiol 16:45–55

Machado D, Couto I, Perdigao J, Rodrigues L, Portugal I, Baptista P, Veigas B, Amaral L, Viveiros M (2012) Contribution of efflux to the emergence of isoniazid and multidrug resistance in *Mycobacterium tuberculosis*. PLoS One 7:e34538. doi:10.1371/journal.pone.0034538

Mandell GL, Bennett JE, Dolin R (2010) Mandell, Douglas, and Bennett's principles and practice of infectious diseases, 7th edn. Churchill Livingstone/Elsevier, Philadelphia

Marquez B (2005) Bacterial efflux systems and efflux pumps inhibitors. Biochimie 87: 1137–1147. doi:10.1016/j.biochi.2005.04.012

Martinez A, Lin J (2006) Effect of an efflux pump inhibitor on the function of the multidrug efflux pump CmeABC and antimicrobial resistance in campylobacter. Foodborne Pathog Dis 3: 393–402. doi:10.1089/fpd.2006.3.393

Martins M, Viveiros M, Amaral L (2008) Inhibitors of Ca2 + and K + transport enhance intracellular killing of *M. tuberculosis* by non-killing macrophages. In vivo 22:69–75

McCune RM Jr, Tompsett R (1956) Fate of *Mycobacterium tuberculosis* in mouse tissues as determined by the microbial enumeration technique. I. The persistence of drug-susceptible tubercle bacilli in the tissues despite prolonged antimicrobial therapy. J Exp Med 104: 737–762

McMurry L, Petrucci RE Jr, Levy SB (1980) Active efflux of tetracycline encoded by four genetically different tetracycline resistance determinants in Escherichia coli. Proc Natl Acad Sci U S A 77:3974–3977

Medjahed H, Gaillard JL, Reyrat JM (2010) *Mycobacterium abscessus*: a new player in the mycobacterial field. Trends Microbiol 18:117–123. doi:10.1016/j.tim.2009.12.007

Mitchison D, Davies G (2012) The chemotherapy of tuberculosis: past, present and future. Int J Tuberc Lung Dis: Off J Int Union Against Tuberc Lung Dis 16:724–732. doi:10.5588/ijtld.12.0083

Molle V, Soulat D, Jault JM, Grangeasse C, Cozzone AJ, Prost JF (2004) Two FHA domains on an ABC transporter, Rv1747, mediate its phosphorylation by PknF, a Ser/Thr protein kinase from *Mycobacterium tuberculosis*. FEMS Microbiol Lett 234:215–223. doi:10.1016/j.femsle.2004.03.033

Morris RP, Nguyen L, Gatfield J (2005) Ancestral antibiotic resistance in Mycobacterium tuberculosis. Proc Natl Acad Sci USA 102(34):12200–12205

Nessar R, Cambau E, Reyrat JM, Murray A, Gicquel B (2012) *Mycobacterium abscessus*: a new antibiotic nightmare. J Antimicrob Chemother 67:810–818. doi:10.1093/jac/dkr578

Neyfakh AA (2002) Mystery of multidrug transporters: the answer can be simple. Mol Microbiol 44:1123–1130

Nishino K, Latifi T, Groisman EA (2006) Virulence and drug resistance roles of multidrug efflux systems of *Salmonella enterica* serovar typhimurium. Mol Microbiol 59:126–141. doi:10.1111/j.1365-2958.2005.04940.x

Ordonez E, Letek M, Valbuena N, Gil JA, Mateos LM (2005) Analysis of genes involved in arsenic resistance in *Corynebacterium glutamicum* ATCC 13032. Appl Environ Microbiol 71:6206–6215. doi:10.1128/AEM.71.10.6206-6215.2005

Padilla E, Llobet E, Domenech-Sanchez A, Martinez–Martinez L, Bengoechea JA, Alberti S (2010) *Klebsiella pneumoniae* AcrAB efflux pump contributes to antimicrobial resistance and virulence. Antimicrob Agents Chemother 54:177–183. doi:10.1128/AAC.00715-09

Pages JM, Amaral L (2009) Mechanisms of drug efflux and strategies to combat them: challenging the efflux pump of Gram-negative bacteria. Biochim Biophys Acta 1794: 826–833. doi:10.1016/j.bbapap.2008.12.011

Parker D, Prince A (2011) Innate immunity in the respiratory epithelium. Am J Respir Cell Mol Biol 45:189–201. doi:10.1165/rcmb.2011-0011RT

Pasca MR, Guglierame P, Arcesi F, Bellinzoni M, De Rossi E, Riccardi G (2004) Rv2686c-Rv2687c-Rv2688c, an ABC fluoroquinolone efflux pump in *Mycobacterium Tuberculosis*. Antimicrob Agents Chemother 48:3175–3178. doi:10.1128/AAC.48.8.3175-3178.2004

Pasca MR, Guglierame P, De Rossi E, Zara F, Riccardi G (2005) mmpL7 gene of *Mycobacterium tuberculosis* is responsible for isoniazid efflux in *Mycobacterium smegmatis*. Antimicrob Agents Chemother 49:4775–4777. doi:10.1128/AAC.49.11.4775-4777.2005

Perez A, Poza M, Fernandez A, Fernandez Mdel C, Mallo S, Merino M, Rumbo-Feal S, Cabral MP, Bou G (2012) Involvement of the AcrAB-TolC efflux pump in the resistance, fitness, and virulence of *Enterobacter cloacae*. Antimicrob Agents Chemother 56:2084–2090. doi:10.1128/AAC.05509-11

Piddock LJ (2006a) Clinically relevant chromosomally encoded multidrug resistance efflux pumps in bacteria. Clin Microbiol Rev 19:382–402. doi:10.1128/CMR.19.2.382-402.2006

Piddock LJ (2006b) Multidrug-resistance efflux pumps—not just for resistance. Nat Rev Microbiol 4:629–636. doi:10.1038/nrmicro1464

Pierre-Audigier C, Jouanguy E, Lamhamedi S, Altare F, Rauzier J, Vincent V, Canioni D, Emile JF, Fischer A, Blanche S, Gaillard JL, Casanova JL (1997) Fatal disseminated *Mycobacterium smegmatis* infection in a child with inherited interferon gamma receptor deficiency. Clinical infectious diseases : an official publication of the Infectious Diseases Society of America 24:982–984

Platz GJ, Bublitz DC, Mena P, Benach JL, Furie MB, Thanassi DG (2010) A tolC mutant of *Francisella tularensis* is hypercytotoxic compared to the wild type and elicits increased proinflammatory responses from host cells. Infect Immun 78:1022–1031. doi:10.1128/IAI.00992-09

Posadas DM, Martin FA, Sabio y Garcia JV, Spera JM, Delpino MV, Baldi P, Campos E, Cravero SL, Zorreguieta A (2007) The TolC homologue of *Brucella suis* is involved in resistance to antimicrobial compounds and virulence. Infect Immun 75:379–389. doi:10.1128/IAI.01349-06

Quillin SJ, Schwartz KT, Leber JH (2011) The novel *Listeria monocytogenes* bile sensor BrtA controls expression of the cholic acid efflux pump MdrT. Mol Microbiol 81:129–142. doi:10.1111/j.1365-2958.2011.07683.x

Ramakrishnan L, Federspiel NA, Falkow S (2000) Granuloma-specific expression of *Mycobacterium virulence* proteins from the glycine-rich PE-PGRS family. Science 288:1436–1439

Ramon-Garcia S, Martin C, Thompson CJ, Ainsa JA (2009) Role of the *Mycobacterium tuberculosis* P55 efflux pump in intrinsic drug resistance, oxidative stress responses, and growth. Antimicrob Agents Chemother 53:3675–3682. doi:10.1128/AAC.00550-09

Reddy TB, Riley R, Wymore F, Montgomery P, DeCaprio D, Engels R, Gellesch M, Hubble J, Jen D, Jin H, Koehrsen M, Larson L, Mao M, Nitzberg M, Sisk P, Stolte C, Weiner B, White J, Zachariah ZK, Sherlock G, Galagan JE, Ball CA, Schoolnik GK (2009) TB database: an integrated platform for tuberculosis research. Nucleic Acids Res 37:D499–D508. doi:10.1093/nar/gkn652

Rengarajan J, Bloom BR, Rubin EJ (2005) Genome-wide requirements for *Mycobacterium tuberculosis* adaptation and survival in macrophages. Proc Natl Acad Sci USA 102: 8327–8332. doi:10.1073/pnas.0503272102

Richter E, Rusch-Gerdes S, Hillemann D (2007) First linezolid-resistant clinical isolates of *Mycobacterium tuberculosis*. Antimicrob Agents Chemother 51:1534–1536. doi:10.1128/AAC.01113-06

Ripoll F, Pasek S, Schenowitz C, Dossat C, Barbe V, Rottman M, Macheras E, Heym B, Herrmann JL, Daffe M, Brosch R, Risler JL, Gaillard JL (2009) Non mycobacterial virulence genes in the genome of the emerging pathogen *Mycobacterium abscessus*. PLoS ONE 4:e5660. doi:10.1371/journal.pone.0005660

Rodrigues L, Ainsa JA, Amaral L, Viveiros M (2011a) Inhibition of drug efflux in mycobacteria with phenothiazines and other putative efflux inhibitors. Recent Pat Anti-Infect Drug Discovery 6:118–127

Rodrigues L, Machado D, Couto I, Amaral L, Viveiros M (2011b) Contribution of efflux activity to isoniazid resistance in the *Mycobacterium tuberculosis* complex. Infection, genetics and evolution : journal of molecular epidemiology and evolutionary genetics in infectious diseases. doi:10.1016/j.meegid.2011.08.009

Rodrigues L, Machado D, Couto I, Amaral L, Viveiros M (2012) Contribution of efflux activity to isoniazid resistance in the *Mycobacterium tuberculosis* complex. Infection, genetics and evolution : journal of molecular epidemiology and evolutionary genetics in infectious diseases 12:695–700. doi:10.1016/j.meegid.2011.08.009

Rodrigues LC, Lockwood D (2011) Leprosy now: epidemiology, progress, challenges, and research gaps. Lancet Infect Dis 11:464–470. doi:10.1016/S1473-3099(11)70006-8

Rohde KH, Veiga DF, Caldwell S, Balazsi G, Russell DG (2012) Linking the transcriptional profiles and the physiological states of *Mycobacterium tuberculosis* during an extended intracellular infection. PLoS Pathog 8:e1002769. doi:10.1371/journal.ppat.1002769

Rosenberg EY, Bertenthal D, Nilles ML, Bertrand KP, Nikaido H (2003) Bile salts and fatty acids induce the expression of Escherichia coli AcrAB multidrug efflux pump through their interaction with rob regulatory protein. Mol Microbiol 48:1609–1619

Saier MH Jr, Yen MR, Noto K, Tamang DG, Elkan C (2009) The transporter classification database: recent advances. Nucleic Acids Res 37:D274–D278. doi:10.1093/nar/gkn862

Sassetti CM, Rubin EJ (2003) Genetic requirements for mycobacterial survival during infection. Proc Natl Acad Sci USA 100:12989–12994. doi:10.1073/pnas.2134250100

Schaefer WB (1954) The effect of isoniazid on growing and resting tubercle bacilli. Am Rev Tuberc 69:125–127

Schnappinger D, Ehrt S, Voskuil MI, Liu Y, Mangan JA, Monahan IM, Dolganov G, Efron B, Butcher PD, Nathan C, Schoolnik GK (2003) Transcriptional adaptation of *Mycobacterium tuberculosis* within macrophages: insights into the phagosomal environment. J Exp Med 198:693–704. doi:10.1084/jem.20030846 jem.20030846 [pii]

Shafer WM, Balthazar JT, Hagman KE, Morse SA (1995) Missense mutations that alter the DNA-binding domain of the MtrR protein occur frequently in rectal isolates of *Neisseria gonorrhoeae* that are resistant to faecal lipids. Microbiology 141(Pt 4):907–911

Shafer WM, Qu X, Waring AJ, Lehrer RI (1998) Modulation of *Neisseria gonorrhoeae* susceptibility to vertebrate antibacterial peptides due to a member of the resistance/nodulation/division efflux pump family. Proc Natl Acad Sci USA 95:1829–1833

Sharma S, Kumar M, Nargotra A, Koul S, Khan IA (2010) Piperine as an inhibitor of Rv1258c, a putative multidrug efflux pump of *Mycobacterium tuberculosis*. J Antimicrob Chemother 65:1694–1701. doi:10.1093/jac/dkq186

Shepard CC (1957) Growth characteristics of tubercle bacilli and certain other mycobacteria in HeLa cells. J Exp Med 105:39–48

Siddiqi N, Das R, Pathak N, Banerjee S, Ahmed N, Katoch VM, Hasnain SE (2004) *Mycobacterium tuberculosis* isolate with a distinct genomic identity overexpresses a tap-like efflux pump. Infection 32:109–111. doi:10.1007/s15010-004-3097-x

Simeone R, Bobard A, Lippmann J, Bitter W, Majlessi L, Brosch R, Enninga J (2012) Phagosomal rupture by *Mycobacterium tuberculosis* results in toxicity and host cell death. PLoS Pathog 8:e1002507. doi:10.1371/journal.ppat.1002507

Singh M, Jadaun GP, Ramdas, Srivastava K, Chauhan V, Mishra R, Gupta K, Nair S, Chauhan DS, Sharma VD, Venkatesan K, Katoch VM (2011) Effect of efflux pump inhibitors on drug susceptibility of ofloxacin resistant *Mycobacterium tuberculosis* isolates. Indian J Med Res 133:535–540

Singh P, Cole ST (2011) *Mycobacterium leprae*: genes, pseudogenes and genetic diversity. Future Microbiol 6:57–71. doi:10.2217/fmb.10.153

Spies FS, da Silva PE, Ribeiro MO, Rossetti ML, Zaha A (2008) Identification of mutations related to streptomycin resistance in clinical isolates of *Mycobacterium tuberculosis* and possible involvement of efflux mechanism. Antimicrob Agents Chemother 52:2947–2949. doi:10.1128/AAC.01570-07

Spivey VL, Molle V, Whalan RH, Rodgers A, Leiba J, Stach L, Walker KB, Smerdon SJ, Buxton RS (2011) Forkhead-associated (FHA) domain containing ABC transporter Rv1747 is positively regulated by Ser/Thr phosphorylation in *Mycobacterium tuberculosis*. J Biol Chem 286:26198–26209. doi:10.1074/jbc.M111.246132

Srinivasan K (2007) Black pepper and its pungent principle-piperine: a review of diverse physiological effects. Crit Rev Food Sci Nutr 47:735–748. doi:10.1080/10408390601062054

Srivastava S, Musuka S, Sherman C, Meek C, Leff R, Gumbo T (2010) Efflux-pump-derived multiple drug resistance to ethambutol monotherapy in *Mycobacterium tuberculosis* and the pharmacokinetics and pharmacodynamics of ethambutol. J Infect Dis 201:1225–1231. doi:10.1086/651377

Stamm LM, Morisaki JH, Gao LY, Jeng RL, McDonald KL, Roth R, Takeshita S, Heuser J, Welch MD, Brown EJ (2003) *Mycobacterium marinum* escapes from phagosomes and is propelled by actin-based motility. J Exp Med 198:1361–1368. doi:10.1084/jem.20031072

Stinear TP, Seemann T, Harrison PF, Jenkin GA, Davies JK, Johnson PD, Abdellah Z, Arrowsmith C, Chillingworth T, Churcher C, Clarke K, Cronin A, Davis P, Goodhead I, Holroyd N, Jagels K, Lord A, Moule S, Mungall K, Norbertczak H, Quail MA, Rabbinowitsch E, Walker D, White B, Whitehead S, Small PL, Brosch R, Ramakrishnan L, Fischbach MA, Parkhill J, Cole ST (2008) Insights from the complete genome sequence of *Mycobacterium marinum* on the evolution of *Mycobacterium tuberculosis*. Genome Res 18:729–741. doi:10.1101/gr.075069.107

Stone BJ, Miller VL (1995) *Salmonella enteritidis* has a homologue of tolC that is required for virulence in BALB/c mice. Mol Microbiol 17:701–712

Taylor DL, Bina XR, Bina JE (2012) *Vibrio cholerae* VexH encodes a multiple drug efflux pump that contributes to the production of cholera toxin and the toxin co-regulated pilus. PLoS One 7:e38208. doi:10.1371/journal.pone.0038208

Tekaia F, Gordon SV, Garnier T, Brosch R, Barrell BG, Cole ST (1999) Analysis of the proteome of *Mycobacterium tuberculosis* in silico. Tuber Lung Dis 79:329–342

Thanassi DG, Cheng LW, Nikaido H (1997) Active efflux of bile salts by Escherichia coli. J Bacteriol 179:2512–2518

Tsukamura M (1976) Properties of *Mycobacterium smegmatis* freshly isolated from soil. Japan J Microbiol 20:355–356

Tzeng YL, Ambrose KD, Zughaier S, Zhou X, Miller YK, Shafer WM, Stephens DS (2005) Cationic antimicrobial peptide resistance in *Neisseria meningitidis*. J Bacteriol 187: 5387–5396. doi:10.1128/JB.187.15.5387-5396.2005

van der Wel N, Hava D, Houben D, Fluitsma D, van Zon M, Pierson J, Brenner M, Peters PJ (2007) *M. tuberculosis* and *M. leprae* translocate from the phagolysosome to the cytosol in myeloid cells. Cell 129:1287–1298. doi:10.1016/j.cell.2007.05.059

Viveiros M, Portugal I, Bettencourt R, Victor TC, Jordaan AM, Leandro C, Ordway D, Amaral L (2002) Isoniazid-induced transient high-level resistance in *Mycobacterium tuberculosis*. Antimicrob Agents Chemother 46:2804–2810

Wallace RJ Jr, Nash DR, Tsukamura M, Blacklock ZM, Silcox VA (1988) Human disease due to *Mycobacterium smegmatis*. J Infect Dis 158:52–59

Wallis RS, Patil S, Cheon SH, Edmonds K, Phillips M, Perkins MD, Joloba M, Namale A, Johnson JL, Teixeira L, Dietze R, Siddiqi S, Mugerwa RD, Eisenach K, Ellner JJ (1999) Drug tolerance in *Mycobacterium tuberculosis*. Antimicrob Agents Chemother 43:2600–2606

Wansbrough-Jones M, Phillips R (2006) Buruli ulcer: emerging from obscurity. Lancet 367:1849–1858. doi:10.1016/S0140-6736(06)68807-7

Ward SK, Abomoelak B, Hoye EA, Steinberg H, Talaat AM (2010) CtpV: a putative copper exporter required for full virulence of *Mycobacterium tuberculosis*. Mol Microbiol 77: 1096–1110. doi:10.1111/j.1365-2958.2010.07273.x

Warner DM, Levy SB (2010) Different effects of transcriptional regulators MarA, SoxS and Rob on susceptibility of Escherichia coli to cationic antimicrobial peptides (CAMPs): Rob-dependent CAMP induction of the marRAB operon. Microbiology 156:570–578. doi:10.1099/mic.0.033415-0

Warner DM, Shafer WM, Jerse AE (2008) Clinically relevant mutations that cause derepression of the *Neisseria gonorrhoeae* MtrC-MtrD-MtrE efflux pump system confer different levels of antimicrobial resistance and in vivo fitness. Mol Microbiol 70:462–478. doi:10.1111/j.1365-2958.2008.06424.x

Woodward JJ, Iavarone AT, Portnoy DA (2010) c-di-AMP secreted by intracellular *Listeria monocytogenes* activates a host type I interferon response. Science 328:1703–1705. doi:10.1126/science.1189801

Woolridge DP, Vazquez-Laslop N, Markham PN, Chevalier MS, Gerner EW, Neyfakh AA (1997) Efflux of the natural polyamine spermidine facilitated by the *Bacillus subtilis* multidrug transporter Blt. The Journal of biological chemistry 272:8864–8866

Wu Y, Vulic M, Keren I, Lewis K (2012) Role of oxidative stress in persister tolerance. Antimicrob Agents Chemother. doi:10.1128/AAC.00921-12

Yamazaki Y, Danelishvili L, Wu M, Hidaka E, Katsuyama T, Stang B, Petrofsky M, Bildfell R, Bermudez LE (2006) The ability to form biofilm influences *Mycobacterium avium* invasion and translocation of bronchial epithelial cells. Cell Microbiol 8:806–814. doi:10.1111/j.1462-5822.2005.00667.x

Yanong RP, Pouder DB, Falkinham JO 3rd (2010) Association of mycobacteria in recirculating aquaculture systems and mycobacterial disease in fish. J Aquat Anim Health 22:219–223. doi:10.1577/H10-009.1

Zahner D, Zhou X, Chancey ST, Pohl J, Shafer WM, Stephens DS (2010) Human antimicrobial peptide LL-37 induces MefE/Mel-mediated macrolide resistance in *Streptococcus pneumoniae*. Antimicrob Agents Chemother 54:3516–3519. doi:10.1128/AAC.01756-09

Zhang M, Yue J, Yang YP, Zhang HM, Lei JQ, Jin RL, Zhang XL, Wang HH (2005) Detection of mutations associated with isoniazid resistance in *Mycobacterium tuberculosis* isolates from China. J Clin Microbiol 43:5477–5482. doi:10.1128/JCM.43.11.5477-5482.2005

Zierski M, Bek E, Long MW, Snider DE Jr (1980) Short-course (6 month) cooperative tuberculosis study in Poland: results 18 months after completion of treatment. Am Rev Respir Dis 122:879–889

Getting Across the Cell Envelope: Mycobacterial Protein Secretion

Aniek D. van der Woude, Joen Luirink and Wilbert Bitter

Abstract Protein secretion is an essential determinant of mycobacterial virulence. *Mycobacterium tuberculosis* has a unique cell envelope consisting of two lipid bilayers, which requires dedicated protein secretion pathways. The conserved general Sec and Tat translocation systems are responsible for protein transport across the inner membrane and are both essential. Additionally, the accessory Sec pathway specifically contributes to virulence. How transport of Sec/Tat substrates across the outer membrane is accomplished is currently an enigma. In addition to these pathways, *M. tuberculosis* also developed specialized secretion systems for protein transport across both membranes, the type VII or ESX secretion systems. Here, we discuss our current knowledge about the mechanisms and substrates of these different protein translocation systems and their role in mycobacterial physiology and virulence.

Contents

A. D. van der Woude · J. Luirink · W. Bitter
Department of Molecular Microbiology, Institute of Molecular Cell Biology, VU University, Amsterdam, The Netherlands

A. D. van der Woude · J. Luirink · W. Bitter (✉)
Department of Medical Microbiology and Infection Control, VU University Medical Center, Amsterdam, The Netherlands
e-mail: w.bitter@vumc.nl

Current Topics in Microbiology and Immunology (2012) 374: 109–134
DOI: 10.1007/82_2012_298

Published Online: 14 December 2012

1 Introduction

Once classified as acid-fast Gram-positive bacteria, mycobacteria have been long considered to be enveloped by a single membrane. However, the mycobacterial cell envelope is in fact even more impermeable for hydrophilic compounds than the cell envelope of Gram-negative bacteria (Jarlier and Nikaido 1990). This characteristic has been attributed to the presence of a "waxy coat", composed of unique mycobacterial (glyco)lipids. Only recently it has been unambiguously shown by cryo-electron tomography that the waxy coat actually consists of an atypical outer membrane surrounded by a loosely associated capsule (Hoffmann et al. 2008; Zuber et al. 2008; Sani et al. 2010). The inner leaflet of this mycolate outer membrane is mainly formed by mycolic acids, long acyl chains that are covalently linked to an arabinogalactan-peptidoglycan polymer. The outer leaflet contains different unusual (glyco)lipids, including again mycolic acids that are now linked to the disaccharide trehalose to form trehalose dimycolate (TDM or cord factor). At the surface of the bacterium is a capsule layer composed of α-glucan, glycolipids and proteins (Sani et al. 2010). Together, the cell envelope represents a thick hydrophobic barrier, which is a major contributor to the high antibiotic resistance of *Mycobacterium tuberculosis*. The fact that their cell envelope comprises two cellular membranes also means that mycobacteria have a periplasm-like compartment. To transport proteins to the different locations in the cell envelope or into the supernatant, mycobacteria have developed dedicated protein secretion pathways, which will be discussed in this chapter.

Protein secretion studies started with extensive (proteomic) analyses to identify proteins of immunological relevance in culture filtrates of *M. tuberculosis* (Nagai et al. 1991; Sonnenberg and Belisle 1997; Rosenkrands et al. 2000; Malen et al. 2007). Among the most abundant culture filtrate proteins are Mpt64, the Antigen 85 complex proteins (Ag85ABC), 19 kDa protein, SodA, EsxA, and EsxB. A number of these proteins are produced with signal sequences and are transferred across the inner membrane via the Sec and the Tat translocation systems. These two systems are omnipresent in the bacterial kingdom and mycobacteria are no exception. The fact that several Sec and Tat substrates have been detected in the culture filtrate, implies that they also cross the special outer membrane. In Gram-negative bacteria the Sec and Tat pathways are often coupled to a secondary secretion mechanism to accomplish outer membrane transport, such as the type II or type V secretion systems. However, since mycobacteria do not possess any of these secretion systems, transport across the outer membrane must be organized differently. In addition to the signal-sequence-dependent proteins, some proteins also seem to be secreted without N-terminal processing. These proteins are secreted by a novel pathway, which has been termed ESX or type VII secretion (T7S). *M. tuberculosis* has five of these T7S systems, which seem to have different functions. Each of the protein transport pathways mentioned above has a specific role in mycobacterial physiology and virulence and will be discussed in the following sections.

2 Translocation Across the Inner Membrane

The ubiquitous Sec and Tat pathways translocate proteins across the inner membrane. These pathways have been extensively studied in a number of bacterial species, most notably *Escherichia coli.* Although a large number of mycobacterial Sec and Tat substrates have been identified, studies dedicated specifically to the working of these translocation systems in mycobacteria are limited, with the exception of the accessory Sec system described in Sects. 2.2 and 2.3. The description of the Sec and Tat secretion systems hereafter will therefore be largely based on their conservation and similarity to the systems of other species.

2.1 Sec Translocation Machinery

The Sec pathway is the major pathway for transport over the cytoplasmic membrane as well as for insertion of integral membrane proteins. Most of our knowledge of this translocation system stems from studies in *E. coli*, but the system is conserved in all bacteria and a homologous system functions for protein import in the endoplasmic reticulum of eukaryotes (recently reviewed by du Plessis et al. (2011), Luirink et al. (2012) and Park and Rapoport (2012)). In bacteria, the core components of this system are three integral membrane proteins SecY, SecE and SecG, which together form a trimeric protein complex: the Sec translocon. Different structure solutions of the Sec translocon show that the protein-conducting channel is formed by the ten transmembrane regions of SecY. Together they form an hour-glass shaped pore, which is enwrapped and stabilized by SecE with SecG associated more peripherally (Breyton et al. 2002; Van den Berg et al. 2004). While SecY and SecE are essential for growth, SecG only increases the transport efficiency and is not essential for translocation per se. An additional auxiliary membrane complex is formed by SecDFYajC, which associates with the translocon. The large periplasmic loops of SecD and SecF are required for efficient protein translocation, but the underlying mechanism is unknown (Nouwen et al. 2005). This auxiliary complex is not

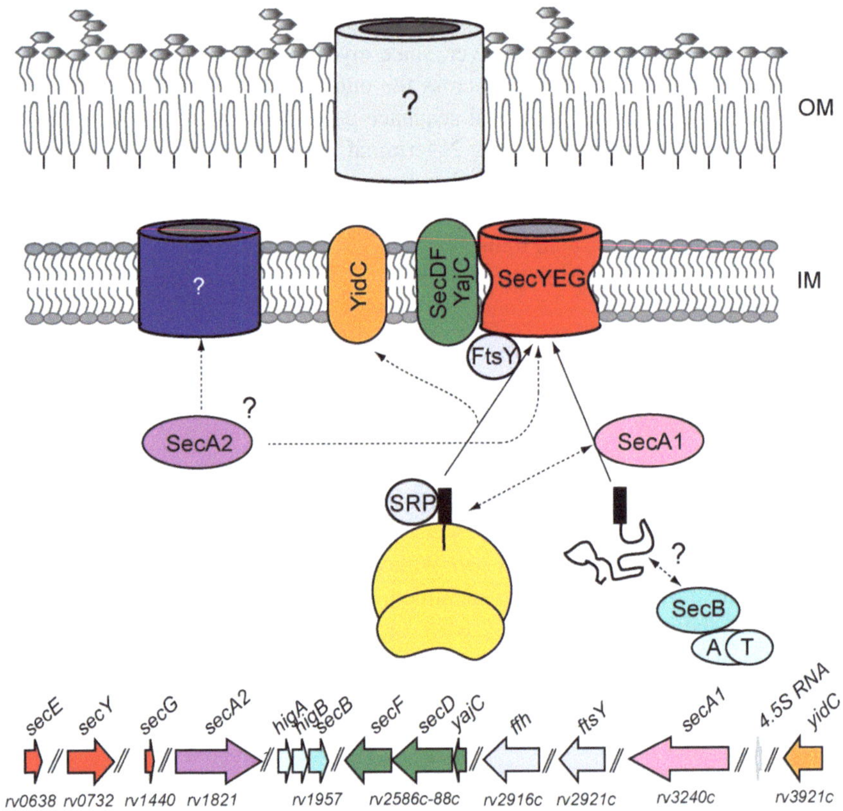

Fig. 1 Model for the general and accessory Sec pathway. Genetic representation and schematic model of the Sec translocation system which targets and transports unfolded preproteins with an N-terminal signal sequence (*in black*). Efficient posttranslational transport through the SecYEG translocon depends on the ATPase SecA1 and the accessory membrane complex SecDFYajC. SecB, which binds to antitoxin (*A*) neutralizing toxin (*T*), might also function as a general chaperone to prevent premature folding of the preprotein. Co-translational targeting of nascent integral inner membrane proteins involves interaction with the SRP that is located near the exit site of the ribosomal tunnel, and FtsY located at the Sec translocon. The role of SecA1 in cotranslational targeting is unclear but it is probably required to energize the insertion of membrane proteins with large translocated domains. YidC is thought to assist in the lateral transfer of transmembrane segments from the Sec translocon into the lipid bilayer. In addition, YidC probably functions as an independent insertase for a subset of small membrane proteins. Secretion of proteins dependent on SecA2 might occur through the canonical SecYEG or an unknown translocon (*in dark purple*). The machinery and mechanism of translocation across the outer membrane (*OM*) is currently completely unclear

absolutely required for Sec transport, but deletion greatly affects bacterial viability. None of the subunits of the Sec trans-locon have been studied specifically for mycobacteria, but they are expected to function in a similar fashion (Fig. 1). High-density transposon mutagenesis in *M. tuberculosis* indicated that all three subunits of the translocon are required for in vitro growth (Griffin et al. 2011), suggesting a difference between mycobacteria and *E. coli* with respect to the role of SecG. The requirement for the auxiliary complex is comparable; SecD and SecF were also scored as essential, whereas YajC appeared nonessential.

The cytosolic ATPase SecA is involved in delivery of preproteins to the tran-slocon and also acts as a motor protein, providing energy for the translocation process. SecA is present in tenfold excess as compared to the other Sec proteins. It has been found to interact with a multitude of molecules, including itself, the ribosome, nascent substrate peptide chain, full length precursor substrate protein, SecB (discussed below), SecY and different phospholipids. Consequently, SecA is found both in the cytosol and associated with the inner membrane. Interestingly, mycobacteria contain two homologs of SecA. Only one of these copies is essential for growth and is considered the 'housekeeping' SecA1. Overexpression of SecA2 does not complement this defect, indicating each SecA performs a specific role in transport (Braunstein et al. 2001). SecA2 is not essential, but is required for full virulence and will be discussed separately in Sect. 2.2. SecA1 has been crystallized as a dimer and the individual subunit structure is similar to other bacterial SecA structures (Sharma et al. 2003). Two nucleotide binding domains (2–221 and 360–587 aa) form the motor domain, hydrolyzing ATP. The translocation domain is composed of a substrate specificity domain (residues 222–359), which binds the preprotein, and a C-terminal helical core domain. Full length *M. tuberculosis* SecA1 does not complement SecA deficiency in *E. coli*, but a chimera with only the N-terminal motor domain of *M. tuberculosis* SecA1 does. This indicates that the substrate specificity domain differs significantly between these species (Owens et al. 2002). SecA has been detected both in monomeric and dimeric form and there is a controversy which conformation is functional. Soluble SecA is usually detected as a homo-dimer, but there is great diversity in dimerization states between different species (Kusters and Driessen 2011). The interaction interface of the dimer structures is sometimes, e.g. for *M. tuberculosis* SecA1, overlapping with the SecA-SecYEG binding region, suggesting that a dimeric form does not reflect the functional state of SecA. Perhaps SecA functions as a monomer during translocation, although lack of definite proof for this mechanism fuels ongoing discussions.

Proteins that are transported via the Sec translocon are synthesized with an N-terminal signal sequence that is removed upon transport. Furthermore, transport through the Sec translocon mainly proceeds in an unfolded state and transport is usually posttranslational. In contrast, insertion of integral membrane proteins via the Sec system is cotranslational and involves the ribosome docking to the Sec-YEG complex via the signal recognition particle (SRP) targeting pathway. Most bacterial SRP's consist of the 4.5 S RNA molecule and the GTPase Ffh, which are both conserved in mycobacteria (Palaniyandi et al. 2012). The SRP binds to ribosomal proteins near the nascent chain exit site and scans nascent chains for the presence of a signal anchor sequence. The complex of ribosome, SRP and nascent membrane protein is then targeted to the SRP receptor FtsY that provides the connection with the Sec translocon to mediate co-translational membrane insertion (Fig. 1). During the insertion process, the Sec translocon opens laterally to allow partitioning of transmembrane segments into the lipid bilayer. The membrane protein YidC has been found to assist the Sec translocon in this process. Furthermore, SecA is required for the translocation of sizeable periplasmic loops of membrane proteins. Of note, a small number of integral inner membrane proteins require only YidC for membrane insertion in a process that is apparently completely independent of the Sec translocon (Luirink et al. 2012). Mycobacteria also have a homologue of the YidC protein, which is essential for growth, but it is unknown whether it has similar functions in membrane protein insertion.

Posttranslational transport of secreted substrates is facilitated by chaperones that keep the preprotein in an unfolded state. In Gram-negative bacteria, this function is usually performed by SecB, which also targets the preprotein to SecA. A SecB-like protein, encoded by *rv1957* in *M. tuberculosis*, can substitute for SecB functions in *E. coli* and thus seems to have a similar chaperone function. Interestingly, this SecB is not generally present in other mycobacteria and is transcribed as part of a toxin-antitoxin operon that is induced under various stress conditions. The unusual SecB was shown to be important for stabilization of the antitoxin protein, thereby neutralizing the toxin (Bordes et al. 2011). Whether this SecB protein has a more general role in the mycobacterial Sec pathway still needs to be determined.

Substrates translocated via the Sec machinery are commonly characterized by a tripartite N-terminal signal sequence, which is cleaved upon transport. The signal sequence consists of a positively charged N terminus followed by a hydrophobic helical region and a polar region containing the cleavage site (including the consensus sequence Ala–X–Ala). Lipoproteins, which are in general also transported by the Sec translocon, contain an additional lipobox motif including a cysteine at the +1 position that is lipid modified prior to cleavage by a specific signal peptidase. Several bioinformatic signal sequence prediction programs have been developed, but none of them seem to be ideal for mycobacteria (Leversen et al. 2009). Experimental approaches using (putative) signal sequences fused to β-lactamase or alkaline phostatase reporter proteins (see also Sect. 2.4) have indicated a large number of proteins with functional signal sequences (Gomez et al. 2000; McCann et al. 2011). However, these methods currently are not specific for Sec substrates and will also identify Tat signal sequences.

2.2 The Accessory Sec Pathway

The mycobacterial genome contains a second copy of SecA (Cole et al. 1998). This SecA2 shares 50 % similarity with the 'housekeeping' SecA1, and is important for mycobacterial virulence (Braunstein et al. 2003), which suggests it is responsible for the secretion of a specific set of substrates involved in virulence. This secretion route, known as the *accessory Sec pathway,* is also present in other mycolic-acid containing bacteria and in many Gram-positive bacteria. Two basic types of accessory Sec systems have been described: the SecA2/SecY2 system and the SecA2 only system. The first type, described for *Streptococcus* sp. and *Staphylococcus* sp., consists not only of SecA2 and SecY2, but also of other proteins specifically involved in protein transport and glycosylation that are all encoded in the same operon (Bensing and Sullam 2002; Siboo et al. 2008). The complete system is involved in transport of a serine-rich glycoprotein that is encoded within the SecA2/SecY2 cluster. This substrate protein, which is secreted independent of the canonical Sec pathway, contains a specific motif behind its relatively long signal sequence that is believed to target it to the accessory Sec pathway (Bensing and Sullam 2010). Interestingly, this complete accessory system can also be functionally expressed in *E. coli* (Bensing et al. 2012). In contrast, substrates of the SecA2 only systems are proposed to codepend on the general Sec pathway. SecA2 only systems are found in different species of mycolates and Gram-positive bacteria, including *Listeria* sp.

How SecA2 functions is still largely unknown. First of all, (how) does SecA2 work in conjunction with the general secretion pathway? SecA2 is roughly 15 kDa smaller than SecA1 and misses large parts of the C terminus that are important for binding to SecB and phospholipids (Breukink et al. 1995; Fekkes et al. 1997; Braunstein et al. 2001). Localization of SecA1 and SecA2 also appears to be different, as shown by studies using *Mycobacterium smegmatis*. Whereas SecA1 is distributed evenly between cell envelope and cytosol, SecA2 localizes primarily in the soluble fraction (Rigel et al. 2009). Similar to generic SecA, the ATPase activity of SecA2 appears important for its function (Hou et al. 2008; Rigel et al. 2009). Disruption of ATP binding even leads to a dominant negative phenotype, demonstrating the importance of ATP hydrolysis for SecA2 function. Interestingly, the ATPase-dead mutant of SecA2 is also affected in localization and is found to be mostly cell envelope-associated (Rigel et al. 2009). As stated before, SecA1 and SecA2 are not functionally redundant. However, depletion of SecA1 in *M. smegmatis* also negatively affected SecA2-mediated secretion, indicating a functional link between the two SecA homologs (Rigel et al. 2009). It therefore seems likely that SecA2 functions in conjunction with the general Sec pathway, and perhaps also communicates with SecA1, possibly through dimerization. Alternatively, a different, currently unidentified translocon is involved in the accessory Sec pathway (Fig. 1). In depth interaction studies might elucidate the mechanism by which SecA2 aids in secretion of specific substrate proteins.

2.3 The Role of SecA2 in Virulence

Deletion of *secA2* in *M. tuberculosis* results in attenuated growth in both cultured macrophages and in murine infection models (Braunstein et al. 2003; Kurtz et al. 2006). However, knowledge about mycobacterial SecA2 substrates is limited. In contrast to the *secA2* mutant of *Listeria monocytogenes*, which lacks a significant amount of its secreted and surface proteins (Lenz and Portnoy 2002; Lenz et al. 2003), only minor differences were found in the culture filtrate of a *secA2* mutant in *M. tuberculosis*. Comparative proteomic analysis identified Superoxide dismutase A (SodA) as SecA2-dependent (Braunstein et al. 2003). Curiously, this putative substrate lacks a signal sequence and is also abundant in the cytosol. Determination of SecA2-dependent substrates in the nonpathogenic species *M. smegmatis* did reveal two specific substrates, both of which contain a predicted N-terminal lipoprotein sequence, Msmeg1704 and Msmeg1712. When cultured in low glucose and glycerol conditions, these homologous lipoproteins were detected in the cell wall, but only when SecA2 and their signal sequences were not perturbed (Gibbons et al. 2007). No homologues of these lipoproteins are present in *M. tuberculosis*, although heterologous expression of Msmeg1704 in *M. tuberculosis* does show moderate SecA2-dependency (Gibbons et al. 2007). Of note, although *M. smegmatis* contains two homologs of SodA, it is unclear whether these are dependent on SecA2.

The function of SodA is to neutralize radical oxygen species, which protects intracellular bacteria during oxidative-burst in macrophages. Therefore, the secretion defect of SodA could be responsible for the observed reduction in virulence of the *secA2* mutant. To test this, SodA was fused to an N-terminal signal sequence, which restored release of SodA into the culture filtrate of the *secA2* mutant, suggesting rerouting to the general Sec system (Hinchey et al. 2007). Although rerouting SodA suppresses the pro-apoptotic phenotype of the *secA2* mutant, it did not restore intracellular growth in macrophages (Hinchey et al. 2007; Sullivan et al. 2012). In addition, the *secA2* mutant is still defective in intracellular growth of oxidative-burst deficient macrophages, further indicating that the absence of SodA secretion is not (solely) responsible for the attenuation of the *secA2* mutant (Kurtz et al. 2006). A specific defect of the *secA2* mutant in phagosome acidification suggests that the accessory Sec pathway also transports effector proteins involved in phagosome maturation arrest (Sullivan et al. 2012). Further proteomic analysis should identify the SecA2 substrates that define its role in virulence.

Another unsolved question is how substrates are targeted to the accessory pathway. Especially SodA presents an enigma, due to its lack of a recognizable signal sequence. Interestingly, also in *L. monocytogenes* several SecA2 substrates, including a homolog of SodA, were shown to lack a signal sequence (Lenz and Portnoy 2002; Archambaud et al. 2006). Although it is currently unknown whether the SecA2-dependency for SodA transport is direct or indirect, increasing evidence indicates that SodA is secreted via the Sec pathway. In *Rhizobium leguminosarum* bv. *viciae* 3841 targeting of SodA to the Sec system was demonstrated to depend on its ten N-terminal amino acids (Krehenbrink et al. 2011). Perhaps this region contains a specific pattern that is sufficient for interaction with/targeting to SecA(2), instead of a classical signal sequence. Because the *M. tuberculosis secA2* mutant is attenuated and defective in apoptosis inhibition, this mutant, combined with an auxotrophic mutation, has been recently tested for vaccine development (Jensen et al. 2012).

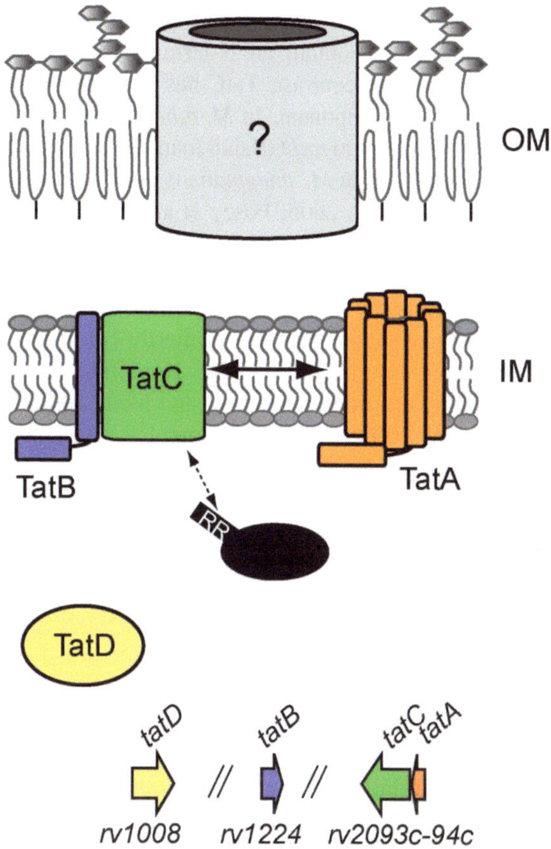

Fig. 2 Tat pathway Genetic representation and schematic model of the Tat translocation system. Folded substrate (*in black*) binds to TatBC depending on an N-terminal signal sequence with a twin-arginine motif (*RR*). Subsequently, TatA subunits are recruited to form a fitting pore for translocation of the preprotein across the inner membrane (*IM*). The role of TatD is unknown. An unidentified outer membrane (*OM*) channel is hypothesized for complete secretion into the extracellular environment

2.4 The Tat Pathway: Transport of Folded Substrates

Many bacteria, including mycobacteria, use the twin-arginine translocation (Tat) system to transport proteins across the inner membrane. The major feature of the Tat system is that it allows secretion of (partially) folded proteins and even oligomers. Like the Sec pathway, the Tat system is best studied in *E. coli*, which was recently reviewed (Palmer and Berks 2012). It consists of three major inner membrane components, TatA, TatB and TatC, which are expressed from an operon in *E. coli*. This operon is also coding for TatD, a cytoplasmic protein with DNase activity. Although this protein is not essential for Tat-mediated transport, it does play a role in turnover of misfolded substrates and thereby contributes to optimal transport (Wexler et al. 2000; Matos et al. 2009). In contrast, most Gram-positive bacteria only have the TatA and TatC component with TatA functionally replacing TatB, and some bacteria have multiple copies of these proteins (Dilks et al. 2003). Mycobacteria have a *tatAC* operon, whereas *tatB* and *tatD* are both separately located elsewhere in the genome (Fig. 2) (McDonough et al. 2005;

Posey et al. 2006). TatA and TatB are small proteins that share limited sequence homology. Both contain an N-terminal transmembrane domain and an amphipathic α-helix. In contrast, TatC has six transmembrane domains and is the most conserved Tat component. In *M. tuberculosis tatA, tatB* and *tatC* are essential for growth in contrast to *tatD* (Saint-Joanis et al. 2006; Griffin et al. 2011). Deletion of *tatA, tatB* or *tatC* in *M. smegmatis* is possible, but results in severe growth defects (McDonough et al. 2005; Posey et al. 2006).

The exact mechanism of translocation via the Tat pathway is unknown. Tat components have been found to form two different membrane complexes. One complex consists mainly of TatC and TatB, whereas the other complex is an oligomer of variable size that primarily consists of TatA subunits (Bolhuis et al. 2001; Mangels et al. 2005). A low resolution structure of the TatBC complex was obtained by electron microscopy and likely consists of 6-8 copies of each protein (Tarry et al. 2009). The TatBC complex, and especially TatC, has been shown to recognize the Tat signal sequence discussed below (Cline and Mori 2001; Tarry et al. 2009). Upon substrate binding, TatA oligomers are recruited, which probably form the translocation channel (Fig. 2). Tat-mediated translocation is energized by the proton-motive force. The reversible association of the Tat subcomplexes allows the system to be modified and activated according to its specific substrates.

Similar to the Sec pathway, Tat substrates are recognized via an N-terminal tripartite signal sequence. The Tat signal sequence is distinguished from the Sec signal sequence by a consensus motif of (S/T)-R-R-x-F-L-K (or, more generally, R-R-x-ϕ-ϕ with ϕ representing an uncharged residue) between the positively charged N terminus and the central hydrophobic domain. The signature arginine pair (twin-arginine) gave rise to the name of the pathway. The two arginine residues are crucial for Tat-mediated transport; mutation of these residues into lysines usually aborts secretion. In addition to the twin-arginine motif, Tat signal sequences have a central domain of lower hydrophobicity and sometimes contain a basic residue in the polar C-terminal region, which was proposed to function as a "Sec-avoidance" motif. Several bioinformatic programs are available to predict Tat substrates based on their signal sequences, including TatP, Tatfind, and Tat-Pred (Rose et al. 2002; Bendtsen et al. 2005; Taylor et al. 2006). A number of these programs have also been used to predict Tat substrates in *M. tuberculosis*, but the overlap in substrate identification using these programs was limited and the programs also suffered from false positive hits (Posey et al. 2006; McDonough et al. 2008). Therefore, mycobacterial Tat substrates will need to be validated experimentally.

Substrates of the Tat pathway in *E. coli* are often proteins that bind cofactor molecules in the cytoplasm, and thus need to be folded prior to export. Using the possibility to create viable *tat* deletion mutants in *M. smegmatis,* several mycobacterial Tat substrates have been identified, including two major β-lactamases, BlaS and BlaC. Consequently, these *tat* mutants are much more sensitive to β-lactam antibiotics (McDonough et al. 2005; Posey et al. 2006). As for all known Tat substrates, export of BlaC was abrogated if the Tat signal sequence was removed or the RR motif mutated. Interestingly, although BlaC does not seem to be associated with a cofactor, its signal sequence could not be functionally replaced by a Sec signal sequence (McDonough et al. 2005), indicating that this protein has to be partially folded in the cytosol. Using truncated BlaC and β-lactam resistance as a reporter for Tat export, a screening method was employed to experimentally identify novel Tat substrates by screening random genomic fusion constructs. This study resulted in 13 novel Tat substrates in *M. tuberculosis,* including four different phospholipases C (McDonough et al. 2008). Additional Tat substrates, including secreted antigenic proteins Apa and Ag85A, were identified in a similar screening method using alkaline phosphatase (PhoA) fusion technology in *E. coli* (Marrichi et al. 2008). One of the identified substrates, phospholipase Rv2525c, was affected in growth rate and β-lactam susceptibility in vitro, but showed enhanced virulence in infection models (Saint-Joanis et al. 2006). Interestingly, a different phospholipase, PlcB, was shown to be Tat-dependent in *M. tuberculosis,* but was not secreted in *M. smegmatis.* This suggests that *M. tuberculosis* could have an additional, yet unidentified chaperone that assist in Tat-mediated translocation (McDonough et al. 2008).

Although the currently described Tat substrates have been shown to influence mycobacterial growth and virulence, none of these Tat substrates are essential for viability. Since the Tat system itself is essential in *M. tuberculosis*, this means that either additional indispensable Tat substrates remain to be identified or that several Tat substrates have an overlapping but essential function.

3 Type VII Secretion

Between 1908 and 1919 at the Institut Pasteur in Lille, Calmette and Guérin cultured a virulent *M. bovis* strain by 230 serial passages on glycerinated bile potato medium until it lost its virulence potential. This attenuation of *M. bovis* resulted in a strain that could be used as a vaccine for tuberculosis, the bacille Calmette-Guérin (BCG). The distribution of this strain to several countries during the 1920s and subsequent subculturing resulted in a number of different BCG vaccine strains. Compared to the original *M. bovis* strain, all BCG vaccine strains consistently lacked one common genomic locus, known as 'region of difference 1'(RD1) (Mahairas et al. 1996; Behr et al. 1999). Deletion of the RD1 locus is one of the main events causing attenuation of the *M. bovis* BCG strain. (Pym et al. 2002; Lewis et al. 2003). The RD1 locus comprises nine genes, including the genes coding for two small secreted proteins, EsxA and EsxB (also known as ESAT-6 and CFP-10, respectively) (Sorensen et al. 1995; Harboe et al. 1996; Berthet et al. 1998). These small secreted proteins are cotranscribed and lack an apparent signal sequence (Berthet et al. 1998).

Completion of the *M. tuberculosis* H37Rv genome sequence revealed the presence of 23 homologs of the *esxAB* genes in 11 different loci (Cole et al. 1998), which were all named *esx* genes. The encoded Esx proteins are generally characterized by a length of $\sim$100 aa and a central WXG motif, and therefore also referred to as WXG100 family proteins (Pallen 2002). Inspection of the genes flanking the *esx* gene couples, showed a specific functional genomic organization for five duplications, which were proposed to form a dedicated secretion apparatus (Tekaia et al. 1999; Gey Van Pittius et al. 2001). This theory was proven in an elegant approach, in which different parts of the RD1 locus were re-introduced in *M. bovis* BCG. Indeed, only the complete RD1 locus was sufficient to restore functional secretion of EsxAB (Pym et al. 2003). This RD1-encoded dedicated secretion system, together with the other four homologous regions were renamed ESX- or type VII secretion (T7S) systems (Abdallah et al. 2007).

3.1 Type VII Secretion Components

The five T7S or ESX systems of *M. tuberculosis* are likely derived from gene duplication events, in the order ESX-4, ESX-1, ESX-3, ESX-2, and finally ESX-5 (Fig. 3d) (Gey Van Pittius et al. 2001). Until now, only the ESX-1, ESX-3, and ESX-5 clusters have been experimentally shown to encode functional secretion systems. Although activity of the ancestral ESX-4 system remains to be shown, it is part of the SigM regulon, which suggests it has a function in vivo (Raman et al. 2006). Also the ESX-2 system does not seem to be expressed under laboratory conditions, but is specifically upregulated during infection (Talaat et al. 2004). Each ESX system appears to have a different role in mycobacterial virulence, but the genetic composition and organization of the five ESX clusters is similar (Fig. 3d). The ESX clusters each contain two *esx* genes surrounded by genes encoding ESX conserved components EccA to EccE and MycP (Bitter et al. 2009), although some of two of these components (EccA and EccE) are lacking in the ancestral ESX-4.

Five ESX components are predicted to be associated with the inner membrane. The first membrane component is EccB, which has one predicted transmembrane domain and a large periplasmic domain. EccC contains 3 GTP/ATP binding sites and 2 N-terminal transmembrane domains. This protein belongs to the family of FtsK/SpoIIIE ATPases and resembles the DNA-coupling protein VirD4, an essential component of the type IV secretion pathway. In some ESX systems *eccC* is split in two different genes (Fig. 3d). The integral membrane protein EccD contains up to twelve transmembrane domains and has been proposed to form the transport channel. EccE has two predicted N-terminal membrane domains and is not found in the ancestral ESX-4 system, which indicates it is the most recent addition to the machinery. Finally, the mycosin (MycP) is a membrane-anchored protease. The catalytic triad domain of MycP, typical for a subtilisin-like serine protease, is located between an N-terminal signal sequence and a C-terminal domain resembling a transmembrane domain, which suggests the active site is located to the periplasm (Brown et al. 2000; Dave et al. 2002). Interestingly, a mutation in the catalytic domain of MycP1 resulted in an attenuated strain but actually enhanced secretion of EsxA (Ohol et al. 2010).

Although studies of the ESX-1 secretion system have demonstrated that these membrane-associated components are each essential for translocation (Stanley et al. 2003; Converse and Cox 2005; Ohol et al. 2010; Champion et al. 2012) it is still unknown how they interact to form a translocation machinery. Preliminary studies of the ESX-5 system have shown that the EccBCDE components form a large multimeric complex of > 1000 kDa, which does not include MycP (Houben et al. 2012b). This indicates that a large translocon is formed by multiple copies of each component, which is likely energized by EccC (Fig. 3a). Whether this large translocon is sufficient to accomplish transport across both membranes of the mycobacterial cell envelope or an additional (unknown) outer membrane component is required (Fig. 3a), remains to be determined.

The ESX systems, except again for ESX-4, also contain the cytosolic ATPase EccA, which belongs to the AAA+ class. Studies of heterologous EccA1 have shown that it forms a hexameric complex with ATPase activity (Luthra et al. 2008). A similar complex was observed for the AAA + ATPases of the type VI secretion pathway, which are crucial for the assembly of system machinery (Bönemann et al. 2010). Disruption of the *eccA* gene of the ESX-1 and ESX-5 systems affects their respective secretion pathways (Gao et al. 2004; Abdallah et al. 2006; Bottai et al. 2012). Yeast-2-hybrid interaction studies suggest that the EccA1 ATPase interacts with some, but not all, ESX-1 substrates (DiGiuseppe Champion et al. 2009). This substrate specificity also correlates with the observed instability of specific ESX-1 substrates upon disruption of *eccA1* (Brodin et al. 2006; McLaughlin et al. 2007; Carlsson et al. 2009).

Although T7S is best studied in mycobacteria, it is not restricted to this genus. In fact, it is found among a large number of Gram-positive bacteria (Pallen 2002), in which it frequently has a role in virulence (Burts et al. 2005; Abdallah et al. 2007). Many GC-rich Gram-positive bacteria, such as *Corynebacterium diphteriae* and *Streptomyces coelicolor*, contain a T7S cluster homologous to the mycobacterial ESX-4 cluster, which suggests that these are derived from a common ancestor (Gey Van Pittius et al. 2001). T7S-like systems of low-GC Gram positives are evolutionary more distant from those in mycobacteria and in fact only contain as conserved components the WXG100 family proteins and a homologue of the FtsK/SpoIII ATPase EccC. A growing number of these type VII secretion systems have been demonstrated to function in transport of their respective Esx proteins (Burts et al. 2005; Garufi et al. 2008; Akpe San Roman et al. 2010).

3.2 Esx Proteins and Other Substrates

The Esx proteins are the most recognizable T7S substrates. Most of the 23 members of this protein family found in *M. tuberculosis* H37Rv, are either encoded as a pair inside one of the *esx* clusters, or are direct duplications of one of these pairs (Gey Van Pittius et al. 2001). EsxA and EsxB, substrates of the ESX-1 system, are the best studied Esx proteins. These proteins form a tight 1:1

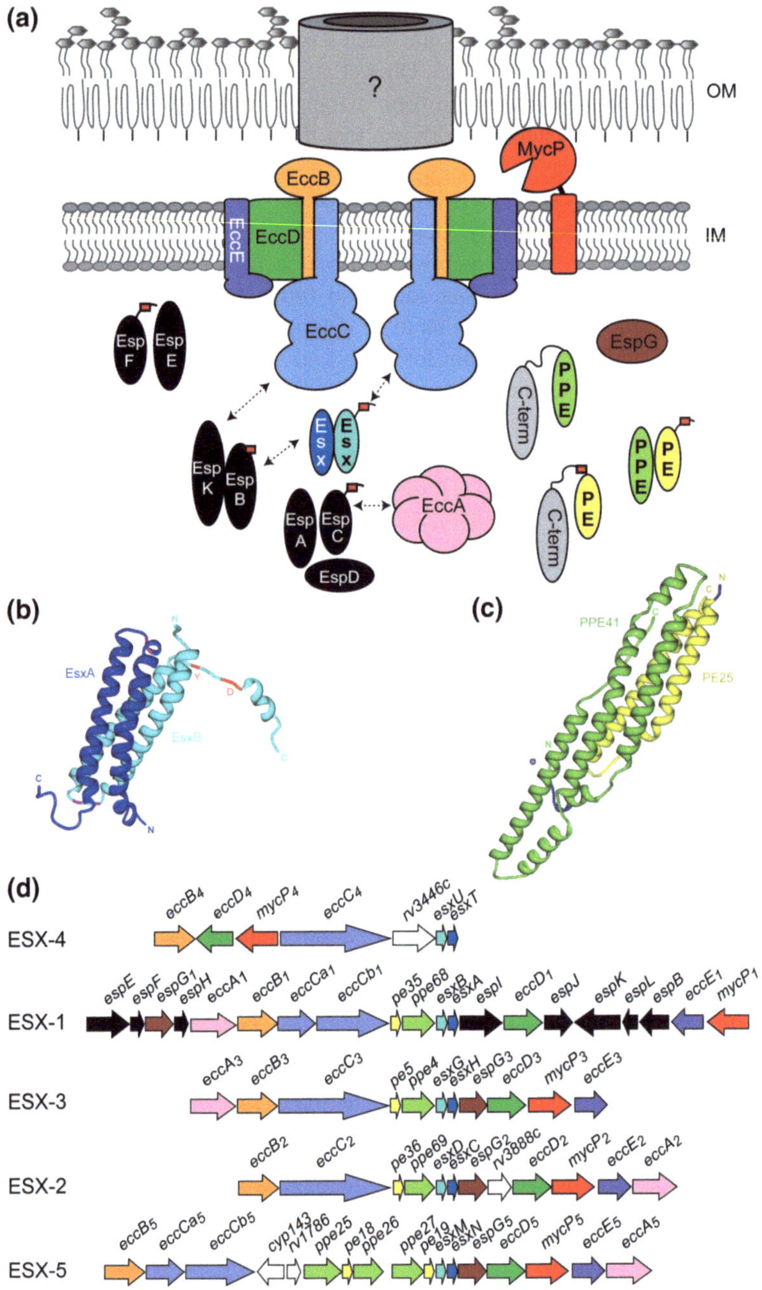
(a)
?
OM
MycP
EccB
IM
EccE
EccD
EccC
Esp F
Esp E
EspG
PPE
C-term
Esx
Esx
Esp K
Esp B
PPE
PE
Esp A
Esp C
EccA
PE
C-term
EspD
(b)
EsxA
EsxB
(c)
PPE41
PE25
(d)
ESX-4
ESX-1
ESX-3
ESX-2
ESX-5

◀ **Fig. 3** Mycobacterial type VII secretion systems. **a** Tentative model for type VII secretion, including conserved cytosolic and membrane components, and a number of ESX-1 and ESX-5-secreted substrates, modified from (Bitter et al. 2009; Houben et al. 2012b). The secretion motif YxxxD/E is indicated with a red box, if present. Plausible interactions between substrates and components are indicated by arrows. Complex formation of inner membrane (*IM*) components is as described in text; Channel present in the outer membrane (*OM*) represents a possible unidentified component. **b** Solution structure of the EsxA/B complex (Renshaw et al. 2005). The WXG loop is marked in magenta and the C-terminal type VII secretion motif YxxxD is marked in red. **c** Solution structure of the PPE41/PE25 complex (Strong et al. 2006). PE/PPE residues are marked in blue. Unstructured C and N termini, including the YxxxE motif present on PE25, are lacking in this structure. Structures were prepared using RCSB Protein Workshop. **d** Genomic organization of the five ESX clusters

heterodimer (Renshaw et al. 2002). The complex consists of two antiparallel positioned proteins forming a four helical-bundle, both attaining a similar helix-turn-helix structure with the WXG motif producing the crucial turn. The unstructured C- and N-terminal tails of these proteins are presented as flexible loops (Fig. 3b) (Renshaw et al. 2005). The formation of a heterodimeric complex is a common feature for all Esx pairs, since EsxG/H and EsxR/S were also found to form a 1:1 complex, although of slightly different structure (Lightbody et al. 2004; Ilghari et al. 2011). Interestingly, the small WXG100 proteins of both *Staphylococcus aureus* and *Streptococcus agalactiae* also form helical-bundle structures, comparable to those of *M. tuberculosis*, but they do so as homodimers (Sundaramoorthy et al. 2008; Shukla et al. 2010). The main difference with the mycobacterial structures is that one of the long α-helices has a kink, which is probably required for homodimerization. The available evidence (see also below) indicates that EsxA and EsxB are secreted as a dimer, which would implicate that T7S facilitates the export of (partially) folded proteins.

Although Esx proteins are the best studied T7S substrates, these proteins are certainly not the only T7S substrates. In recent years, thorough analysis of ESX-1 and ESX-5 has shown that each ESX system secretes a specific set of additional substrates. The ESX-1 secretion system is responsible for secretion of a number of so-called ESX-1 secretion-associated proteins (Esp), most of which are encoded within the ESX-1 cluster (Fig. 3a/d) (Fortune et al. 2005; MacGurn et al. 2005; McLaughlin et al. 2007; Bitter et al. 2009; Carlsson et al. 2009; Sani et al. 2010). Intriguingly, some secreted Esp's, including EspACD and EspB, are essential for EsxA/B secretion, and vice versa (Fortune et al. 2005; MacGurn et al. 2005; McLaughlin et al. 2007; Stoop et al. 2011; Chen et al. 2012). The cause of this mutual interdependence is currently unknown, but it might reflect an interaction between these proteins before transport.

Also PE and PPE proteins were found to be T7S substrates (Fortune et al. 2005; Abdallah et al. 2006, 2009; Bottai et al. 2012; Daleke et al. 2012). A number of PE/PPEs is encoded within the ESX clusters (Fig. 3d), but the majority is found at different locations. The large, mycolate-specific families of *pe* and *ppe* genes were one of the major surprises in the *M. tuberculosis* H37Rv genome sequence with their 99 and 69 members, respectively (Cole et al. 1998). The names of PE and PPEs are derived from an N-terminal signature motif, which consists of Proline (P) and Glutamic acid (E). The relatively conserved N termini, referred to as PE and PPE domains, are ~110 and ~180 aa in length. Conversely, the variable C-terminal domains of these protein families are the basis of division into several subfamilies, of which the PE_PGRS (with polymorphic GC-rich sequence) is the largest. The ESX-5 system is the major secretion pathway for PE/PPEs, but some have also been found to be secreted via ESX-1 (Fortune et al. 2005; Abdallah et al. 2006, 2009; Sani et al. 2010; Daleke et al. 2011; Bottai et al. 2012; Daleke et al. 2012). Interestingly, disruption of specific *pe* and *ppe* genes can also affect the ESX secretion pathways (Bottai et al. 2012). Similar to Esx proteins, PE/PPE proteins are believed to specifically interact with their cognate partner (Riley et al. 2008). One PE/PPE pair, consisting of the small PE/PPE-domain-only PPE41 and PE25, has been crystallized and forms a heterodimeric complex of helix-turn-helix structure similar to the EsxA/B complex (Fig. 3c) (Strong et al. 2006). Most PPE proteins also have a WXG motif producing a turn between the two α-helices pairing with the PE protein (Daleke et al. 2012).

In depth analysis of the secretion requirements for the PPE41/PE25 complex has identified a short specific secretion motif at the C terminus of the PE protein: YxxxD/E. Mutation of the tyrosine or the negatively charged residue, or changing the spacing between these two residues completely disrupts secretion of both PPE41 and PE25. Significantly, this secretion motif is conserved at the end of all PE domains and is essential for secretion, irrespective of the secretion system that the PE protein is targeted to. A similar YxxxD/E secretion motif is also detected on EsxB, EspB, and various other Esp proteins (Fig. 3a/b) and is also important for secretion via ESX-1 for these different classes of substrates (Daleke et al. 2012). The secretion motif partly corresponds to the specific C-terminal secretion sequence that was previously described to be important for secretion of EsxB (Champion et al. 2006). Exchange of this motif between different substrates resulted in functional secretion, but did not alter the secretion route to a different ESX system (Daleke et al. 2012). Therefore, YxxxD/E is a general T7S motif, which does not provide system specificity. Likely, an additional signal is necessary for targeting to a specific ESX system. The identification of a general T7S motif greatly contributes to the identification of novel T7S substrates. Bioinformatic analysis of the coding sequence of *M. tuberculosis* for the secretion motif has already indicated a number of novel ESX substrates (Daleke et al. 2012). Further experiments are required to establish how extensive the set of T7S substrates really is.

3.3 ESX-1: An Interplay Between Substrates Essential for Virulence

Since the deletion of RD1 was shown to greatly affect mycobacterial virulence, the associated secretion system ESX-1 has been the object of many studies. A number of studies now indicate that the ESX-1 secretion system is associated with membrane damage/rupture. For example, wildtype *Mycobacterium marinum* bacteria, but not ESX-1 mutants, can cause lysis of red blood cells upon interaction (Gao et al. 2004). Moreover, a functional ESX-1 system is essential for escape of both *M. tuberculosis* and *M. marinum* from the phagosome to the cytosol (Gao et al. 2004; van der Wel et al. 2007; Smith et al. 2008; Simeone et al. 2012). This escape is likely facilitated by rupture of the phagosomal membrane by one of the ESX-1 substrates. EsxA has been postulated to be responsible for this, since purified EsxA was shown to destabilize artificial lipid membranes (de Jonge et al. 2007; Smith et al. 2008). However this theory is not undisputed, because several (non-pathogenic) mycobacterial species do secrete EsxA, but do not escape the phagolysosomal compartment (Houben et al. 2012a). A complicating factor is that any EsxA/B secretion defect in mycobacteria always coincides with impaired secretion of Esp's, so observed effects could also be indirect. It has been shown that abrogation of EspA secretion specifically without affecting EsxAB secretion results in serious attenuation in macrophage infections (Garces et al. 2010). It is thus well possible that the phagosomal membrane rupture associated with ESX-1 secretion is actually caused by one of the Esp's or a combination of ESX-1 substrates. Further research should identify the true ESX-1 substrate necessary for phagolysosomal escape.

Interestingly, in *M. smegmatis* ESX-1 is additionally involved in a completely different process, i.e. specific conjugal DNA transfer mechanism. Disruption of ESX-1 in the donor strain resulted in hyperconjugative mutants (Flint et al. 2004), whereas in the recipient strain a functional ESX-1 system seems to be crucial for conjugation (Coros et al. 2008). Whether specific ESX-1 substrates are involved in this process remains to be determined.

Since ESX-1 secretion was shown to be crucial in a specific step of the mycobacterial infection pathway, it is no surprise that expression of this system seems to be strictly regulated. One important ESX-1 regulator is the two-component system PhoP/R. This two-component systems regulates directly the *espACD* operon (Walters et al. 2006). As EspA secretion is essential for ESX-1 functioning, PhoP/R therefore also indirectly regulate EsxAB secretion (Frigui et al. 2008; Gonzalo-Asensio et al. 2008).

The *espACD* operon is also regulated by the repressor Lsr2 (Gordon et al. 2010) and activated by Rv3849. Since this latter protein was suspected to be a secreted substrate of ESX-1, it was named EspR (Raghavan et al. 2008). However, it has been shown recently that EspR is not secreted and actually binds and regulates a large number of genomic loci. A consensus binding motif of EspR was determined, which also corresponds to the region upstream of the *espACD* operon (Blasco et al. 2012). It therefore seems likely that EspR is a nucleoid-associated protein that performs a more general role in regulation.

3.4 ESX-3: A role in Iron Uptake

The ESX-3 system is essential for in vitro growth in *M. tuberculosis*, as shown by whole genome transposon screens (Griffin et al. 2011). This suggests that ESX-3 has an important physiological role. A role in metal metabolism was suspected, since the complete ESX-3 cluster is regulated by zinc uptake regulator Zur, as well as iron-dependent transcriptional repressor IdeR (Rodriguez et al. 2002; Maciag et al. 2007, 2009). This link to iron and zinc was further established by studies with conditional ESX-3 mutants in *M. tuberculosis.* Although these mutations are lethal in normal culture conditions, the supplemental addition of either iron or zinc to the culture medium allowed ESX-3 mutations (Serafini et al. 2009). Remarkably, ESX-3 mutations are well-tolerated in *M. smegmatis*, although not under iron-limiting conditions (Siegrist et al. 2009). Currently, the only known substrates of this system are its encoded WXG100 family proteins EsxG/H (Siegrist et al. 2009; Rosenberger et al. 2012). Structure solution of the EsxG/H complex revealed a Zinc binding site (Ilghari et al. 2011), which could suggest that this substrate has a role in zinc uptake. Iron uptake in *M. tuberculosis* is facilitated by the siderophore mycobactin, which suggests that a substrate of the ESX-3 system is involved in the acquisition and utilization of iron bound to the secreted siderophore mycobactin (Siegrist et al. 2009). Whether this unknown ESX-3 substrate is a secreted mycobactin-binding protein or an iron uptake system in the cell wall, remains to be determined.

3.5 ESX-5: Major Pathway for PE/PPE Proteins

The most recently evolved type VII secretion system, ESX-5, is restricted to the genomes of the slow-growing mycobacterial species, including all the major pathogens (Gey Van Pittius et al. 2001). This could imply that ESX-5 contributes to either the slow-growing phenotype and/or mycobacterial virulence. Interestingly, the introduction of the genome cluster encoding ESX-5 precedes the large expansion of *pe* and *ppe* genes (Gey van Pittius et al. 2006). In line with this observation, the ESX-5 system was demonstrated to be the major secretion pathway for the PE and PPE proteins, including the PE_PGRSs (Abdallah et al. 2006, 2009; Bottai et al. 2012). Several studies have shown that the PE and PPE domains of these proteins are essential for their secretion (Cascioferro et al. 2011; Daleke et al. 2011), which can also be explained by the presence of the T7S motif on the PE domain as described above (Daleke et al. 2012). The function of PE/PPE proteins is largely unknown, although a number have been shown to be involved in different steps of mycobacterial virulence (recently reviewed by (Sampson 2011)). For instance, LipY, a PE protein containing a C-terminal lipase, has an important role in hydrolyzing triacylglycerols upon nutrient starvation (Deb et al. 2006). LipY was also shown to be secreted via ESX-5 (Daleke et al. 2011).

Infection studies with several ESX-5 mutants of both *M. tuberculosis* and *M. marinum* have shown that they are attenuated in macrophage infection experiments, although this effect is not observed for every gene disruption of the ESX-5 cluster (Abdallah et al. 2006, 2011; Bottai et al. 2012). Similar to ESX-1 mutants, mutations in ESX-5 affect macrophage cell death, but for this secretion system the escape of the bacteria from the phagolysosome into the cytosol is not affected (Abdallah et al. 2008, 2011). In striking contrast with these macrophage infection experiments, the same *M. marinum* ESX-5 mutant showed a marked hypervirulence phenotype in adult zebrafish, which appeared independent of the adaptive immune system. This ESX-5 mutant has a specific growth advantage in adult zebrafish, which indicates that ESX-5 substrates might be involved in establishing a moderate and persistent infection in its natural host (Weerdenburg et al. 2012). For *M. tuberculosis* removal of *eccD5* or the five *pe/ppe* genes of the ESX-5 cluster resulted in an attenuated phenotype in both immunocompetent and SCID mice (Bottai et al. 2012; Sayes et al. 2012). Interestingly, this last mutant also showed vaccine potential (Sayes et al. 2012). The inconsistency in virulence defects observed for *M. tuberculosis* and *M. marinum* ESX-5 mutants might be explained by differences in ESX-5 substrates, since the *M. marinum* genome contains many more *pe/ppe* genes, especially *pe_pgrs,* as compared to *M. tuberculosis* (Stinear et al. 2008). Future experiments will have to show the exact role of ESX-5 secreted proteins.

4 Conclusion

Protein secretion pathways are essential elements of mycobacterial virulence. Similar to other bacteria, the Sec and the Tat pathway play a central role in translocation across the inner membrane and insertion of proteins into the inner membrane. In addition to these systems, mycobacteria also use the accessory SecA2 protein to participate in the transport of specific virulence factors. Both the mechanism of routing via SecA2 and the subset of virulence factors that depend on this pathway for efficient transport, need to be examined in more detail. Similarly, the substrate specificity of the Tat pathway need to be analyzed to explain why this pathway is essential for growth in *M. tuberculosis* unlike the situation in *E. coli.*

Of note, a number of proteins that contain a signal sequence, *e.g.* Apa, Mpt64, and the antigen 85 proteins are cell surface exposed or secreted into the extracellular environment. While the Sec and Tat translocons are likely responsible for translocation of these proteins across the inner membrane, the mechanism and machinery for translocation across the outer membrane is completely unknown. Given the unique chemical composition of this membrane and the absence of obvious homologs of other secretion systems active in Gram-negative bacteria, a yet undiscovered protein conducting channel likely exists. Future studies in this direction would benefit from the elucidation of the proteome of this intriguing mycomembrane.

In contrast, the recently discovered mycobacterial T7S systems are hypothesized to transport proteins that lack a classical signal peptide across both membranes. The key players of this system are currently pieced together, but defining the molecular basis for protein transfer will prove a challenging task. At present, it is not even clear whether translocation across the inner membrane and mycomembrane occurs in one synchronized mechanism or in two consecutive steps. The five different T7S systems of *M. tuberculosis* each could be responsible for secretion of different virulence factors and therefore also play a role in different steps of the tuberculosis infection cycle. Interestingly, disruption of either of the ESX-1, ESX-3, or ESX-5 system leads to serious attenuation and even results in strains that could potentially be used as (novel) vaccines for tuberculosis (Sweeney et al. 2011; Sayes et al. 2012). Many of the effector proteins involved have been identified. Future challenges include the identification of novel T7S substrates and their role in virulence, the composition and the structure of the secretion machinery in both the inner membrane and the mycomembrane and the role of this system in other bacteria.

References

Abdallah AM, Verboom T, Hannes F et al (2006) A specific secretion system mediates PPE41 transport in pathogenic mycobacteria. Mol Microbiol 62:667–679

Abdallah AM, Gey van Pittius NC, Champion PA et al (2007) Type VII secretion–mycobacteria show the way. Nat Rev Microbiol 5:883–891

Abdallah AM, Savage ND, van Zon M et al (2008) The ESX-5 secretion system of *Mycobacterium marinum* modulates the macrophage response. J Immunol 181:7166–7175

Abdallah AM, Verboom T, Weerdenburg EM et al (2009) PPE and PE_PGRS proteins of *Mycobacterium marinum* are transported via the type VII secretion system ESX-5. Mol Microbiol 73:329–340

Abdallah AM, Bestebroer J, Savage ND et al (2011) Mycobacterial secretion systems ESX-1 and ESX-5 play distinct roles in host cell death and inflammasome activation. J Immunol 187:4744–4753

Akpe San Roman S, Facey PD, Fernandez-Martinez L et al (2010) A heterodimer of EsxA and EsxB is involved in sporulation and is secreted by a type VII secretion system in *Streptomyces coelicolor*. Microbiology 156:1719–1729

Archambaud C, Nahori MA, Pizarro-Cerda J et al (2006) Control of Listeria superoxide dismutase by phosphorylation. J Biol Chem 281:31812–31822

Behr MA, Wilson MA, Gill WP et al (1999) Comparative genomics of BCG vaccines by whole-genome DNA microarray. Science 284:1520–1523

Bendtsen JD, Nielsen H, Widdick D et al (2005) Prediction of twin-arginine signal peptides. BMC Bioinform 6:167

Bensing BA, Sullam PM (2002) An accessory sec locus of *Streptococcus gordonii* is required for export of the surface protein GspB and for normal levels of binding to human platelets. Mol Microbiol 44:1081–1094

Bensing BA, Sullam PM (2010) Transport of preproteins by the accessory Sec system requires a specific domain adjacent to the signal peptide. J Bacteriol 192:4223–4232

Bensing BA, Yen YT, Seepersaud R et al (2012) A specific interaction between SecA2 and a region of the preprotein adjacent to the signal peptide occurs during transport via the accessory Sec system. J Biol Chem 287:24438–24447

Berthet FX, Rasmussen PB, Rosenkrands I et al (1998) A *Mycobacterium tuberculosis* operon encoding ESAT-6 and a novel low-molecular-mass culture filtrate protein (CFP-10). Microbiology 144(Pt 11):3195–3203

Bitter W, Houben EN, Bottai D et al (2009) Systematic genetic nomenclature for type VII secretion systems. PLoS Pathog 5:e1000507

Blasco B, Chen JM, Hartkoorn R et al (2012) Virulence regulator EspR of *Mycobacterium tuberculosis* is a nucleoid-associated protein. PLoS Pathog 8:e1002621

Bolhuis A, Mathers JE, Thomas JD et al (2001) TatB and TatC form a functional and structural unit of the twin-arginine translocase from *Escherichia coli*. J Biol Chem 276:20213–20219

Bönemann G, Pietrosiuk A, Mogk A (2010) Tubules and donuts: a type VI secretion story. Mol Microbiol 76:815–821

Bordes P, Cirinesi AM, Ummels R et al (2011) SecB-like chaperone controls a toxin-antitoxin stress-responsive system in *Mycobacterium tuberculosis*. Proc Natl Acad Sci U S A 108:8438–8443

Bottai D, Di Luca M, Majlessi L et al (2012) Disruption of the ESX-5 system of *Mycobacterium tuberculosis* causes loss of PPE protein secretion, reduction of cell wall integrity and strong attenuation. Mol Microbiol 83:1195–1209

Braunstein M, Brown AM, Kurtz S et al (2001) Two nonredundant SecA homologues function in mycobacteria. J Bacteriol 183:6979–6990

Braunstein M, Espinosa BJ, Chan J et al (2003) SecA2 functions in the secretion of superoxide dismutase A and in the virulence of *Mycobacterium tuberculosis*. Mol Microbiol 48:453–464

Breukink E, Nouwen N, van Raalte A et al (1995) The C terminus of SecA is involved in both lipid binding and SecB binding. J Biol Chem 270:7902–7907

Breyton C, Haase W, Rapoport TA et al (2002) Three-dimensional structure of the bacterial protein-translocation complex SecYEG. Nature 418:662–665

Brodin P, Majlessi L, Marsollier L et al (2006) Dissection of ESAT-6 system 1 of *Mycobacterium tuberculosis* and impact on immunogenicity and virulence. Infect Immun 74:88–98

Brown GD, Dave JA, Gey van Pittius NC et al (2000) The mycosins of *Mycobacterium tuberculosis* H37Rv: a family of subtilisin-like serine proteases. Gene 254:147–155

Burts ML, Williams WA, DeBord K et al (2005) EsxA and EsxB are secreted by an ESAT-6-like system that is required for the pathogenesis of *Staphylococcus aureus* infections. Proc Natl Acad Sci U S A 102:1169–1174

Carlsson F, Joshi SA, Rangell L et al (2009) Polar localization of virulence-related Esx-1 secretion in mycobacteria. PLoS Pathog 5:e1000285

Cascioferro A, Daleke MH, Ventura M et al (2011) Functional dissection of the PE domain responsible for translocation of PE_PGRS33 across the mycobacterial cell wall. PLoS One 6:e27713

Champion PA, Stanley SA, Champion MM et al (2006) C-terminal signal sequence promotes virulence factor secretion in *Mycobacterium tuberculosis*. Science 313:1632–1636

Champion MM, Williams EA, Kennedy GM et al (2012) Direct detection of bacterial protein secretion using whole colony proteomics. Mol Cell Proteomics 11:596–604

Chen JM, Boy-Rottger S, Dhar N et al (2012) EspD is critical for the virulence-mediating ESX-1 secretion system in *Mycobacterium tuberculosis*. J Bacteriol 194:884–893

Cline K, Mori H (2001) Thylakoid DeltapH-dependent precursor proteins bind to a cpTatC-Hcf106 complex before Tha4-dependent transport. J Cell Biol 154:719–729

Cole ST, Brosch R, Parkhill J et al (1998) Deciphering the biology of *Mycobacterium tuberculosis* from the complete genome sequence. Nature 393:537–544

Converse SE, Cox JS (2005) A protein secretion pathway critical for *Mycobacterium tuberculosis* virulence is conserved and functional in *Mycobacterium smegmatis*. J Bacteriol 187:1238–1245

Coros A, Callahan B, Battaglioli E et al (2008) The specialized secretory apparatus ESX-1 is essential for DNA transfer in *Mycobacterium smegmatis*. Mol Microbiol 69:794–808

Daleke MH, Cascioferro A, de Punder K et al (2011) Conserved Pro-Glu (PE) and Pro–Pro-Glu (PPE) protein domains target LipY lipases of pathogenic mycobacteria to the cell surface via the ESX-5 pathway. J Biol Chem 286:19024–19034

Daleke MH, Ummels R, Bawono P et al (2012) General secretion signal for the mycobacterial type VII secretion pathway. Proc Natl Acad Sci U S A 109:11342–11347

Dave JA, Gey van Pittius NC, Beyers AD et al (2002) Mycosin-1, a subtilisin-like serine protease of *Mycobacterium tuberculosis*, is cell wall-associated and expressed during infection of macrophages. BMC Microbiol 2:30

de Jonge MI, Pehau-Arnaudet G, Fretz MM et al (2007) ESAT-6 from *Mycobacterium tuberculosis* dissociates from its putative chaperone CFP-10 under acidic conditions and exhibits membrane-lysing activity. J Bacteriol 189:6028–6034

Deb C, Daniel J, Sirakova TD et al (2006) A novel lipase belonging to the hormone-sensitive lipase family induced under starvation to utilize stored triacylglycerol in *Mycobacterium tuberculosis*. J Biol Chem 281:3866–3875

DiGiuseppe Champion PA, Champion MM, Manzanillo P et al (2009) ESX-1 secreted virulence factors are recognized by multiple cytosolic AAA ATPases in pathogenic mycobacteria. Mol Microbiol 73:950–962

Dilks K, Rose RW, Hartmann E et al (2003) Prokaryotic utilization of the twin-arginine translocation pathway: a genomic survey. J Bacteriol 185:1478–1483

du Plessis DJ, Nouwen N, Driessen AJ (2011) The Sec translocase. Biochim Biophys Acta 1808:851–865

Fekkes P, van der Does C, Driessen AJ (1997) The molecular chaperone SecB is released from the carboxy-terminus of SecA during initiation of precursor protein translocation. EMBO J 16:6105–6113

Flint JL, Kowalski JC, Karnati PK et al (2004) The RD1 virulence locus of *Mycobacterium tuberculosis* regulates DNA transfer in *Mycobacterium smegmatis*. Proc Natl Acad Sci U S A 101:12598–12603

Fortune SM, Jaeger A, Sarracino DA et al (2005) Mutually dependent secretion of proteins required for mycobacterial virulence. Proc Natl Acad Sci U S A 102:10676–10681

Frigui W, Bottai D, Majlessi L et al (2008) Control of *M. tuberculosis* ESAT-6 secretion and specific T cell recognition by PhoP. PLoS Pathog 4:e33

Gao LY, Guo S, McLaughlin B et al (2004) A mycobacterial virulence gene cluster extending RD1 is required for cytolysis, bacterial spreading and ESAT-6 secretion. Mol Microbiol 53:1677–1693

Garces A, Atmakuri K, Chase MR et al (2010) EspA acts as a critical mediator of ESX1-dependent virulence in *Mycobacterium tuberculosis* by affecting bacterial cell wall integrity. PLoS Pathog 6:e1000957

Garufi G, Butler E, Missiakas D (2008) ESAT-6-like protein secretion in *Bacillus anthracis*. J Bacteriol 190:7004–7011

Gey Van Pittius NC, Gamieldien J, Hide W et al (2001) The ESAT-6 gene cluster of *Mycobacterium tuberculosis* and other high G + C Gram-positive bacteria. Genome Biol 2:RESEARCH0044

Gey van Pittius NC, Sampson SL, Lee H et al (2006) Evolution and expansion of the *Mycobacterium tuberculosis* PE and PPE multigene families and their association with the duplication of the ESAT-6 (esx) gene cluster regions. BMC Evol Biol 6:95

Gibbons HS, Wolschendorf F, Abshire M et al (2007) Identification of two *Mycobacterium smegmatis* lipoproteins exported by a SecA2-dependent pathway. J Bacteriol 189:5090–5100

Gomez M, Johnson S, Gennaro ML (2000) Identification of secreted proteins of *Mycobacterium tuberculosis* by a bioinformatic approach. Infect Immun 68:2323–2327

Gonzalo-Asensio J, Mostowy S, Harders-Westerveen J et al (2008) PhoP: a missing piece in the intricate puzzle of *Mycobacterium tuberculosis* virulence. PLoS ONE 3:e3496

Gordon BR, Li Y, Wang L et al (2010) Lsr2 is a nucleoid-associated protein that targets AT-rich sequences and virulence genes in *Mycobacterium tuberculosis*. Proc Natl Acad Sci U S A 107:5154–5159

Griffin JE, Gawronski JD, Dejesus MA et al (2011) High-resolution phenotypic profiling defines genes essential for mycobacterial growth and cholesterol catabolism. PLoS Pathog 7:e1002251

Harboe M, Oettinger T, Wiker HG et al (1996) Evidence for occurrence of the ESAT-6 protein in *Mycobacterium tuberculosis* and virulent *Mycobacterium bovis* and for its absence in *Mycobacterium bovis* BCG. Infect Immun 64:16–22

Hinchey J, Lee S, Jeon BY et al (2007) Enhanced priming of adaptive immunity by a proapoptotic mutant of *Mycobacterium tuberculosis*. J Clin Invest 117:2279–2288

Hoffmann C, Leis A, Niederweis M et al (2008) Disclosure of the mycobacterial outer membrane: cryo-electron tomography and vitreous sections reveal the lipid bilayer structure. Proc Natl Acad Sci U S A 105:3963–3967

Hou JM, D'Lima NG, Rigel NW et al (2008) ATPase activity of *Mycobacterium tuberculosis* SecA1 and SecA2 proteins and its importance for SecA2 function in macrophages. J Bacteriol 190:4880–4887

Houben D, Demangel C, van Ingen J et al (2012a) ESX-1 Mediated Translocation to the Cytosol controls Virulence of Mycobacteria. Cell Microbiol 14:1287–1298

Houben EN, Bestebroer J, Ummels R et al (2012b) Composition of the type VII secretion system membrane complex. Mol Microbiol 86:472–484

Ilghari D, Lightbody KL, Veverka V et al (2011) Solution structure of the *Mycobacterium tuberculosis* EsxG.EsxH complex: functional implications and comparisons with other *M. tuberculosis* Esx family complexes. J Biol Chem 286:29993–30002

Jarlier V, Nikaido H (1990) Permeability barrier to hydrophilic solutes in *Mycobacterium chelonei*. J Bacteriol 172:1418–1423

Jensen K, Ranganathan UD, Van Rompay KK et al (2012) A recombinant attenuated *Mycobacterium tuberculosis* vaccine strain is safe in immunosuppressed SIV-infected infant macaques. Clin Vaccine Immunol 19:1170–1181

Krehenbrink M, Edwards A, Downie JA (2011) The superoxide dismutase SodA is targeted to the periplasm in a SecA-dependent manner by a novel mechanism. Mol Microbiol 82:164–179

Kurtz S, McKinnon KP, Runge MS et al (2006) The SecA2 secretion factor of *Mycobacterium tuberculosis* promotes growth in macrophages and inhibits the host immune response. Infect Immun 74:6855–6864

Kusters I, Driessen AJ (2011) SecA, a remarkable nanomachine. Cell Mol Life Sci 68:2053–2066

Lenz LL, Portnoy DA (2002) Identification of a second Listeria secA gene associated with protein secretion and the rough phenotype. Mol Microbiol 45:1043–1056

Lenz LL, Mohammadi S, Geissler A et al (2003) SecA2-dependent secretion of autolytic enzymes promotes Listeria monocytogenes pathogenesis. Proc Natl Acad Sci U S A 100:12432–12437

Leversen NA, de Souza GA, Malen H et al (2009) Evaluation of signal peptide prediction algorithms for identification of mycobacterial signal peptides using sequence data from proteomic methods. Microbiology 155:2375–2383

Lewis KN, Liao R, Guinn KM et al (2003) Deletion of RD1 from *Mycobacterium tuberculosis* mimics Bacille Calmette-Guerin attenuation. J Infect Dis 187:117–123

Lightbody KL, Renshaw PS, Collins ML et al (2004) Characterisation of complex formation between members of the *Mycobacterium tuberculosis* complex CFP-10/ESAT-6 protein family: towards an understanding of the rules governing complex formation and thereby functional flexibility. FEMS Microbiol Lett 238:255–262

Luirink J, Yu Z, Wagner S et al (2012) Biogenesis of inner membrane proteins in *Escherichia coli*. Biochim Biophys Acta 1817:965–976

Luthra A, Mahmood A, Arora A et al (2008) Characterization of Rv3868, an essential hypothetical protein of the ESX-1 secretion system in *Mycobacterium tuberculosis*. J Biol Chem 283:36532–36541

MacGurn JA, Raghavan S, Stanley SA et al (2005) A non-RD1 gene cluster is required for Snm secretion in *Mycobacterium tuberculosis*. Mol Microbiol 57:1653–1663

Maciag A, Dainese E, Rodriguez GM et al (2007) Global analysis of the *Mycobacterium tuberculosis* Zur (FurB) regulon. J Bacteriol 189:730–740

Maciag A, Piazza A, Riccardi G et al (2009) Transcriptional analysis of ESAT-6 cluster 3 in *Mycobacterium smegmatis*. BMC Microbiol 9:48

Mahairas GG, Sabo PJ, Hickey MJ et al (1996) Molecular analysis of genetic differences between *Mycobacterium bovis* BCG and virulent *M. bovis*. J Bacteriol 178:1274–1282

Malen H, Berven FS, Fladmark KE et al (2007) Comprehensive analysis of exported proteins from *Mycobacterium tuberculosis* H37Rv. Proteomics 7:1702–1718

Mangels D, Mathers J, Bolhuis A et al (2005) The core TatABC complex of the twin-arginine translocase in *Escherichia coli*: TatC drives assembly whereas TatA is essential for stability. J Mol Biol 345:415–423

Marrichi M, Camacho L, Russell DG et al (2008) Genetic toggling of alkaline phosphatase folding reveals signal peptides for all major modes of transport across the inner membrane of bacteria. J Biol Chem 283:35223–35235

Matos CF, Di Cola A, Robinson C (2009) TatD is a central component of a Tat translocon-initiated quality control system for exported FeS proteins in *Escherichia coli*. EMBO Rep 10:474–479

McCann JR, McDonough JA, Sullivan JT et al (2011) Genome-wide identification of *Mycobacterium tuberculosis* exported proteins with roles in intracellular growth. J Bacteriol 193:854–861

McDonough JA, Hacker KE, Flores AR et al (2005) The twin-arginine translocation pathway of *Mycobacterium smegmatis* is functional and required for the export of mycobacterial beta-lactamases. J Bacteriol 187:7667–7679

McDonough JA, McCann JR, Tekippe EM et al (2008) Identification of functional Tat signal sequences in *Mycobacterium tuberculosis* proteins. J Bacteriol 190:6428–6438

McLaughlin B, Chon JS, MacGurn JA et al (2007) A mycobacterium ESX-1-secreted virulence factor with unique requirements for export. PLoS Pathog 3:e105

Nagai S, Wiker HG, Harboe M et al (1991) Isolation and partial characterization of major protein antigens in the culture fluid of *Mycobacterium tuberculosis*. Infect Immun 59:372–382

Nouwen N, Piwowarek M, Berrelkamp G et al (2005) The large first periplasmic loop of SecD and SecF plays an important role in SecDF functioning. J Bacteriol 187:5857–5860

Ohol YM, Goetz DH, Chan K et al (2010) *Mycobacterium tuberculosis* MycP1 protease plays a dual role in regulation of ESX-1 secretion and virulence. Cell Host Microbe 7:210–220

Owens MU, Swords WE, Schmidt MG et al (2002) Cloning, expression, and functional characterization of the *Mycobacterium tuberculosis* secA gene. FEMS Microbiol Lett 211:133–141

Palaniyandi K, Veerasamy M, Narayanan S (2012) Characterization of Ffh of *Mycobacterium tuberculosis* and its interaction with 4.5 S RNA. Microbiol Res 167:520–525

Pallen MJ (2002) The ESAT-6/WXG100 superfamily—and a new Gram-positive secretion system? Trends Microbiol 10:209–212

Palmer T, Berks BC (2012) The twin-arginine translocation (Tat) protein export pathway. Nat Rev Microbiol 10:483–496

Park E, Rapoport TA (2012) Mechanisms of Sec61/SecY-mediated protein translocation across membranes. Annu Rev Biophys 41:21–40

Posey JE, Shinnick TM, Quinn FD (2006) Characterization of the twin-arginine translocase secretion system of *Mycobacterium smegmatis*. J Bacteriol 188:1332–1340

Pym AS, Brodin P, Brosch R et al (2002) Loss of RD1 contributed to the attenuation of the live tuberculosis vaccines *Mycobacterium bovis* BCG and *Mycobacterium microti*. Mol Microbiol 46:709–717

Pym AS, Brodin P, Majlessi L et al (2003) Recombinant BCG exporting ESAT-6 confers enhanced protection against tuberculosis. Nat Med 9:533–539

Raghavan S, Manzanillo P, Chan K et al (2008) Secreted transcription factor controls *Mycobacterium tuberculosis* virulence. Nature 454:717–721

Raman S, Puyang X, Cheng TY et al (2006) Mycobacterium tuberculosis SigM positively regulates Esx secreted protein and nonribosomal peptide synthetase genes and down regulates virulence-associated surface lipid synthesis. J Bacteriol 188:8460–8468

Renshaw PS, Panagiotidou P, Whelan A et al (2002) Conclusive evidence that the major T-cell antigens of the *Mycobacterium tuberculosis* complex ESAT-6 and CFP-10 form a tight, 1:1 complex and characterization of the structural properties of ESAT-6, CFP-10, and the ESAT-6*CFP-10 complex. Implications for pathogenesis and virulence. J Biol Chem 277:21598–21603

Renshaw PS, Lightbody KL, Veverka V et al (2005) Structure and function of the complex formed by the tuberculosis virulence factors CFP-10 and ESAT-6. EMBO J 24:2491–2498

Rigel NW, Gibbons HS, McCann JR et al (2009) The accessory SecA2 system of mycobacteria requires ATP binding and the canonical SecA1. J Biol Chem 284:9927–9936

Riley R, Pellegrini M, Eisenberg D (2008) Identifying cognate binding pairs among a large set of paralogs: the case of PE/PPE proteins of *Mycobacterium tuberculosis*. PLoS Comput Biol 4:e1000174

Rodriguez GM, Voskuil MI, Gold B et al (2002) IdeR, an essential gene in *Mycobacterium tuberculosis*: role of IdeR in iron-dependent gene expression, iron metabolism, and oxidative stress response. Infect Immun 70:3371–3381

Rose RW, Bruser T, Kissinger JC et al (2002) Adaptation of protein secretion to extremely high-salt conditions by extensive use of the twin-arginine translocation pathway. Mol Microbiol 45:943–950

Rosenberger T, Brulle JK, Sander P (2012) A beta-lactamase based reporter system for ESX dependent protein translocation in mycobacteria. PLoS One 7:e35453

Rosenkrands I, King A, Weldingh K et al (2000) Towards the proteome of *Mycobacterium tuberculosis*. Electrophoresis 21:3740–3756

Saint-Joanis B, Demangel C, Jackson M et al (2006) Inactivation of Rv2525c, a substrate of the twin arginine translocation (Tat) system of Mycobacterium tuberculosis, increases beta-lactam susceptibility and virulence. J Bacteriol 188:6669–6679

Sampson SL (2011) Mycobacterial PE/PPE proteins at the host-pathogen interface. Clin Dev Immunol 2011:497203

Sani M, Houben EN, Geurtsen J et al (2010) Direct visualization by cryo-EM of the mycobacterial capsular layer: a labile structure containing ESX-1-secreted proteins. PLoS Pathog 6:e1000794

Sayes F, Sun L, Di Luca M et al (2012) Strong immunogenicity and cross-reactivity of *Mycobacterium tuberculosis* ESX-5 type VII secretion -encoded PE-PPE proteins predicts vaccine potential. Cell Host Microbe 11:352–363

Serafini A, Boldrin F, Palu G et al (2009) Characterization of a *Mycobacterium tuberculosis* ESX-3 conditional mutant: essentiality and rescue by iron and zinc. J Bacteriol 191:6340–6344

Sharma V, Arockiasamy A, Ronning DR et al (2003) Crystal structure of *Mycobacterium tuberculosis* SecA, a preprotein translocating ATPase. Proc Natl Acad Sci U S A 100:2243–2248

Shukla A, Pallen M, Anthony M et al (2010) The homodimeric GBS1074 from Streptococcus agalactiae. Acta Crystallogr Sect F: Struct Biol Cryst Commun 66:1421–1425

Siboo IR, Chaffin DO, Rubens CE et al (2008) Characterization of the accessory Sec system of *Staphylococcus aureus*. J Bacteriol 190:6188–6196

Siegrist MS, Unnikrishnan M, McConnell MJ et al (2009) Mycobacterial Esx-3 is required for mycobactin-mediated iron acquisition. Proc Natl Acad Sci U S A 106:18792–18797

Simeone R, Bobard A, Lippmann J et al (2012) Phagosomal rupture by *Mycobacterium tuberculosis* results in toxicity and host cell death. PLoS Pathog 8:e1002507

Smith J, Manoranjan J, Pan M et al (2008) Evidence for pore formation in host cell membranes by ESX-1-secreted ESAT-6 and its role in *Mycobacterium marinum* escape from the vacuole. Infect Immun 76:5478–5487

Sonnenberg MG, Belisle JT (1997) Definition of *Mycobacterium tuberculosis* culture filtrate proteins by two-dimensional polyacrylamide gel electrophoresis, N-terminal amino acid sequencing, and electrospray mass spectrometry. Infect Immun 65:4515–4524

Sorensen AL, Nagai S, Houen G et al (1995) Purification and characterization of a low-molecular-mass T-cell antigen secreted by *Mycobacterium tuberculosis*. Infect Immun 63:1710–1717

Stanley SA, Raghavan S, Hwang WW et al (2003) Acute infection and macrophage subversion by *Mycobacterium tuberculosis* require a specialized secretion system. Proc Natl Acad Sci U S A 100:13001–13006

Stinear TP, Seemann T, Harrison PF et al (2008) Insights from the complete genome sequence of *Mycobacterium marinum* on the evolution of *Mycobacterium tuberculosis*. Genome Res 18:729–741

Stoop EJ, Schipper T, Huber SK et al (2011) Zebrafish embryo screen for mycobacterial genes involved in the initiation of granuloma formation reveals a newly identified ESX-1 component. Dis Model Mech 4:526–536

Strong M, Sawaya MR, Wang S et al (2006) Toward the structural genomics of complexes: crystal structure of a PE/PPE protein complex from *Mycobacterium tuberculosis*. Proc Natl Acad Sci U S A 103:8060–8065

Sullivan JT, Young EF, McCann JR et al (2012) The *Mycobacterium tuberculosis* SecA2 system subverts phagosome maturation to promote growth in macrophages. Infect Immun 80:996–1006

Sundaramoorthy R, Fyfe PK, Hunter WN (2008) Structure of *Staphylococcus aureus* EsxA suggests a contribution to virulence by action as a transport chaperone and/or adaptor protein. J Mol Biol 383:603–614

Sweeney KA, Dao DN, Goldberg MF et al (2011) A recombinant *Mycobacterium smegmatis* induces potent bactericidal immunity against *Mycobacterium tuberculosis*. Nat Med 17:1261–1268

Talaat AM, Lyons R, Howard ST et al (2004) The temporal expression profile of *Mycobacterium tuberculosis* infection in mice. Proc Natl Acad Sci U S A 101:4602–4607

Tarry MJ, Schafer E, Chen S et al (2009) Structural analysis of substrate binding by the TatBC component of the twin-arginine protein transport system. Proc Natl Acad Sci U S A 106:13284–13289

Taylor PD, Toseland CP, Attwood TK et al (2006) TATPred: a Bayesian method for the identification of twin arginine translocation pathway signal sequences. Bioinformation 1:184–187

Tekaia F, Gordon SV, Garnier T et al (1999) Analysis of the proteome of *Mycobacterium tuberculosis* in silico. Tuber Lung Dis 79:329–342

Van den Berg B, Clemons WM Jr, Collinson I et al (2004) X-ray structure of a protein-conducting channel. Nature 427:36–44

van der Wel N, Hava D, Houben D et al (2007) *M. tuberculosis* and *M. leprae* translocate from the phagolysosome to the cytosol in myeloid cells. Cell 129:1287–1298

Walters SB, Dubnau E, Kolesnikova I et al (2006) The *Mycobacterium tuberculosis* PhoPR two-component system regulates genes essential for virulence and complex lipid biosynthesis. Mol Microbiol 60:312–330

Weerdenburg EM, Abdallah AM, Mitra S et al (2012) ESX-5-deficient *Mycobacterium marinum* is hypervirulent in adult zebrafish. Cell Microbiol 14:728–739

Wexler M, Sargent F, Jack RL et al (2000) TatD is a cytoplasmic protein with DNase activity. No requirement for TatD family proteins in sec-independent protein export. J Biol Chem 275:16717–16722

Zuber B, Chami M, Houssin C et al (2008) Direct visualization of the outer membrane of mycobacteria and corynebacteria in their native state. J Bacteriol 190:5672–5680

A Single-Cell Perspective on Non-Growing but Metabolically Active (NGMA) Bacteria

Giulia Manina and John D. McKinney

Abstract A long-standing and fundamental problem in microbiology is the non-trivial discrimination between live and dead cells. The existence of physically intact and possibly viable bacterial cells that fail to replicate during a more or less protracted period of observation, despite environmental conditions that are ostensibly propitious for growth, has been extensively documented in many different organisms. In clinical settings, non-culturable cells may contribute to non-apparent infections capable of reactivating after months or years of clinical latency, a phenomenon that has been well documented in the specific case of *Mycobacterium tuberculosis*. The prevalence of these silent but potentially problematic bacterial reservoirs has been highlighted by classical approaches such as limiting culture dilution till extinction of growing cells, followed by resuscitation of apparently "viable but non-culturable" (VBNC) subpopulations. Although these assays are useful to demonstrate the presence of VBNC cells in a population, they are effectively retrospective and are not well suited to the analysis of non-replicating cells per se. Here, we argue that research on a closely related problem, which we shall refer to as the "non-growing but metabolically active" state, is poised to advance rapidly thanks to the recent development of novel technologies and methods for real-time single-cell analysis. In particular, the combination of fluorescent reporter dyes and strains, microfluidic and microelectromechanical systems, and time-lapse fluorescence microscopy offers tremendous and largely untapped potential for future exploration of the physiology of non-replicating cells.

G. Manina (✉) · J. D. McKinney
School of Life Sciences, Swiss Federal Institute of Technology in Lausanne (EPFL), 1015 Lausanne, Switzerland
e-mail: giulia.manina@epfl.ch

Current Topics in Microbiology and Immunology (2013) 374: 135–161

DOI: 10.1007/82_2013_333

Published Online: 21 June 2013

Contents

1 Introduction

The second half of the nineteenth century marked the beginning of a new era in microbiology. A number of techniques were introduced to select and maintain pure bacterial cultures, resulting in the emergence of the "germ theory" of disease causation followed by the establishment of a system of proofs to link a specific microbe to a specific diseases state known as "Koch's postulates." It was during this period that microbes like *Mycobacterium tuberculosis* and *Vibrio cholerae* were identified as the etiological agents of specific communicable diseases. At the same time, the introduction of differential-staining techniques dramatically improved the resolution of features that could be achieved by optical microscopy, allowing the user to distinguish the physical and morphological characteristics of cells in order to assess the viability and death of individual bacteria. In subsequent years, fluorescent dyes—for example, dyes that specifically stain nucleic acids, were broadly used in conjunction with epifluorescence microscopy or flow cytometry to assess total bacterial counts or specific metabolic activities, such as respiration or intracellular enzyme activities (Roszak and Colwell 1987). More recently, microbiologists have relied on commercial kits based on combinations of fluorescent reagents to differentiate between live and dead cells. Despite the widespread application of such "live/dead" reagents, it is important to note that these reagents provide only an indirect assessment of cellular viability, as they typically measure cellular physiological parameters (e.g., specific enzyme activities, cell envelope permeability, etc.) that are assumed but not necessarily proven to correlate with live/dead cell states.

The application of fluorescent reagents and single-cell measurements, often involving optical microscopy or flow cytometry, led to the realization that bacterial populations are phenotypically heterogeneous. This realization raised difficult questions about the relationship among the metabolic state, growth rate, and viability of individual cells. Many techniques for counting the number of live cells in a bacterial population, such as the standard method for plate counts of colony-forming units (CFU), rely on bacterial growth for detection. Although growth-dependent techniques unequivocally detect viable cells and exclude dead cells (which cannot grow, by definition), they may detect only the most active subpopulations of cells within a culture and may underestimate the actual number of living cells, because some viable cells may fail to grow during the time window of observation. Furthermore, these techniques typically provide only a retrospective view of the physiology of cells within a population—for example, by counting colonies of *M. tuberculosis* that do not become visible until several weeks after plating. These issues became increasingly problematic when investigators realized that bacterial populations often contain large numbers of cells that appear to be viable, inasmuch as they are physically intact and may even display metabolic activity, yet which fail to grow and divide during the period of observation. Despite decades of intensive research, the significance of these so-called "viable but non-culturable" (VBNC) cells remains elusive (Mukamolova et al. 2003; Oliver 2010; Trevors 2011), and in only a few cases has it been shown unequivocally and reproducibly that VBNC cells can be converted to actively growing cells by the application of a specific stimulus.

When bacteria do not grow under conditions that are apparently favorable for growth, a number of questions arise. Is a non-growing cell truly dead, in the sense that it has lost all capacity to resume growth in future? What is the metabolic state of a cell that remains intact but fails to grow? Can we consider a cell to be "alive" in some meaningful sense if it is metabolically active but non-replicating? For how long can such a non-replicating cell remain capable of resuming growth, and what stimuli might be effective in triggering the resumption of growth? Finally, related to that last question, for how long should we monitor a cell before concluding that it has entered a non-replicating but possibly still-viable state? These questions are not merely of academic interest, as VBNC cells have been implicated in the ability of bacteria to persist in diverse environments, including animal and human hosts.

In general, we can consider a bacterial cell to be "viable" if its genetic information is intact and if it carries out essential metabolic processes, generates ATP, and maintains a metabolic potential required to sustain chromosome replication, RNA transcription, protein translation, balanced increase in biomass, surface synthesis, partition of cytoplasmic macromolecules, and finally division into two daughter cells (Fig. 1). Thus, assessing the dynamics of the molecular events that are necessary to sustain cell survival and proliferation can serve as reliable evidence of cell viability. However, the time scales on which these processes unwind may differ quite substantially between individual cells of the same species. According to this perspective, in order to reach a more thorough and accurate understanding of bacterial physiology, bacteria should be studied not only at the population level but also at the level of individual cells. As most biological assays are based on population-averaged measurements, the latter goal necessitates the development and application of new investigative tools for real-time single-cell analysis of cell physiology and metabolism.

In the following chapter, we will address issues relating to the definition of life and death of individual bacteria, focusing on the still-enigmatic state of

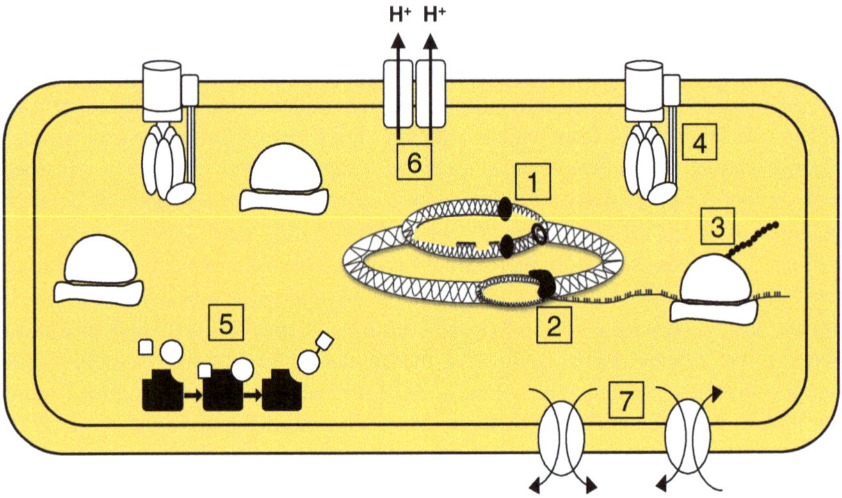

Fig. 1 Schematic illustration of fundamental cellular processes. Cells carry out a number of processes necessary to their survival, including: DNA replication (*1*), transcription (*2*), translation (*3*), ATP synthesis (*4*), enzymatic reactions involved in metabolic processes (*5*), respiration (*6*), symport and antiport of various molecules (*7*). The use of fluorescent reporter-tagging methods to visualize one or more of these intracellular processes, in combination with microfluidic culture systems and time-lapse fluorescence microscopy, can provide direct evidence of ongoing metabolic activity occurring in replicating cells as well as non-replicating NGMA cells

non-culturability of intact and potentially viable cells. We will describe some documented examples of VBNC organisms, and attempt to relate this literature to *M. tuberculosis*, a naturally slow-growing species that may also be capable of long-term survival in a VBNC-like state. We will describe recent progress in the field of mycobacteriology with the aim of understanding the physiology of the dormant and persistent states, both in vitro and in vivo, and we will highlight some classic experiments as well as new technologies. Lastly, we will discuss the use of fluorescent markers for direct, real-time measurements of physiological processes inside individual living cells by time-lapse fluorescence microscopy, and the specific utility of these techniques for the study of non-replicating bacteria.

2 Non-Growing but Metabolically Active (NGMA) Bacteria

2.1 Viable, Dead, or Just "Non-Culturable"? Not a Trivial Question…

The visible processes of bacterial growth and division giving rise to two daughter cells not only ensures the perpetuation of the genetic material but also provides compelling evidence that the bacterium is alive. On the other hand, demonstrating unequivocally that a bacterium is *not* alive is less straightforward. Ascertaining the metabolic and growth states of individual cells within a microbial community is a highly relevant albeit complex and challenging problem in microbiology that has important implications for microbial ecology and human health. Bacterial populations, even those that are isogenic, are characterized by substantial phenotypic heterogeneity (Balaban 2011; Davidson and Surette 2008; Dhar and McKinney 2007; Lidstrom and Konopka 2010; Walling and Shepard 2011), which makes the task of discerning the physiological status of a cell population using markers or reporters a real challenge. An additional layer of complexity is the further modulation of this phenotypic heterogeneity by environmental inputs.

It is thought that microorganisms may sometimes enter into a VBNC state as a survival strategy in response to unfavorable growth conditions or stressful environmental assaults. The term VBNC was coined by Colwell and collaborators in their early studies on the detection of *Escherichia coli* and *V. cholerae* in aquatic environments, in which they noted the importance to public health of measuring the true viable counts of water-borne pathogens (reviewed in Colwell 2000). They observed a markedly smaller number of culturable cells, compared to direct enumeration by microscopy-based techniques, suggesting the presence of subpopulations of cells that were non-culturable using standard culture methods yet could potentially serve as a reservoir of organisms retaining pathogenic potential (Colwell 2000). In the ensuing years, many non-sporulating bacterial species, including some pathogenic species, have been shown to form VBNC cells. Typically, the VBNC state is induced by exposure of bacteria to nutrient starvation or physicochemical stresses, resulting in diversified cellular phenotypes (Oliver 2010).

Traditional bulk-cell techniques, such as incorporation of radiolabeled precursor molecules (Rahman et al. 1994; Pawlowski et al. 2011), determination of intracellular ATP levels (Gengenbacher et al. 2010), real-time PCR (Asakura et al. 2006; Lahtinen et al. 2008; Lothigius et al. 2010; Trevors 2011), microarrays (Liu et al. 2008), and protein gel electrophoresis (Heim et al. 2002; Muela et al. 2008) can provide strong evidence of ongoing metabolic activity, including active transcription and translation (Table 1). However, these techniques are not optimal for the study of the VBNC phenomenon. Bulk-cell techniques provide population-averaged measurements, which do not necessarily provide reliable information about the status of individual cells or minor subpopulations (discussed in Dhar and McKinney 2007; Lidstrom and Konopka 2010; Walling and Shepard 2011; Yin and Marshall 2012). Thus, even the earliest studies of the VBNC state used bulk-cell techniques in conjunction with the single-cell techniques that were available at the time, especially epifluorescence microscopy and flow cytometry.

Total cell counts can be determined by epifluorescence microscopy of cells stained with dyes like acridine orange and 4′,6-diamidino-2-phenylindole, which stain nucleic acids (Roszak and Colwell 1987). Similarly, cellular "viability" can be assessed using microscopy-based techniques to measure cellular phenotypes that are thought to be characteristic of live or dead cells. Kogure's "direct viable counting" (DVC) method relies on the microscopic detection of cell elongation

Table 1 Old and new techniques for population-averaged and single-cell studies

Invasive bulk-averaged	Invasive single-cell	Live single-cell	Single-cell recovery
Radiolabel incorporation	Fluorescent dyes	GFP fusions	Flow cytometry
Intracellular ATP content	Fluorescence microscopy	MS2-tagging, PCA	FACS-based cell sorting
Real-time PCR	FISH	Imaging in microfluidics	Optical tweezers
Microarrays	RING-FISH	FLIP, FRAP	Dielectrophoresis (DEP)
Mass spectrometry	Flow cytometry	FRET	Microfluidic cell sorting

over time (Kogure et al. 1987); thus, it is not capable of distinguishing between dead cells and viable but non-growing cells. Respiratory activity of single cells can be detected by respiration-dependent reduction of 5-cyano-2,3-ditolyl tetrazolium chloride, a water-soluble non-fluorescent dye, to the corresponding formazan product, which is water-insoluble and red-fluorescent (Lew et al. 2010). Intracellular enzymatic activity can be detected by cellular uptake of non-fluorescent fluorescein diacetate (FDA) and its hydrolytic conversion to fluorescent FDA by intracellular esterases (Hamid Salim et al. 2006). Energization of the bacterial cytoplasmic membrane can be detected by staining cells with Rhodamine 123, a red-fluorescent dye that preferentially accumulates in energized membranes (Kell et al. 1998). Maintenance of the cell envelope permeability barrier can be assessed with the so-called "LIVE/DEAD" BacLight kit (Molecular Probes), which combines two nucleic acid stains: membrane-permeable SYTO9 (green) and membrane-impermeable propidium iodide (red), to identify "live" cells (which stain green) versus "dead" cells (which stain red).

Although these reagents can provide useful insights into the physiological status of single cells when used in conjunction with epifluorescence microscopy or flow cytometry (Davey et al. 2004), at least two drawbacks should be mentioned. First, the assays themselves are potentially toxic for bacterial cells, which may result in under-estimation of the number of "viable" cells in a population. Second, because the assays measure phenotypes that are only indirectly related to cell viability, they may result in incorrect assignment of cells to the "live" or "dead" category—for example, a cell that appears be dead because it can be stained with propidium iodide may nonetheless be capable of repairing its cell envelope and resuming active growth and division (Davey and Hexley 2011).

Fluorescent in situ hybridization (FISH) has also been used to study the VBNC phenomenon, especially the dynamics of fecal bacteria in river water. A combination of DVC and FISH has been used to assess the VBNC status of environmental *Helicobacter pylori*, in the presence or absence of different DNA-gyrase inhibitors (Piqueres et al. 2006), and Recognition of Individual Gene-FISH (RING-FISH) has been used to target low copy chromosomal DNA (Griffitt et al. 2011). However, one could argue that DNA is too stable to be a reliable indicator of cell viability, and could potentially persist long after the cells had actually died; moreover, FISH requires sample fixation so it cannot be used on living cells (Table 1).

An important innovation allowing the real-time detection of cellular activities in unperturbed (unstained and unfixed) samples was the introduction of methods for engineering cells to express one or more chromatic variants of green fluorescent protein (GFP). By tagging cells with the expressed GFP marker, non-destructive measurements of single-cell fluorescence variation could be used to distinguish between different cell states—active, dormant, live, dead, VBNC, etc. (Lowder et al. 2000). The introduction of fluorescent expressed reporter tags opened a promising new avenue for the investigation of non-culturable cells that has only partly been exploited (Table 1).

2.2 Interpretations of the VBNC Phenomenon

In their natural environments, bacterial communities frequently encounter different types of stresses, such as nutrient starvation, natural antimicrobials, irradiation, temperature shifts, host immune mechanisms, and competition with other microorganisms. Besides genetic mutations, phenotypic heterogeneity due to non-genetic variability, arising from multiple sources such as stochastic fluctuations in gene expression and asymmetric partitioning of cell components during cell division, may allow microbes to adapt quickly to new conditions, guaranteeing the survival of at least a fraction of the original population (Balaban 2011; Davidson and Surette 2008; Dhar and McKinney 2007; Lidstrom and Konopka 2010; Walling and Shepard 2011). Not only single-cell variation in expression of specific genes that modulate the fitness of individual cells but also variation in single-cell growth rates could contribute to survival in fluctuating environments. Spontaneous entry of individual cells into a non-growing and potentially non-culturable state could potentially serve as an advantageous mechanism to attenuate or modify their metabolic requirements, thereby providing a long-term state of tolerance to multiple external stresses (Balaban et al. 2004; Kussell et al. 2005; Kussell and Leibler 2005).

An extreme example of an adaptive response to unfavorable conditions is the transition from replicating vegetative cell to non-replicating and metabolically dormant spore in sporulating bacteria. Spores are structurally distinct from vegetative cells, and the stimuli required for germination and outgrowth of spores are known. Recently it has been shown that the RNA profile is highly dynamic during and after the sporulation process, with the degradation rates of RNA molecules changing according to temperature variations (Segev et al. 2012). Amino acids and sugars are well-known germinants, and cell wall muropeptides released by growing cells have been shown to play a major role in exit from dormancy in *Bacillus subtilis* spores (Shah et al. 2008). Once triggered, germination is a fast process that is dependent on the ability of dormant spores to sense changes occurring in their external environment. The signals and processes that govern sporulation and germination are understood in considerable detail. In contrast, the physiological basis and dynamics of entry into and exit from a state of non-culturability in non-sporulating bacteria, which may provide some of the same survival advantages as sporulation, are less well understood. In part, this "knowledge gap" stems from the difficulty of obtaining for analysis homogeneous populations of VBNC cells with no contaminating culturable cells (Barcina and Arana 2009).

By comparison, more is known about the ability of non-sporulating bacteria to sense changes in their external environment and to modulate their growth rate accordingly. For example, the stringent response is triggered by various stresses including nutrient (especially amino acids) starvation, leading to intracellular accumulation of the "alarmone" (p)ppGpp, which influences both initiation and elongation of rRNA synthesis and expression of stress-response genes (Jin et al. 2012). Another phenomenon observed in several bacterial species during stationary phase is the so-called "ribosomal hibernation", which is controlled by a set of proteins that promote the transition and dimerization of the active 70S ribosomal subunit into the inactive 100S form. Formation of 100S hibernating ribosomes is a mechanism used by both Gram-negative and Gram-positive bacteria to temporarily interrupt the ribosome cycle under stressful conditions (Ueta et al. 2008, 2010; Williamson et al. 2012).

In this chapter, we will focus on phenotypic states of non-culturability in which the cells retain metabolic activities that can be detected at the single-cell level using a variety of techniques, some of which have already been described. Because the viability (i.e., future replicative potential) of these cells is uncertain, we will largely eschew the term VBNC, which seems to beg the question. As an alternative, we propose the more conservative term NGMA to describe cells in this physiological state, which are indeed something of an *enigma* in the field of microbiology.

The search for stimuli and molecular mechanisms that could trigger entry into and exit from the NGMA state has elicited different interpretations from different groups. Nyström interpreted the phenomenon of bacterial non-culturability, induced by starvation or stress, in two alternative ways (Nyström 2001). The first interpretation evokes direct causation, whereby loss of culturability could be due to cellular deterioration after exposure to essentially any damaging stress. Although metabolic rates are reduced under nutrient starvation, the respiration, and consequent production of reactive oxygen species (ROS) probably remain substantial, as suggested by the observed induction of oxidative stress-response proteins and accumulation of oxidatively-damaged proteins (Nyström 2001; Desnues et al. 2003). Consistent with the idea that ROS might contribute to the non-culturability of nutrient-starved cells, it has been shown that addition of catalase or ROS scavengers to the medium of nutrient-starved cultures increased plating efficiency (i.e., culturability) of the stressed cells (Mizunoe et al. 1999). The second interpretation evokes indirect causation, whereby an active adaptive response allows stressed cells to enter into a non-replicating and non-culturable state as a mechanism to promote their long-term survival in the face of potentially lethal stresses. Consistent with the idea that non-culturability is an actively regulated phenomenon, mutants of *V. cholera* have been isolated that display altered frequencies of entry into the VBNC state (discussed in Nyström 2001). Stress could also arise from a natural process of self-catabolism during long-term nutrient starvation, resulting in depletion of intracellular stores and cellular components and gradual accumulation of intracellular waste products. The ability of subpopulations of aging cells, during prolonged stationary phase, to express stress-response genes and outcompete younger cells was defined as a "Growth Advantage in Stationary Phase" (GASP) phenotype (Finkel 2006; Navarro Llorens et al. 2010). GASP was proven to be due to accumulation of advantageous mutations in aging cultures, thereby increasing the overall fitness of the population under nutrient-limiting conditions.

Focusing on these stages of life in stationary-phase bacterial communities: on the one hand, the high-density population of nutrient-scavenging cells leads to nutrient depletion from the external milieu; on the other hand, death of a fraction of cells releases molecules that can be scavenged by the survivors. The simultaneous occurrence of these two opposing processes could permit the maintenance of a relatively stable number of culturable cells over time due to dynamic and balanced processes of cell death and cell proliferation. It could even be envisaged that natural selection might favor a "division of labor" along these lines, in which starving populations bifurcate into cells that are "eaters" and cells that are "eaten." Consistent with this idea, a model of cannibalism has been reported for *B. subtilis* in which starving populations show marked cell-to-cell variation in the expression of the master regulator Spo0A, which governs exit from vegetative growth and entry into the differentiation pathway leading to spore formation. Individual cells with high levels of Spo0A produce toxins that kill and lyse their non-sporulating siblings, thereby releasing nutrients that delay commitment to sporulation (Gonzàlez-Pastor 2011).

Connecting this concept to the NGMA phenomenon, any bacterial population subjected to environmental challenge or nutrient limitation could potentially release molecules, either to kill or to cause loss of culturability in other members of the population, thereby reducing the overall metabolic rates and nutrient requirements of the population. The presence of so-called "altruistic" and "survivor" members within a bacterial population has been proposed as a successful survival strategy to cope with environmental stresses (Barcina and Arana 2009). The maintenance of "non-demanding" subpopulations of cells could represent an inexpensive bacterial reservoir for future re-expansion when auspicious conditions return. Moreover, slowing down or freezing cellular functions could reduce the physiological impact of target inhibition by antimicrobials.

The "Scout Hypothesis" provides a different perspective in the presence of apparently altruistic cells in a dormant bacterial population (Epstein 2009). According to this hypothesis, individual cells within a dormant population could periodically "wake up" due to stochastic changes in expression of a master regulatory gene, for example. The awakened "scout" cells may resume proliferating if conditions are favorable or may die if conditions are adverse. Such random

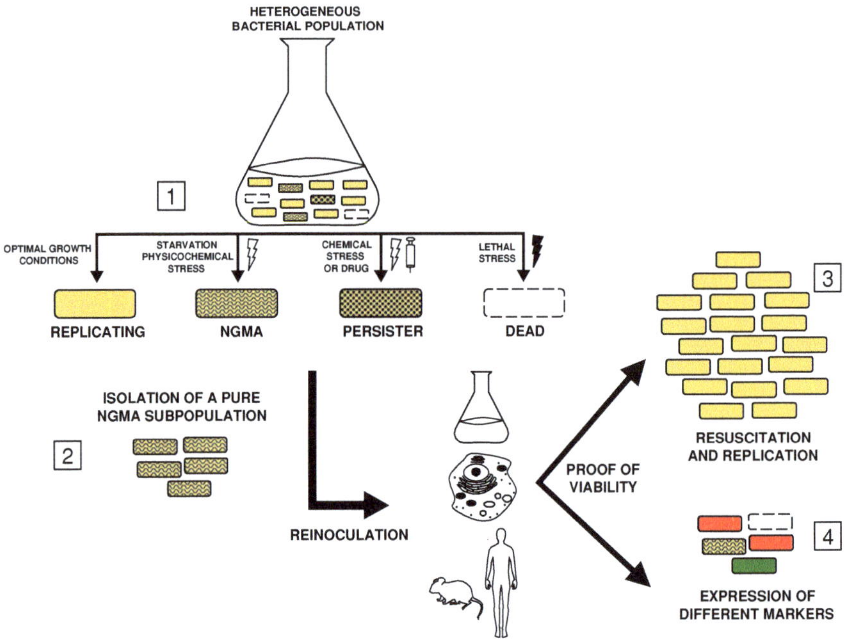

Fig. 2 Bacterial heterogeneity and identification of NGMA subpopulations. Bacterial populations are characterized by the presence of physiologically distinct subsets of cells, due to genetic and non-genetic heterogeneity, dynamic patterns of gene expression, cell–cell communication, environmental fluctuations, and physical stresses (*1*). The classical method of NGMA cells isolation by dilution to extinction of the replicating populations and reinoculation in vitro, ex vivo, or in vivo is commonly used to demonstrate viability through resuscitation of the non-culturable population (*2* and *3*). The use of fluorescent reporter-tagging methods, coupled to cell sorting, optical tweezers, or dielectrophoresis (DEP), represents a valid alternative approach to isolate and purify NGMA supopulations, which can then be subjected to 'omic studies or reinoculated under different environmental conditions to test for resuscitation (*4*)

awakenings have been proposed as a parsimonious strategy for a population of dormant cells to sample the environment, thereby risking only a few individuals to sense the environment and stimulate the revival of their dormant siblings if conditions are favorable for growth. Recently, the scout hypothesis has been investigated in *E. coli* and *Mycobacterium smegmatis*, a non-pathogenic and fast-growing relative of *M. tuberculosis* (Buerger et al. 2012).

The common element linking these studies and interpretations is that bacterial populations are heterogeneous, comprising subsets of cells characterized by different phenotypes and different roles within their community. Long-term single-cell studies are required to better understand the NGMA phenomenon, and to elucidate how non-genetic phenotypic diversity within a population could potentially influence the response and survival of bacterial populations under different stressful conditions (Fig. 2).

2.3 Recovery and Virulence of Non-Culturable Cells

A common challenge in the VBNC field is to perform successful resuscitation experiments, thereby proving that revival is truly due to resuscitation of previously non-culturable cells rather than outgrowth of a small minority of viable and culturable cells still present in the population. This technical problem highlights once again the need for new approaches to obtain pure subpopulations of bacteria from mixed cultures (Fig. 2).

A commonly used method utilized to separate culturable from non-culturable cells is the Most Probable Number (MPN) strategy, which involves serially diluting a mixed culture sufficiently such that each subculture contains, on average, less than one CFU (Roszak and Colwell 1987). Afterward, resuscitation of the diluted cultures is induced by appropriately modifying the culture conditions and the appearance of culturable cells is monitored using appropriate techniques, for example, by counting CFU. If the number of viable counts that appears in the stimulated cultures exceeds the number that could be achieved by simple proliferation of a small number of culturable cells contaminating the original culture, the experimenter can infer that previously non-culturable cells have been resuscitated (i.e., have reverted to a culturable state) by the stimulus.

Experiments with *Micrococcus luteus* have provided one of the most convincing cases of bacterial resuscitation from a state of non-culturable dormancy (Kaprelyants and Kell 1993). Transition of *M. luteus* to a dormant state can be induced by prolonged incubation in spent culture medium. Typically, following entry into stationary phase, the CFU count declines to non-detectable levels while the total cell count decreases by about 40 % and then stabilizes. Entry of cells into the non-culturable state is accompanied by a reduction in cell diameter, respiration rate, rhodamine uptake, and protein amount per cell, whereas the DNA content per cell remains constant. To eliminate any residual culturable cells, bacteria are further treated with penicillin G, which kills replicating cells but spares non-replicating cells. Resuscitation is then triggered by washing the dormant cells and inoculating them into fresh medium at high density. About 20 h later, the revived bacteria show signs of metabolic activity and the viable counts return to the initial level. The increase in CFU following resuscitation can be ascribed to resuscitation of previously non-culturable cells, because the 4-h generation time of *M. luteus* is too long to explain the observed increase in CFU from 1,000 CFU/ml at 30 h of recovery up to 100 million CFU/ml at 50 h. However, expansion of the CFU count by this magnitude could be explained by re-awakening of a large fraction of a VBNC population whose starting total cell number, based on microscopic counts, was about 200 million CFU/ml.

Interestingly, resuscitation of non-culturable cells of *M. luteus* can be enhanced by addition of cell-free supernatant from stationary-phase cultures (Kaprelyants et al. 1994). The extracellular factor responsible for recovery of dormant *M. luteus* has been identified as a secreted protein termed Resuscitation Promoting Factor

(Rpf). Purified *M. luteus* Rpf can also exert a resuscitation effect on heterologous species, including *M. tuberculosis*. It is noteworthy that the *M. tuberculosis* genome encodes five Rpf-like genes (Mukamolova et al. 2002), which are collectively dispensable for growth in liquid medium but required for resuscitation from a non-culturable state; interestingly, this defect can be rescued by addition of wild-type culture supernatant (Kana et al. 2008).

Taking into consideration the evidence suggesting that non-culturable cells of different species can be resuscitated by appropriate stimuli, an important question is the state of virulence of these non-replicating populations. This problem is relevant not only for water-borne pathogens, which have historically been a major focus in the VBNC field, but also for food-borne or air-borne bacteria that cause long-term clinically inapparent infections, such as *M. tuberculosis*. To test the virulence potential of VBNC cells, it is first required to isolate pure sub-populations of non-culturable cells, assess their potential to be resuscitated, and verify the ability to re-initiate an infection (Fig. 2). This procedure in fact recalls Koch's postulates, four criteria proposed to establish a deterministic relation between a disease and its causative microbe. Once a microorganism has been identified and isolated as a pure culture from a diseased host, it should be able to cause the same disease if re-inoculated into a healthy host; it should then be possible to re-isolate the microorganism from the secondarily infected host and confirm that it is identical to the initial causative agent. Even if we accept the concept that in vivo resuscitation of non-culturable bacteria represents a potential risk to human health, it will be difficult to demonstrate that resuscitation and virulence are exclusively due to the reawakening of non-culturable cells, unless and until a clean procedure for single-cell identification and purification becomes available.

In 1995, Oliver and Bokian tested whether non-culturable cells of *Vibrio vulnificus* were able to infect mice (Oliver and Bockian 1995). The VBNC state was induced by incubation of *V. vulnificus* in artificial seawater and exposure to a temperature downshift, causing the viable counts in the culture to drop to levels that were undetectable by Kogure's method. Subsequently, mice were infected with different doses of the VBNC culture and the lethal dose causing death of half the infected mice (LD_{50}) was determined, as well as viable counts (CFU) from various tissues. These experiments established that VBNC cells of *V. vulnificus* retain pathogenicity. However, there was an inverse temporal relationship between time of incubation at low temperature and virulence, which could indicate either a gradual shift into the VBNC state, the presence of mixed populations of culturable and non-culturable cells in the inocula used to infect the animals, or gradual death of VBNC cells.

A similar example of in vivo resuscitation was performed in human volunteers using an attenuated strain of *V. cholerae* O1 that expressed a mutated subunit of cholera toxin (Colwell et al. 1996). The non-culturable state was induced by nutrient starvation and temperature downshift for either 10 or 28 days; total and viable (culturable) counts were measured using standard techniques. Each volunteer ingested about one billion cells, including an estimated ten million VBNC cells, the remainder being simply non-viable. Remarkably, "young" (10 days) inocula of VBNC cells retained the ability to cause infection resulting in clinical signs and symptoms, whereas "old" (28 days) inocula did not.

Interpretation of studies reporting in vivo resuscitation and virulence of VBNC cells, whether in mice (Oliver and Bockian 1995) or in humans (Colwell et al. 1996), is complicated by the difficulty of achieving pure VBNC populations using standard bulk methods. Interestingly, VBNC cells of several species have been shown to exit the non-culturable state when co-cultured with different types of eukaryotic cells (Senoh et al. 2012). Further studies along these lines would greatly benefit from the advent of new non-destructive single-cell techniques to isolate pure subpopulations of non-culturable cells in order to assess the pathogenic potential of non-culturable bacteria both in vivo and ex vivo.

3 *Mycobacterium tuberculosis*

3.1 *Mycobacterium tuberculosis: A Case of Dormancy and Revival*

Bacterial growth is a complex process that can display a broad spectrum of states. In comparison to actively replicating cells, which are the usual object of microbiology research, assessing the physiological states of non-replicating cells can be technically challenging. Dormancy, a reversible state of low metabolic activity used by non-sporulating bacteria to survive for long periods of time under stressful environmental conditions, is often coupled to non-culturability.

M. tuberculosis is a major human pathogen and a prominent example of bacterial dormancy. It is estimated that perhaps one-third of the global human population is latently infected with tubercular bacilli, and it is generally assumed (although not proven) that these bacteria are metabolically sluggish or quiescent. Understanding latent *M. tuberculosis* infection is important for at least two reasons: first, reactivation of latent *M. tuberculosis* infections represents a huge burden of future disease; second, latent *M. tuberculosis* infections are thought to be refractory to antibiotic therapy (Young et al. 2006). Regarding the latter, there is an apparent discrepancy between the refractoriness of nonreplicating *M. tuberculosis* toward drugs that target cell wall biogenesis, such as isoniazid, and the effectiveness of prolonged prophylactic isoniazid therapy in preventing reactivation of latent tuberculosis infection. This discrepancy could indicate that dormant cells of *M. tuberculosis* retain some sensitivity to wall-targeting antibiotics or, alternatively, that dormant cells periodically "wake up" and revert to an active state of drug susceptibility. A related question concerns the true extent of effectiveness of anti-tuberculosis therapy for active disease, and whether some dormant bacteria could persist in the lungs despite prolonged dug therapy and maintain the

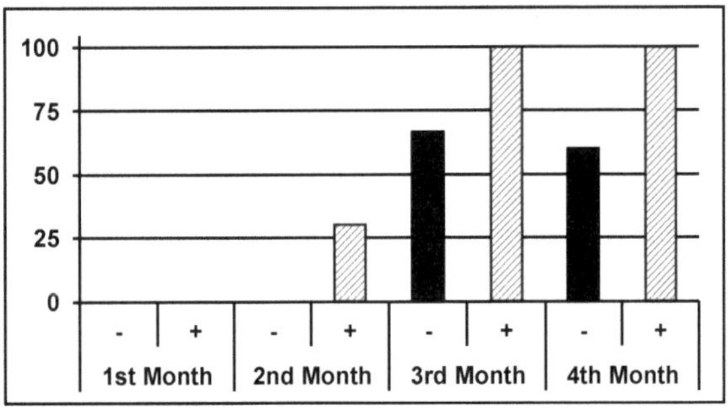

Fig. 3 Percentage of animals exhibiting relapse, after prolonged treatment with combination of drugs. Cortisone was administered (+) in groups of mice, which were monitored over a period of 4 months following anti-tuberculosis drug therapy (Modified from McCune et al. 1966)

potential to reactivate in the future. These open questions underscore the need to elucidate the metabolic state of nonculturable *M. tuberculosis*, a topic that had been investigated intensively over the years but which has not, as yet, yielded much in the way of definitive answers.

In the 1950s and 1960s, McCune and colleagues at Cornell University Medical College performed a series of pioneering experiments focused on *M. tuberculosis* dormancy and antibiotic persistence in a murine model of tuberculosis (McCune et al. 1966). In the so-called "Cornell Model", mice were infected with *M. tuberculosis* and treated extensively with various combinations of anti-tuberculosis drugs. At the conclusion of treatment, the bacilli were apparently cleared ("vanished") from the lungs of mice that received the most effective regimens, inasmuch as no remaining bacteria could be detected using sensitive techniques, such as CFU assays, acid-fast staining of lung sections, and re-inoculation of tissue homogenates into normal and immune-suppressed mice. Nonetheless, it became clear that the apparently "cured" mice still harbored tubercle bacilli, apparently in a non-culturable state, because immune suppression of these animals led to nearly uniform bacteriological relapse, as verified by CFU counts of tissues following immune suppression (Fig. 3). Remarkably, the bacteria that were recovered by plate culture post-relapse retained sensitivity to the drugs that were used to treat the infected animals. In sum, prolonged treatment of infected mice with anti-tuberculosis drugs resulted in "vanishing" of culturable bacteria from the tissues, whilst immune-suppression of the treated animals led to "reappearance" of culturable bacteria in the tissues. These experiments provided strong evidence for reactivation of *M. tuberculosis* from a VBNC-like state in vivo. Similar results have been reported in a guinea pig model of tuberculosis (Obregón-Henao et al. 2012).

Molecular analyses of the "vanishing" phenomenon in the Cornell Model have also been performed (de Wit et al. 1995; Hu et al. 2000). De Wit and colleagues searched for the presence of *M. tuberculosis* DNA in tissue homogenates using quantitative PCR (qPCR). The estimated bacterial count by qPCR in acutely infected mice was similar to the numbers detected by plating for CFU, whereas in chronically infected mice the bacillary count by qPCR exceeded the CFU count, suggesting that some bacteria had died, or, alternatively, entered into a VBNC-like state. Following prolonged drug therapy of infected mice, resulting in the "vanishing" of CFU from the tissues, bacterial DNA could still be detected in the tissues by qPCR; however, this approach was not informative about the metabolic status and potential for resuscitation of bacteria in the tissues of the treated animals and the detected DNA could simply represent dead cells. Inoculation of the apparently sterile (no CFU) tissue homogenates into healthy mice resulted in the induction of infection and disease in a small fraction of the inoculated animals. In these experiments, as in the original experiments by the Cornell group, it is extremely difficult to distinguish between resuscitation of VBNC cells and outgrowth of a very small number of persistent viable (culturable) cells that may have escaped detection. In principle, these interpretations could be distinguished by purification of the nonculturable cell population in order to exclude any contamination by culturable cells prior to inoculation. In practice, this ideal goal seems unachievable using current techniques.

In principle, the relatively shorter half-life of RNA compared to DNA should make RNA a better marker of the metabolic state of dormant and VBNC-like *M. tuberculosis*. Hu et al. (2000) detected the presence of ribosomal RNA and messenger RNA transcripts in the Cornell Model and in an in vitro model of hypoxia-induced dormancy, the so-called "Wayne model" (Wayne 1976). They also demonstrated ongoing transcription in rifampicin-treated cultures of *M. tuberculosis* in vitro by measuring incorporation of tritiated uridine, despite the "vanishing" of detectable CFU in the rifampicin-treated cultures (Hu et al. 2000). In these experiments, it was also shown that inoculation of liquid medium with serially diluted cultures provided a more sensitive measure of residual viable counts compared to inoculation of solid medium and CFU formation. A similar discrepancy between results obtained by inoculation of liquid versus solid media was also reported for experiments with stationary-phase cultures or tissue homogenates from chronically infected mice (Dhillon et al. 2004). These observations underscore the fact that the current distinction between viable and VBNC-like mycobacteria is essentially an operational definition (does the cell form a colony on solid medium? does it replicate in liquid medium?) that provides little insight into the actual physiology of these two contrasting states.

Despite the pioneering work done by the Cornell group nearly half a century ago, in the intervening decades there has been rather slight progress in the understanding of the VBNC-like state in *M. tuberculosis*. The relative stagnation of this area of inquiry, which lags behind the remarkable advances that have been made in other areas of tuberculosis research in recent years, underscores the lack of suitable tools to query the physiology of VBNC-like cells of *M. tuberculosis* at the single-cell level in the absence of small numbers of contaminating and potentially confounding "ordinary" (i.e., viable and culturable) cells.

3.2 Resuscitation in Mycobacteria: From Bulk to Single-Cell Studies

The physiological and molecular bases of mycobacterial entry into and exit from a state of dormancy are still poorly understood, despite the existence of several models in vitro (Hu et al. 2000; Dhillon et al. 2004; Sala et al. 2010; Shleeva et al. 2011), ex vivo (Biketov et al. 2000), and in vivo (McCune et al. 1966; Obregón-Henao et al. 2012). In this section, we will focus on experiments that have attempted to address these questions, and in this context we will argue for the potential importance of single-cell approaches based on new technologies and nondestructive techniques that permit the real-time tracking of individual cells.

Long-term in vitro cultures subjected to medium acidification, oxygen depletion, or nitric oxide exposure are some of the conditions that have been used to induce *M. tuberculosis* dormancy, based on the belief that these conditions mimic the conditions that mycobacteria encounter in vivo in the host tissue environment. After prolonged incubation of *M. tuberculosis* in stationary-phase liquid cultures, the cells become nonculturable and phase-dark ovoid cell bodies accumulate (Shleeva et al. 2011). These morphologically abnormal cells are dwarfish, acid-fast positive, and propidium iodide negative, implying an intact cell envelope barrier; although these cultures are non-culturable in the sense that they do not form colonies on plates, they incorporate tritiated uracil and they are capable of outgrowth when inoculated into liquid medium. Furthermore, using the MPN method described above, these non-culturable cells can be resuscitated by exposure to culture supernatants from actively growing cultures of *M. tuberculosis* or (albeit less efficiently) by exposure to Rpf from *M. luteus* (Shleeva et al. 2011). Similar results have been obtained in experiments demonstrating the ability of phospholipids and peptides isolated from culture supernatants of *M. tuberculosis* to resuscitate very old (6 months) stationary-phase cultures of *M. tuberculosis* (Zhang et al. 2001). Fatty acids, especially oleic acid, can also promote the resuscitation of VBNC-like cells of *M. smegmatis* (Nazarova et al. 2011). It has also been shown that *M. tuberculosis* in the sputum of untreated patients could be cultivated only in the presence of Rpf; however, this Rpf-dependence was eventually lost after sustained in vitro cultivation (Mukamolova et al. 2010). Rpf displays peptidoglycan hydrolytic activity, and it has recently been shown that mycobacterial peptidoglycan fragments, obtained by Rpf hydrolysis, are sufficient to resuscitate VBNC-like mycobacterial cells (Nikitushkin et al. 2013).

The presence of non-culturable forms of *M. tuberculosis* has also been inferred from in vivo studies using a "replication clock" system to assess the replication dynamics of *M. tuberculosis* in chronically infected mice (Gill et al. 2009). This system comprises a strain of *M. tuberculosis* harboring an unstable episomal plasmid that is lost at a rate proportional to the rate of bacterial division. By measuring the rates of plasmid loss during the course of murine infection, in conjunction with mathematical modeling, the authors concluded that the actual numbers of division events occurring in chronically infected mice exceeded the numbers predicted by counting CFUs. These observations suggest that a fraction of the replicating population is either killed or enters into a VBNC-like state, thereby resulting in a deceptively stable number of culturable cells (CFU) over time.

Although the experiments described above provide evidence for the existence of VBNC-like forms of *M. tuberculosis* that can be resuscitated by appropriate stimuli, the conclusions are largely based on indirect assays, such as MPN, and are limited to bulk culture measurements or destructive microscopy techniques. Thus, they provide little insight into the physiology and behavior of individual cells during entry into and exit from the VBNC-like state. Flow cytometry, combined with fluorescent markers and dyes that can report on specific aspects of bacterial physiology—such as gene expression, respiratory activity, membrane potential, etc.—can provide detailed insight into the "instantaneous" phenotype of individual cells (Davey et al. 2004). A major drawback of this technique, however, is that it does not permit the temporal tracking of individual cells and, unless coupled to cell sorting, each cell that is measured is lost to further analysis. Thus, flow cytometry suffers from the same limitation as the techniques described above, viz. it cannot be used to track the behavior of individual cells as they enter into and exit from the VBNC-like state.

A promising technique to separate and recover subpopulations of cells for further analysis is cellular dielectrophoresis (DEP), which can be used to separate intact and damaged cells based on their different dielectric behavior (Table 1). Recently, DEP has been used to study non-culturable populations of *M. smegmatis* and to separate live, dead, and dormant cells (Zhu et al. 2010a, b). In these studies, the authors used exponentially growing, late stationary-phase, and killed cultures of *M. smegmatis*. Each of these cell preparations was characterized by differences in cell morphology, structure, and composition, resulting in different cellular conductivities, which is a *sine qua non* condition for DEP-based separation of dielectric particles. In particular, the calculated conductivities were highest for dead cells (915 ± 15 μS/cm), intermediate for stationary-phase cells (812 ± 10 μS/cm), and lowest for exponentially growing cells (560 ± 20 μS/cm), presumably due, at least in part, to differences in the insulating properties of the cell envelope. By modulating the frequency of the alternating current electric field, crossover frequencies were identified at which cells with different conductivities displayed either positive DEP (particles migrate to regions of high field strength) or negative DEP (particles migrate to regions of low field strength) (Zhu et al. 2010b). Thus, DEP-based separation is a potentially powerful approach to obtain relatively pure preparations of cell subpopulations from mixed cell cultures; if this technique can be scaled up, it might be possible to obtain sufficient numbers of purified cells to perform detailed phenotypic analyses of these distinct cell states. An important caveat is the damage that could result from subjecting cells, even briefly, to a strong electric field, which could lead to damage-related artifacts.

Notwithstanding this potential confounder, DEP-based cell separation looks like a promising approach to purification and analysis of bacterial subpopulations from mixed starting cultures.

4 Tools to Investigate the NGMA Phenomenon at the Single-Cell Level

Microbial phenotypic heterogeneity occurs at different levels of environmental complexity, ranging from axenic cultures in vitro to host tissue environments in vivo, where fluctuations of physico-chemical parameters can induce replicating cells to die or enter a non-replicative state while maintaining a certain level of metabolic activity. A deeper understanding of the NGMA phenomenon at the single-cell level—and, in particular, the ability to track individual cells as they enter and exit the NGMA state—would provide new insights into the survival mechanisms of microorganisms. The metabolism and physiology of individual cells, linked to different growth states (e.g., replicating, non-replicating, and non-culturable), can be characterized by measuring fundamental cellular activities at the single-cell level, such as growth, division, DNA replication, gene expression, biochemical networks, respiratory activity, membrane energization, metabolite fluctuations, etc. (Fig. 1).

Phenotypic heterogeneity can arise from multiple sources, including (but not limited to) stochastic flucutations in gene expression (Golding et al. 2005; Yu et al. 2006; Larson et al. 2009; Suter et al. 2011; Ferguson et al. 2012; Lionnet and Singer 2012), noisy production of proteins (Cai et al. 2006; Bar-Even et al. 2006), and asymmetric partitioning of cell components at division (Huh and Paulsson 2011). A number of studies have demonstrated that both eukarytic and prokaryotic cells are characterized by extremely heterogeneous dynamics in gene expression. These processes can be studied at the single-cell and single-gene level by use of fluorescent reporters, time-lapse microscopy, and quantitative modeling tools.

Two broad categories of gene expression noise have been proposed to influence the transcriptional activity of individual genes (Locke and Elowitz 2009; Lionnet and Singer 2012). "Extrinsic" noise reflects fluctuations at the level of the cell; for example, time-dependent variation in the amount or activity of a regulatory factor. "Intrinsic" noise reflects fluctuations at the level of the gene itself; for example, thermodynamic fluctuations affecting the binding or unbinding of a regulatory factor to its cognate binding site on the DNA. These two sources of gene expression noise obviously intersect, as, for example, intrinsic noise affecting the expression of a regulatory factor could propagate as extrinsic noise affecting expression of the target genes that are regulated by that transcription factor.

It is conceivable that entry into and exit from the NGMA state could be triggered, or at least modulated, by noise-driven fluctuations in gene expression, resulting in transient phenotypic switching between growing and non-growing states. This is an area of investigation that seems ideally suited to real-time, single-cell experimental approaches, such as time-lapse fluorescence microscopy, in conjunction with computational modeling (Locke and Elowitz 2009; Chubb and Liverpool 2010). Within this realm, there is broad scope for application of the rapidly expanding arsenal of biomolecular and microengineered technologies (Rowat, et al. 2009; Tang et al. 2009; Taniguchi et al. 2010; Dalerba et al. 2011; Fidalgo and Maerkl 2011; Gobaa et al. 2011; White et al. 2011; Srigunapalan et al. 2012). For example, highly automated microfluidic tools, micro-patterned with chambers to contain cells, microchannels for delivery of liquid media, and pneumatic valves to permit user-specified opening and closing of the micro-channels, allow rapid and precise control over the cells' microenvironment. These microdevices can be deigned and inexpensively manufactured to match the end-users' specific requirements and can be used in conjunction with time-lapse microscopy for real-time single-cell analysis of microbial responses to fluctuating environments.

The use of fluorescent reporter strains of bacteria, often comprising fusions of sequences encoding a fluorescent protein to promoter sequences (transcriptional fusions) or protein-coding sequences (translational fusions), has revolutionized the study of gene expression in single cells. This approach is particularly powerful when combined with live-cell time-lapse microscopy and microfluidic culture systems that provide dynamic control over the culture environment (Golding et al. 2005; Lionnet and Singer 2012). The development of new fluorescent proteins, such as short-lived (Andersen et al. 1998), "timer" (Terskikh et al. 2000), photoactivatable (Patterson and Lippincott-Schwartz 2002), and photo-switchable variants (Bates et al. 2007), is progressively expanding the spectrum of applications and enabling deeper insights into intracellular processes, time-dependent gene expression patterns, protein localization in space and time, and molecular interactions at the sub-micron scale (Table 1).

In the specific case of the NGMA phenomenon, there is exciting but largely untapped potential in the use of transcriptional and translational fusions of fluorescent proteins to genes controlling processes involved in cell growth, metabolism, ribosome production, chromosome replication, etc., to elucidate the physiology of individual NGMA cells (Fig. 1). By tracking time-dependent changes in the production, degradation, and localization of fluorescently-tagged proteins in single cells, it should be possible to construct a picture of the metabolic activities and potentialities of NGMA cells as compared to replicating cells. The use of conditional gene expression systems to switch on/off the expression of fluorescent reporters should also permit a quantitative analysis of the transcriptional and translational potentialities of NGMA cells compared to replicating cells. Demonstration that individual NGMA cells retain the ability to respond to an external inducing agent by switching on de novo transcription and translation of a fluorescent reporter would go a long way to strengthening the case that these cells might be viable.

New tools have also been introduced for microscopy-based monitoring of time-dependent fluctuations in single-cell gene expression at single-molecule resolution. Examples include the mRNA MS2 tagging system (Golding et al. 2005) and the fluorescent Protein Complementation Assay (PCA) (Valencia-Burton et al. 2007). These techniques can be used to monitor individual gene transcription events in real-time, as a stretch of the newly transcribed RNA, tagged with MS2 arrays or by RNA aptamers (which are bound by copies of split GFP that combine only at the moment of transcription), provide very fine temporal resolution of individual gene expression events (Table 1).

Other relevant methods that could be appropriate for investigating the physiology of individual NGMA cells include the combination of fluorescent reporters with photobleaching techniques, optical nanosensors, and real-time fluorescence microscopy. These tools can be used for single-cell techniques such as Fluorescence Loss in Photobleaching (FLIP), Fluorescence Recovery after Photobleaching (FRAP), and nanosensors based on Förster Resonance Enegy Transfer (FRET) (Lippincott-Schwartz et al. 2001). In both FLIP and FRAP, PB causes the loss of fluorescence in one portion of a cell or the entire cell. In FLIP, the intracellular movement of molecules can be followed through fluorescence diffusion after PB. In FRAP, the resurgence of fluorescence provides a real-time readout of de novo transcription and translation occurring after PB. FRET-based nanosensors employ two distinct fluorescent proteins, a donor molecule and an acceptor molecule, characterized by overlapping emission spectra and coupled to a substrate-binding protein domain. Upon binding of the specific substrate to the substrate-binding protein, conformational changes bring the fluorescent proteins into closer juxtaposition, thereby enhancing resonance energy transfer from donor to acceptor following excitation of the donor. Thus, changes in the fluorescence ratio (donor fluorescence : acceptor fluorescence) provide an estimate of the intracellular substrate/metabolite concentration.

Combining the aforementioned techniques with flow cytometry-based cell sorting, microfluidic cell sorting, DEP-based cell sorting (Zhu et al. 2010a), or single-cell trapping with optical tweezers (Ericsson et al. 2000) could, in principle, allow the isolation and collection of NGMA cells for 'omics-based studies, new targets identification, and virulence assessment through in vivo recovery experiments (Fig. 2). Fluorescent reporter strains can also be used for live-cell studies of host-pathogen interactions, thereby avoiding the artifacts associated with cell fixation (Fig. 4). Microfluidic devices for ex vivo studies (James et al. 2009), in vivo imaging systems (Manicassamy et al. 2010; Nam et al. 2011), and intravital microscopy (Meissner et al. 2009) have become increasingly available to academic researchers. Despite these exciting advances, in vivo imaging is still far from achieving the sensitivity and resolution required to perform studies on individual bacteria residing within infected animals. In principle, infecting cultured mammalian cells or animal models with bacterial reporter strains whose intracellular activities can be monitored through a wide variety of fluorescent reporters would provide the most direct way to visualize growth dynamics, metabolic activity, cellular processes, and virulence potential of both replicating and NGMA bacteria.

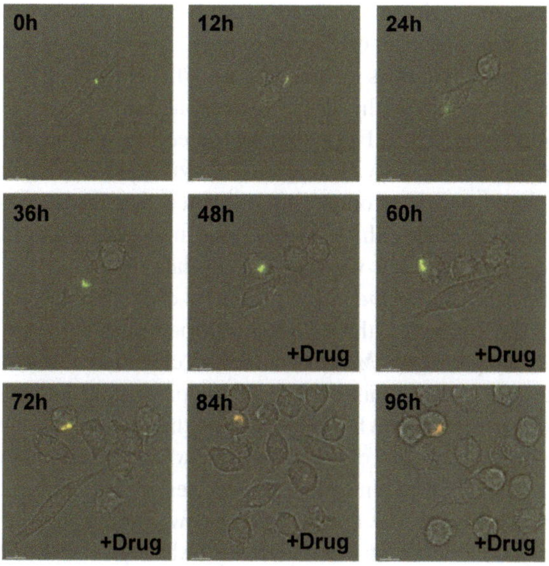

Fig. 4 Microscopy-based real-time analysis of host-pathogen interactions at the single-cell level. RAW macrophages infected with two-color fluorescently tagged *M. tuberculosis* (*GFP* short-lived reporter of metabolic activity, *DsRed* long-lived constitutively expressed reporter). Representative snapshots were taken over 4 days of continuous time-lapse fluorescence microscopy of a single macrophage infected with a single bacterium. The addition of a potent anti-tuberculosis drug to the medium causes a marked decrease in green fluorescence (metabolic reporter) within 24 h

5 Conclusions

The ecological and clinical relevance of VBNC bacteria has been a fascinating and prominent but largely unresolved issue in microbiology for several decades. Due to limitations in standard culturing techniques, artifacts associated with cell fixation, and potential toxicity of reporter dyes, there is a clear need for new methods to study the physiology of non-growing but physically intact and metabolically active bacteria. Ideally, such studies should be focused on the single-cell level. Population-averaged bulk-cell measurements can provide only partial and potentially misleading information on NGMA cells, inasmuch as there is no certainty of isolating pure subpopulations of NGMA cells, which might themselves comprise multiple distinct phenotypic states. Moreover, the standard limiting dilution techniques used to isolate non-growing from growing cells has obvious limitations and resuscitation experiments would benefit from higher-resolution techniques. A promising approach that has not been fully exploited to date is the combination of fluorescent reporter strains, time-lapse fluorescence microscopy, and microfluidic systems for real-time single-cell analysis of NGMA cells. Other techniques for single-cell analysis, some of which have been described in this chapter, have proven their value in single-cell analysis of replicating cells but have not, so far, been applied to the analysis of NGMA cells. Tools to detect and measure molecular events and dynamics within intact and metabolically active cells that have lost (at least temporarily) the ability to grow and divide offer alternative ways to assess cellular “viability”. For example, detection of ongoing transcription, translation, and metabolic activity (e.g., ATP synthesis) within individual non-growing cells would provide compelling evidence that these cells retain fundamental activities associated with “viability” and may also retain the capability of reactivating growth if provided with the appropriate stimulus.

In the specific case of *M. tuberculosis*, there is considerable scope for applying new single-cell approaches to the study of dormant and persistent states in this unusually refractory pathogen. *M. tuberculosis* is considered to be an exemplar of “persistent pathogens” that maintain long-term parasitic relationships with their chronically infected mammalian hosts by slowing their rates of metabolism, growth, and cell division. Reactivation of *M. tuberculosis* after years or even decades of clinically latent infection may, in some respects, resemble the resuscitation of non-culturable microorganisms. The application of new single-cell micro-technologies to the study of mycobacteria is still in its infancy (Aldridge et al. 2012; Golchin et al. 2012; Wakamoto et al. 2013), and to the best of our knowledge the new approaches have not been applied to the specific problem of the NGMA state in mycobacteria. In future, it should be possible to develop co-culture microsystems for real-time single-cell analysis of host-pathogen interactions—for example, bacteria interacting with cultured host cells or bacteria within tissue sections derived from infected animals. These approaches could provide new insights into the NGMA phenomenon during the course of infection and might identify new strategies for targeting this physiologically distinct subpopulation.

References

Aldridge BB, Fernandez-Suarez M, Heller D, Ambravaneswaran V, Irimia D, Toner M, Fortune SM (2012) Asymmetry and aging of mycobacterial cells lead to variable growth and antibiotic susceptibility. Science 335:100–104

Andersen JB, Sternberg C, Poulsen LK, Petersen Bjorn S, Givskov M, Molin S (1998) New unstable variants of green fluorescent protein for studies of transient gene expression in bacteria. Appl Environ Microbiol 64:2240–2246

Asakura H, Ishiwa A, Arakawa E, Makino S, Okada Y, Yamamoto S, Igimi S (2006) Gene expression profile of *Vibrio cholerae* in the cold stress-induced viable but non-culturable state. Environ Microbiol 9:869–879

Balaban NQ (2011) Persistence: mechanisms for triggering and enhancing phenotypic variability. Curr Opin Genet Dev 21:768–775

Balaban NQ, Merrin J, Chait R, Kowalik L, Leibler S (2004) Bacterial persistence as a phenotypic switch. Science 305:1622–1625

Barcina I, Arana I (2009) The viable but nonculturable phenotype: a crossroads in the life-cycle of non-differentiating bacteria? Rev Environ Sci Biotechnol 8:245–255

Bar-Even A, Paulsson J, Maheshri N, Carmi M, O'Shea E, Pilpell Y, Barkai N (2006) Noise in protein expression scales with natural protein abundance. Nat Genet 38:636–643

Bates M, Huang B, Dempsey GT, Zhuang X (2007) Multicolor super-resolution imaging with photo-switchable fluorescent probes. Science 317:1749–1753

Biketov S, Mukamolova GV, Potapov V, Gilenkov E, Vostroknutova g, Kell DB, Young M, Kaprelyants AS (2000) Culturability of *Mycobacterium tuberculosis* cells isolated from murine macrophages: a bacterial growth factor promotes recovery. FEMS Immunol Med Microbiol 29:233–240

Buerger S, Spoering A, Gavrish E, Leslin C, Ling L, Epstein SS (2012) Microbial scout hypothesis, stochastic exit from dormancy, and the nature of slow growers. Appl Environ Microbiol 78:3221–3228

Cai L, Friedman N, Xie XS (2006) Stochastic protein expression in individual cells at the single molecule level. Nature 440:358–362

Chubb JR, Liverpool TB (2010) Bursts and pulses: insights from single cell studies into transcriptional mechanisms. Curr Opin Genet Develop 20:478–484

Colwell RR (2000) Viable but nonculturable bacteria: a survival strategy. J Infect Chemother 6:121–125

Colwell RR, Brayton P, Herrington D, Tall B, Huq A, Levine MM (1996) Viable but non-culturable *Vibrio cholerae* 01 revert to a cultivable state in the human intestine. World J Microbiol Biotechnol 12:28–31

Dalerba P, Kalisky T, Sahoo D, Rajendran PS, Rothenberg ME, Leyrat AA, Sim S, Okamoto J, Johnston DM, Qian D, Zabala M, Bueno J, Neff NF, Wang J, Shelton AA, Visser B, Hisamori S, Shimono Y, van de Wetering M, Clevers H, Clarke MF, Quake SR (2011) Single-cell dissection of transcriptional heterogeneity in human colon tumors. Nat Biotechnol 29:1120–1127

Davey HM, Hexley P (2011) Red but not dead? Membranes of stressed *Saccharomyces cerevisiae* are permeable to propidium iodide. Environ Microbiol 13:163–171

Davey HM, Kell DB, Weichart DH, Kaprelyants AS (2004) Estimation of microbial viability using flow cytometry. Curr Protoc Cytom 11(3):1–21

Davidson CJ, Surette MG (2008) Individuality in bacteria. Annu Rev Genet 42:253–268

de Wit D, Wootton M, Dhillont J, Mitchisont DA (1995) The bacterial DNA content of mouse organs in the Cornell model of dormant tuberculosis. Tuber Lung Dis 76:555–562

Desnues B, Cuny C, Grégori G, Dukan S, Aguilaniu H, Nyström T (2003) Differential oxidative damage and expression of stress defence regulons in culturable and non-culturable *Escherichia coli* cells. EMBO Rep 4:400–404

Dhar N, McKinney JD (2007) Microbial phenotypic heterogeneity and antibiotic tolerance. Curr Opin Micriobiol 10:30–38

Dhillon J, Lowrie DB, Mitchison DA (2004) *Mycobacterium tuberculosis* from chronic murine infections that grows in liquid but not on solid medium. BMC Infect Dis 4:1–4

Epstein SS (2009) Microbial awakenings. Nature 457:1083

Ericsson M, Hanstorp D, Hagberg P, Enger J, Nyström T (2000) Sorting out bacterial viability with optical tweezers. J Bacteriol 182:5551–5555

Ferguson ML, Le Coq D, Jules M, Aymerich S, Radulescu O, Declerck N, Royera CA (2012) Reconciling molecular regulatory mechanisms with noise patterns of bacterial metabolic promoters in induced and repressed states. Proc Natl Acad Sci U S A 109:155–160

Fidalgo LM, Maerkl SJ (2011) A software-programmable microfluidic device for automated biology. Lab Chip 11:1612–1619

Finkel SE (2006) Long-term survival during stationary phase: evolution and the GASP phenotype. Nat Rev Microbiol 4:113–120

Gengenbacher M, Rao SPS, Pethe K, Dick T (2010) Nutrient-starved, non-replicating *Mycobacterium tuberculosis* requires respiration, ATP synthase and isocitrate lyase for maintenance of ATP homeostasis and viability. Microbiology 156:81–87

Gill WP, Harik NS, Whiddon MR, Liao RP, Mittler JE, Sherman DR (2009) A replication clock for *Mycobacterium tuberculosis*. Nat Med 15:211–214

Gobaa S, Hoehnel S, Roccio M, Negro A, Kobel S, Lutolf MP (2011) Artificial niche microarrays for probing single stem cell fate in high throughput. Nat Methods 8:949–955
Golchin SA, Stratford J, Curry RJ, McFadden J (2012) A microfluidic system for long-term time-lapse microscopy studies of mycobacteria. Tuberculosis (Edinb.) 92:489–496
Golding I, Paulsson J, Zawilski SM, Cox EC (2005) Real-time kinetics of gene activity in individual bacteria. Cell 123:1025–1036
Gonzàlez-Pastor JE (2011) Cannibalism: a social behavior in sporulating *Bacillus subtilis*. FEMS Microbiol Rev 35:415–424
Griffitt KJ, Noriea NF, Johnson CN, Grimes DJ (2011) Enumeration of *Vibrio parahaemolyticus* in the viable but nonculturable state using direct plate counts and recognition of individual gene fluorescence in situ hybridization. J Microbiol Methods 85:114–118
Hamid Salim A, Aung KJ, Hossain MA, Van Deun A (2006) Early and rapid microscopy-based diagnosis of true treatment failure and MDR-TB. Int J Tuberc Lung Dis 10:1248–1254
Heim S, Lleo MdM, Bonato B, Guzman CA, Canepari P (2002) The viable but nonculturable state and starvation are different stress responses of *Enterococcus faecalis*, as determined by proteome analysis. J Bacteriol 184:6739–6745
Hu Y, Mangan JA, Dhillon J, Sole KM, Mitchison DA, Butcher PD, Coates ARM (2000) Detection of mRNA transcripts and active transcription in persistent *Mycobacterium tuberculosis* induced by exposure to rifampin or pyrazinamide. J Bacteriol 182:6358–6365
Huh D, Paulsson J (2011) Non-genetic heterogeneity from stochastic partitioning at cell division. Nat Genet 43:95–100
James CD, Moorman MW, Carson BD, Branda CS, Lantz JW, Manginell RP, Martino A, Singh AK (2009) Nuclear translocation kinetics of NF-κB in macrophages challenged with pathogens in a microfluidic platform. Biomed Microdevices 11:693–700
Jin DJ, Cagliero C, Zhou YN (2012) Growth rate regulation in *Escherichia coli*. FEMS Microbiol Rev 36:269–287
Kana BD, Gordhan BG, Downing KJ, Sung N, Vostroktunova G, Machowski EE, Tsenova L, Young M, Kaprelyants A, Kaplan G, Mizrahi V (2008) The resuscitation-promoting factors of *Mycobacterium tuberculosis* are required for virulence and resuscitation from dormancy but are collectively dispensable for growth in vitro. Mol Microbiol 67:672–684
Kaprelyants AS, Kell DB (1993) Dormancy in stationary-phase cultures of *Micrococcus luteus*: flow cytometric analysis of starvation and resuscitation. Appl Environ Microbiol 59:3187–3196
Kaprelyants AS, Mukamolova GV, Kell DB (1994) Estimation of dormant *Micrococcus luteus* cells by penicillin lysis and by resuscitation in cell-free spent culture medium at high dilution. FEMS Microbiol Lett 115:347–352
Kell DB, Kaprelyants AS, Weichart DH, Harwood CR, Barer MR (1998) Viability and activity in readily culturable bacteria: a review and discussion of the practical issues. Antonie van Leeuwenhoek 73:169–187
Kogure K, Simidu U, Taga N, Colwell RR (1987) Correlation of direct viable counts with heterotrophic activity for marine bacteria. Appl Environ Microbiol 53:2322–2327
Kussell E, Leibler S (2005) Phenotypic diversity, population growth, and information in fluctuating environments. Science 309:2075–2078
Kussell E, Kishony R, Balaban NQ, Leibler S (2005) Bacterial persistence: a model of survival in changing environments. Genetics 169:1807–1814
Lahtinen SJ, Ahokoski H, Reinikainen JP, Gueimonde M, Nurmi J, Ouwehand AC, Salminen SJ (2008) Degradation of 16S rRNA and attributes of viability of viable but nonculturable probiotic bacteria. Lett Appl Microbiol 46:693–698
Larson DR, Singer RH, Zenklusen D (2009) A single molecule view of gene expression. Trends Cell Biol 19:630–637
Lew S, Lew M, Mieszczynski T, Szarek J (2010) Selected fluorescent techniques for identification of the physiological state of individual water and soil bacterial cells. Folia Microbiol (Praha) 55:107–118

Lidstrom ME, Konopka MC (2010) The role of physiological heterogeneity in microbial population behavior. Nat Chem Biol 6:705–712

Lionnet T, Singer RH (2012) Transcription goes digital. EMBO Rep 13:313–321

Lippincott-Schwartz J, Snapp E, Kenworthy A (2001) Studying protein dynamics in living cells. Nat Methods 2:444–456

Liu Y, Gilchrist A, Zhang J, Li XF (2008) Detection of viable but nonculturable *Escherichia coli* O157:H7 bacteria in drinking water and river water. Appl Environ Microbiol 74:1502–1507

Locke JCW, Elowitz MB (2009) Using movies to analyze gene circuit dynamics in single cells. Nat Rev Microbiol 7:383–392

Lothigius A, Sjöling A, Svennerholm AM, Bölin I (2010) Survival and gene expression of enterotoxigenic *Escherichia coli* during long-term incubation in sea water and freshwater. J Appl Microbiol 108:1441–1449

Lowder M, Unge A, Maraha N, Jansson JK, Swiggett J, Oliver JD (2000) Effect of starvation and the viable-but-nonculturable state on green fluorescent protein (GFP) fluorescence in GFP-tagged *Pseudomonas fluorescens* A506. Appl Environ Microbiol 66:3160–3165

Manicassamy B, Manicassamy S, Belicha-Villanueva A, Pisanelli G, Pulendran B, García-Sastre A (2010) Analysis of in vivo dynamics of influenza virus infection in mice using a GFP reporter virus. Proc Natl Acad Sci U S A 107:11531–11536

McCune RM, Feldmann FM, Lambert HP, McDermott W (1966) Microbial persistence. I. The capacity of tubercle bacilli to survive sterilization in mouse tissue. J Exp Med 123:445–468

Meissner S, Knels L, Krueger A, Koch T, Koch E (2009) Simultaneous three-dimensional optical coherence tomography and intravital microscopy for imaging subpleural pulmonary alveoli in isolated rabbit lungs. J Biomed Opt 14:054020

Mizunoe Y, Wai SN, Takade A, Yoshida S (1999) Restoration of culturability of starvation-stressed and low-temperature-stressed *Escherichia coli* O157 cells by using H_2O_2-degrading compounds. Arch Microbiol 172:63–67

Muela A, Seco C, Camafeita E, Arana I, Orruno M, Lopez JA, Barcina I (2008) Changes in *Escherichia coli* outer membrane subproteome under environmental conditions inducing the viable but nonculturable state. FEMS Microbiol Lett 64:28–36

Mukamolova GV, Turapov OA, Young DI, Kaprelyants AS, Kell DB, Young M (2002) A family of autocrine growth factors in *Mycobacterium tuberculosis*. Mol Microbiol 46:623–635

Mukamolova GV, Kaprelyants AS, Kell DB, Young M (2003) Adoption of the transiently non-culturable state—a bacterial survival strategy? Adv Microb Physiol 47:65–129

Mukamolova GV, Turapov O, Malkin J, Woltmann G, Barer MR (2010) Resuscitation-promoting factors reveal an occult population of tubercle bacilli in sputum. Am J Respir Crit Care Med 181:174–180

Nam SW, Chen X, Lim J, Kim SH, Kim ST, Cho YH, Yoon J, Park S (2011) *In vivo* fluorescence imaging of bacteriogenic cyanide in the lungs of live mice infected with cystic fibrosis pathogens. PLoS One 6:e21387

Navarro Llorens JM, Tormo A, Martínez-García E (2010) Stationary phase in gram-negative bacteria. FEMS Microbiol Rev 34:476–495

Nazarova EV, Shleeva MO, Morozova NS, Kudykina YK, Vostroknutova GN, Ruzhitsky AO, Selishcheva AA, Sorokoumova GM, Shvets VI, Kaprelyants AS (2011) Role of lipid components in formation and reactivation of *Mycobacterium smegmatis* "nonculturable" cells. Biochemistry (Mosc) 76:636–644

Nikitushkin VD, Demina GR, Shleeva MO, Kaprelyants AS (2013) Peptidoglycan fragments stimulate resuscitation of "non-culturable" mycobacteria. Antonie van Leeuwenhoek 103:37–46

Nyström T (2001) Not quite dead enough: on bacterial life, culturability, senescence, and death. Arch Microbiol 176:159–164

Obregón-Henao A, Shanley CA, Shang S, Caraway ML, Basaraba RJ, Duncan CG, Ordway DJ, Orme IM (2012) Cortisone-forced reactivation of weakly acid fast positive *Mycobacterium tuberculosis* in guinea pigs previously treated with chemotherapy. Mycobac Dis 2:1–5

Oliver JD (2010) Recent findings on the viable but nonculturable state in pathogenic bacteria. FEMS Microbiol Rev 34:415–425
Oliver JD, Bockian R (1995) *In vivo* resuscitation, and virulence towards mice, of viable but nonculturable cells of *Vibrio vulnificus*. Appl Environ Microbiol 61:2620–2623
Patterson GH, Lippincott-Schwartz J (2002) A photoactivatable GFP for selective photolabeling of proteins and cells. Science 297:1873–1877
Pawlowski DR, Metzger DJ, Raslawsky A, Howlett A, Siebert G, Karalus RJ, Garrett S, Whitehouse CA (2011) Entry of *Yersinia pestis* into the viable but nonculturable state in a low-temperature tap water microcosm. PLoS One 6:e17585
Piqueres P, Moreno Y, Alonso JL, Ferrús MA (2006) A combination of direct viable count and fluorescent in situ hybridization for estimating *Helicobacter pylori* cell viability. Res Microbiol 157:345–349
Rahman I, Shahamat M, Kirchman PA, Russek-Cohen E, Colwell RR (1994) Methionine uptake and cytopathogenicity of viable but nonculturable *Shigella dysenteriae* Type 1. Appl Environ Microbiol 60:3573–3578
Roszak DB, Colwell RR (1987) Survival strategies of bacteria in the natural environment. Microbiol Rev 51:365–379
Rowat AC, Bird JC, Agresti JJ, Rando OJ, Weitza DA (2009) Tracking lineages of single cells in lines using a microfluidic device. Proc Natl Acad Sci U S A 106:18149–18154
Sala C, Dhar N, Hartkoorn RC, Zhang M, Ha YH, Schneider P, Cole ST (2010) Simple model for testing drugs against nonreplicating *Mycobacterium tuberculosis*. Antimicrob Agents Chemother 54:4150–4158
Segev E, Smith Y, Ben-Yehuda S (2012) RNA dynamics in aging bacterial spores. Cell 148:139–149
Senoh M, Ghosh-Banerjee J, Ramamurthy T, Colwell RR, Miyoshi S, Nair GB, Takeda Y (2012) Conversion of viable but nonculturable enteric bacteria to culturable by co-culture with eukaryotic cells. Microbiol Immunol 56:342–345
Shah IM, Laaberki MH, Popham DL, Dworkin J (2008) A eukaryotic-like Ser/Thr kinase signals bacteria to exit dormancy in response to peptidoglycan fragments. Cell 135:486–496
Shleeva MO, Kudykina YK, Vostroknutova GN, Suzina NE, Mulyukin AL, Kaprelyants AS (2011) Dormant ovoid cells of *Mycobacterium tuberculosis* are formed in response to gradual external acidification. Tuberculosis (Edinb) 91:146–154
Srigunapalan S, Eydelnant IA, Simmons CA, Wheeler AR (2012) A digital microfluidic platform for primary cell culture and analysis. Lab Chip 12:369–375
Suter DM, Molina N, Gatfield D, Schneider K, Schibler U, Naef F (2011) Mammalian genes are transcribed with widely different bursting kinetics. Science 332:472–474
Tang F, Barbacioru C, Wang Y, Nordman E, Lee C, Xu N, Wang X, Bodeau J, Tuch BB, Siddiqui A, Lao K, Surani MA (2009) mRNA-Seq whole-transcriptome analysis of a single cell. Nat Methods 6:377–382
Taniguchi Y, Choi PJ, Li GW, Chen H, Babu M, Hearn J, Emili A, Xie XS (2010) Quantifying *E. coli* proteome and transcriptome with single-molecule sensitivity in single cells. Science 329:533–537
Terskikh A, Fradkov A, Ermakova G, Zaraisky A, Tan P, Kajava AV, Zhao X, Lukyanov S, Matz M, Kim S, Weissman I, Siebert P (2000) "Fluorescent timer": protein that changes color with time. Science 290:1585–1588
Trevors JT (2011) Viable but non-culturable (VBNC) bacteria: gene expression in planktonic and biofilm cells. J Microbiol Methods 86:266–273
Ueta M, Ohniwa RL, Yoshida H, Maki Y, Wada C, Wada A (2008) Role of HPF (hibernation promoting factor) in translational activity in *Escherichia coli*. J Biochem 143:425–433
Ueta M, Wada C, Wada A (2010) Formation of 100S ribosomes in *Staphylococcus aureus* by the hibernation promoting factor homolog SaHPF. Genes Cells 15:43–58
Valencia-Burton M, McCullough RM, Cantor CR, Broude NE (2007) RNA visualization in live bacterial cells using fluorescent protein complementation. Nat Methods 4:421–427

Wakamoto Y, Dhar N, Chait R, Schneider K, Signorino-Gelo F, Leibler S, McKinney JD (2013) Dynamic persistence of antibiotic-stressed mycobacteria. Science 339:91–95

Walling MA, Shepard JRE (2011) Cellular heterogeneity and live cell arrays. Chem Soc Rev 40:4049–4076

Wayne LG (1976) Dynamics of submerged growth of *Mycobacterium tuberculosis* under aerobic and microaerophilic conditions. Am Rev Respir Dis 114:807–811

White AK, VanInsberghe M, Petriv OI, Hamidi M, Sikorski D, Marra MA, Piret J, Aparicio S, Hansen CL (2011) High-throughput microfluidic single-cell RT-qPCR. Proc Natl Acad Sci U S A 108:13999–14004

Williamson KS, Richards LA, Perez-Osorio AC, Pitts B, McInnerney K, Stewart PS, Franklin MJ (2012) Heterogeneity in *Pseudomonas aeruginosa* biofilms includes expression of ribosome hibernation factors in the antibiotic-tolerant subpopulation and hypoxia-induced stress response in the metabolically active population. J Bacteriol 194(8):2062–2073

Yin H, Marshall D (2012) Microfluidics for single cell analysis. Curr Opin Biotechnol 23:110–119

Young DB, Gideon HP, Wilkinson RJ (2006) Eliminating latent tuberculosis. Trends Microbiol 17:183–188

Yu J, Xiao J, Ren X, Lao X, Xie XS (2006) Probing gene expression in live cells, one protein molecule at a time. Science 311:1600–1603

Zhang Y, Yang Y, Woods A, Cotter RJ, Sun Z (2001) Resuscitation of dormant *Mycobacterium tuberculosis* by phospholipids or specific peptides. Biochem Biophys Res Comm 284:542–547

Zhu K, Kaprelyants AS, Salina EG, Markx GH (2010a) Separation by dielectrophoresis of dormant and nondormant bacterial cells of *Mycobacterium smegmatis*. Biomicrofluidics 4:022809

Zhu K, Kaprelyants AS, Salina EG, Schuler M, Markx GH (2010b) Construction by dielectrophoresis of microbial aggregates for the study of bacterial cell dormancy. Biomicrofluidics 4:022810

Mycobacterium Tuberculosis Metabolism and Host Interaction: Mysteries and Paradoxes

Sabine Ehrt and Kyu Rhee

Abstract Metabolism is a widely recognized facet of all host–pathogen interactions. Knowledge of its roles in pathogenesis, however, remains comparatively incomplete. Existing studies have emphasized metabolism as a cell autonomous property of pathogens used to fuel replication in a quantitative, rather than qualitatively specific, manner. For *Mycobacterium tuberculosis,* however, matters could not be more different. *M. tuberculosis* is a chronic facultative intracellular pathogen that resides in humans as its only known host. Within humans, *M. tuberculosis* resides chiefly within the macrophage phagosome, the cell type, and compartment most committed to its eradication. *M. tuberculosis* has thus evolved its metabolic network to both maintain and propagate its survival as a species within a single host. The specific ways in which its metabolic network serves these distinct, through interdependent, functions, however, remain incompletely defined. Here, we review existing knowledge of the *M. tuberculosis*–host interaction, highlighting the distinct phases of its natural life cycle and the diverse microenvironments encountered therein.

S. Ehrt (✉)
Department of Microbiology and Immunology, Weill Cornell Medical College, New York, NY 10065, USA
e-mail: sae2004@med.cornell.edu

K. Rhee
Department of Medicine, Weill Cornell Medical College, New York, NY 10065, USA
e-mail: kyr9001@med.cornell.edu

Current Topics in Microbiology and Immunology (2012) 374: 163–188 163
DOI: 10.1007/82_2012_299

Published Online: 16 December 2012

Contents

1 Introduction

Most bacterial pathogens reside in polymicrobial niches and seek to replicate as quickly as possible to overtake their host niche. However, humans are the only known host and reservoir of *M. tuberculosis*. *M. tuberculosis* has thus evolved within a niche, in which either unrestrained replication or perfect symbiosis threaten its very existence as a species. Instead, the *M. tuberculosis* lifecycle consists in a prolonged period of replicative quiescence tolerant to host immunity and conventional chemotherapy, and while sometimes lasting decades, if not the lifetime of the host, is followed by an obligatory episode of replication, required for transmission to a new host. *M. tuberculosis* thus both replicates and persists within the same host species.

Interest in metabolic aspects of the *M. tuberculosis*–host interaction dates to Koch's discovery of *M. tuberculosis* and its distinct chemical staining properties among the "heaping nuclei and detritus" of tuberculous lesions in animals that had died from tuberculosis several weeks earlier, suggesting the existence of specific metabolic differences (Sakula 1983). This discovery prompted several decades of human autopsy, experimental animal, bacteriologic, and biochemical studies that focused chiefly on histopathologic descriptions of the host niches occupied by *M. tuberculosis*, its crude biochemical composition, and cell autonomous growth capabilities (Rich 1946; Youmans 1979). The advent of protein biochemistry later shifted attention to descriptions of its enzymatic activities as reported by whole cell lysates and indirect chemical reporters (Murthy et al. 1962; Wheeler and Blanchard 2005). Based on such studies, Segal and Bloch adduced seminal evidence that *M. tuberculosis* grown in vitro and recovered from the lungs of infected animals existed in distinct metabolic states, in the latter of which it was found to preferentially respire on fatty acids rather than on carbohydrates more typically used to cultivate *M. tuberculosis* in vitro (Bloch and Segal 1956). Following publication of the *M. tuberculosis* genome sequence in 1998, studies of the *M. tuberculosis*–host interaction turned heavily, if not exclusively, to the use of comparative genomics, transcriptional profiling, homology-based bioinformatic modeling, and gene knockout approaches (Cole et al. 1998; Sassetti et al. 2001; Rhee et al. 2011; Griffin et al. 2011). These studies afforded unprecedented genome-scale insight into the biology of *M. tuberculosis* with single nucleotide resolution.

Notwithstanding, knowledge of the *M. tuberculosis*–host interaction remains incomplete. Unlike the case for other cellular processes, metabolism consists of reactions whose substrates and products are atomic, rather than molecular, in diversity and often cannot be inherently deduced from genetic information. As a result, homology-based comparison have failed to suggest a function for nearly 40 % of *M. tuberculosis* annotated genes, while up to 30 % of detected enzymatic activities have not been ascribed to a known gene (Cole et al. 1998; Chen 2007). Their levels, connectivities, and directionalities remain similarly undefined due to regulation by multiple pathways and regulatory circuits, many of which are non-genetic in nature. In this regard, efforts to infer the in vivo metabolic state of *M. tuberculosis* based on its transcriptional profile or survival of gene knockouts in animal models of tuberculosis have been hindered by fundamental biochemical ambiguities. While emergent metabolomics technologies have begun to resolve such ambiguities, these same technologies have simultaneously revealed unique properties of its metabolic network that have prompted a re-evaluation current experimental knowledge.

All tools and approaches have inherent strengths and limitations. Classical bacteriology and biochemistry have provided insight into the cell autonomous traits and capabilities of *M. tuberculosis* and its proteins but also been limited by the artificial culture conditions used to study them. Genetic approaches, by contrast, have afforded a singular specificity but relied almost completely on the accuracy and completeness of their accompanying bioinformatic gene annotations. Metabolomics, defined as the simultaneous measurement of all metabolites in a biological system under a given set of conditions, is the youngest of systems level disciplines and begun to deliver unique insights into the intracellular biochemistry of *M. tuberculosis* and its response to perturbation. However, while metabolites represent the integrated product of a cell's genome, proteome and environment, and most direct reporters of a cell's metabolic state, technical limitations have thus far precluded the ability to profile *M. tuberculosis* recovered from host cells or tissues (Rhee et al. 2011).

Technical limitations aside, knowledge of the *M. tuberculosis*–host interaction has been limited by biological dissociations between *M. tuberculosis* life cycle in vitro and in vivo. As a result, experimental studies of *M. tuberculosis* pathogenesis have primarily focused on its ability to persist, while its physiology has been most extensively studied when replicating at maximal growth rates.

Here, we review and assimilate existing knowledge of *M. tuberculosis* physiologic and pathophysiologic traits within the context of its natural life cycle and primary experimental evidence.

2 The *Mycobacterium tuberculosis* Host Niche: Down the Rabbit Hole

Following aerosol transmission to a new host, *M. tuberculosis* first encounters the terminal airspace, or alveolus, of the lung where it is phagocytosed by tissue resident alveolar macrophages and dendritic cells. *M. tuberculosis* is then believed to undergo a period of unrestricted replication, migrate to the local draining lymph node, and spread throughout the bloodstream, infecting additional macrophages and reseeding additional regions of the lung to establish a systemic infection until contained by the onset of cellular immunity. Activation of the adaptive immune response results in the production of interferon-γ (IFN-γ), which enables macrophages to kill or restrain replication of *M. tuberculosis* and marks the onset of a period in which *M. tuberculosis* persists in a non- or slowly-replicating state that often lasts decades, if not the lifetime of the host (Russell 2001; Barry et al. 2009; Philips and Ernst 2012).

Mycobacterium tuberculosis has been reported to infect and/or be found in both phagocytic and nonphagocytic cell types, including myeloid dendritic cells, neutrophils, adipocytes, and epithelial cells (Neyrolles et al. 2006; Wolf et al. 2007; Nathan 2009). However, macrophages constitute the quantitatively largest and the most significant cellular reservoir of *M. tuberculosis* during the establishment and maintenance of a chronic infection. Moreover, not all macrophages, nor their phagosomes, are the same. Though microscopically well defined, phagosomes are both diverse and dynamic. Following internalization of a particle or pathogen, phagosomes undergo a series of rapid and extensive changes in their biochemical composition and function that can be further modulated by immune activation. Nascent phagosomes, for example, while hypoxic and nutrient poor, are generally not competent to kill intracellular bacteria (James et al. 1995; Vieira et al. 2002). However, subsequent fusion with endomembrane compartments, such as lysosomes, leads to a progressive acidification, alteration in ionic composition, and accumulation of lytic enzymes, oxygenated lipids, and reactive oxygen species that can be further intensified with immune activation by cytokines. Cytokines such as IFN-γ can further modify this chemistry with the coordinated production and trafficking of reactive oxygen and nitrogen intermediates, antimicrobial peptides and mediators of autophagy, such as inducible nitric oxide synthase (iNOS), the immunity-related GTPase LRG-47, and guanylate-binding proteins (GBPs) (MacMicking et al. 1997; MacMicking 2003; Gutierrez et al. 2004; Alonso et al. 2007; Purdy and Russell 2007; Kim et al. 2011; Shenoy et al. 2012).

Mycobacterium tuberculosis-containing phagosomes have long been noted to exhibit an exceptional degree of heterogeneity (Russell 2001; Rohde et al. 2007). Unlike the case for pyogenic bacteria, some *M. tuberculosis*-containing phagosomes are remarkably long lived, lasting days, rather than hours, in vitro, and years, rather than days, in vivo, while others succumb to their replication, and yet others undergo apoptosis (Fratazzi et al. 1999). In addition, considerable biochemical and cell biologic evidence has established that *M. tuberculosis* can actively regulate the phagosome, by preventing its fusion with the lysosome and antagonizing its acidification (Sturgill-Koszycki et al. 1994; Schaible et al. 1998). More recent studies have suggested that *M. tuberculosis* may even escape the phagosome to reside in the cytosol (van der Wel et al. 2007; Houben et al. 2012).

Given this extensive heterogeneity, it is likely, if not certain, that *M. tuberculosis* encounters phagosomal environments that vary according to the specific cell type, location, and time of infection. Indeed, bone marrow-derived macrophages, alveolar macrophages, and dendritic cells have all been shown to be capable of producing similar levels of nitric oxide (NO) (the major determinant of immune control of *M. tuberculosis* in mice), upon immune stimulation with IFN-γ (MacMicking et al. 1997; Bodnar et al. 2001; Choi et al. 2002; Roy et al. 2004). Yet, infection of each is associated with a distinct physiologic outcome. For example, alveolar macrophages support sufficient replication to enable *M. tuberculosis* to establish infection, while dendritic cells achieve sufficient control to enable maturation of a T cell response and bone marrow-derived macrophages, which are believed to predominate during the chronic or persistent phase of infection, and characterized by their unique ability to restrict, if not kill, *M. tuberculosis* over prolonged periods of time. Such differences thus highlight key, yet unresolved, differences in the cellular microenvironments associated with the *M. tuberculosis* lifecycle.

From an experimental perspective, it is important to note that biochemical studies of the host milieu of *M. tuberculosis* have relied heavily, if not exclusively, on the in vitro use of murine bone marrow-derived macrophages or immortalized cell lines (Vogt and Nathan 2011) that, when used, have been challenged by methodologic shortfalls associated with identifying their constituents and preserving their concentrations during subcellular fractionation.

Limitations aside, experimental studies of *M. tuberculosis*-infected macrophages of immunocompetent mice or humans have provided direct biochemical (rather than inferred genetic) evidence for: a transient burst of ROI mediated by assembly of the phagocyte oxidase complex accompanying formation of the phagocytic cup; decreased oxygen; a pH of about 6.3 in resting macrophages, due to exclusion of the vacuolar proton-exchanging ATPase, that can be overcome with IFN-γ-stimulation to achieve a pH of approximately 4.5; and NO generated upon IFN-γ-induced production and vesicular localization of iNOS as well as from chemical dismutation of its auto-oxidation product, nitrite (Nathan and Shiloh 2000; MacMicking 2003; Vandal et al. 2008). The relevance of NO-derived species in human TB has been specifically affirmed by the demonstration of co-localizing immunoreactivity of iNOS and nitrotyrosine, a biochemical footprint of NO reactivity (Choi et al. 2002). Transcriptional profiling studies of *M. tuberculosis* recovered from macrophages in vitro and from the lungs of infected mice have been used to infer the availability of fatty acids, and lack carbohydrates, lysine, leucine, and iron in resting and immune-activated phagosomes (Sambandamurthy et al. 2002; Schnappinger et al. 2003; Timm et al. 2003; Voskuil et al. 2003). However, direct biochemical support for such inferences remains lacking. Interestingly, studies of *M. tuberculosis*' bioactive cell surface lipids and secreted proteins have documented trafficking into host cytosol and extracellular vesicles, while electron microscopy has demonstrated co-localization of *M. tuberculosis*-containing phagosomes with host lipid bodies, suggesting a potential biochemical connectivity (Russell 2001; Russell et al. 2009).

Looking more macroscopically, both human patient and animal studies have extended this heterogeneity to include broader changes in the tissue surrounding *M. tuberculosis*-infected macrophages. For example, while the initial infection of alveolar macrophages triggers a localized inflammatory response consisting in tumor necrosis factor (TNF)-α and IFN-γ, the concurrent release of inflammatory chemokines initiates the formation of a granuloma, the pathologic hallmark of tuberculosis (Russell 2006). Initially the granuloma, is an amorphous mass of macrophages, monocytes, and neutrophils, but over time undergoes differentiation giving rise to specialized cell types, including multinucleated giant cells, foamy macrophages and epithelioid macrophages, that organize into a multicellular structure consisting in a macrophage-rich center surrounded by a mantle of lymphocytes and fibrous sheath. This microanatomic heterogeneity is accompanied by further variations in vascular supply and oxygen tension that can range from neovascularized to hypoxic necrotic granulomas within the same host (Via et al. 2008; Russell et al. 2010). Moreover, studies of rabbit and human tuberculous granulomas have shown *M. tuberculosis* to exist in anatomically distinct transcriptional states within a single lesion (Kaplan et al. 2003).

Albeit brief, *M. tuberculosis* also traverses key extracellular environments that are critical to completion of its life cycle. As granulomas progress, their fibrous sheath becomes more marked, while their vasculature wanes and the number of foamy macrophages increases. These changes are accompanied by further increases in hypoxia, as reported by in situ staining with pimonidazole, and presumptive gradients of other blood borne nutrients, that culminate in necrosis, a central caseation that is sometimes surrounded by calcification, and accumulation of extracellular bacilli (Boshoff and Barry 2005; Russell 2006; Via et al. 2008). Biochemical analysis of the major lipid species within this caseum has revealed an abundance of cholesterol ester, cholesterol, and triacylglycerol, and also a high level of lactosylceramide (Kim et al. 2010). Ultimately, such granulomas rupture into the airways, perhaps via an increase in levels of matrix metalloproteinase 9 (MMP9), and spill thousands of viable bacilli, poised to re-enter cell cycle (Russell 2006; Elkington and D'Armiento 2011). Once released into the airways, *M. tuberculosis* completes its lifecycle via infectious sputum aerosol and transmission to a new host. Though generally considered aerated mucus polysaccharides and epithelial cell debris of the lower respiratory tract, cytologic studies of

M. tuberculosis-containing sputum have identified a significant number of expelled bacteria within neutrophils, while bacteriologic and transcriptional profiling studies of *M. tuberculosis* have revealed distinct subpopulations of nonreplicating as well as viable but not culturable subpopulations suggestive of a more complex environment (Garton et al. 2008; Eum et al. 2010; Mukamolova et al. 2010). Sputum aerosol studies have further established a distinct array of physical characteristics associated with droplets of different size that may further contribute to the heterogeneity of this critical niche (Rieder 1999).

3 The *Mycobacterium tuberculosis* Looking Glass: Current Gaps in Knowledge

3.1 Eating and Sleeping on an Atkins Diet?

Looking from the pathogen seminal studies from Segal and Bloch taught that *M. tuberculosis* isolated from the lungs of infected mice preferentially metabolized fatty acids over carbohydrates implicating fatty acids as the predominant carbon source encountered by *M. tuberculosis* in vivo (Bloch and Segal 1956). This view was later enforced with the completion of the *M. tuberculosis* genome and discovery of its extensive duplications of genes encoding enzymes associated with beta-oxidation (Cole et al. 1998). Genome wide transcriptome analyses of *M. tuberculosis* recovered from primary mouse macrophages, mouse lungs and TB patients similarly revealed that genes encoding enzymes required for lipid and fatty acid metabolism were induced in the host niche (Schnappinger et al. 2003; Timm et al. 2003).

The specific lipid and fatty acid substrates accessible to *M. tuberculosis* remain incompletely defined. However, *M. tuberculosis* expresses a number of lipases and phospholipases capable of catalyzing the release of fatty acids from host lipids (Côtes et al. 2008; Singh et al. 2010; Dedieu et al. 2012). Accordingly, an *M. tuberculosis* strain lacking three phospholipase C enzymes failed to persist normally during mouse infection indicating a potential demand for lipid degradation in the later phase of infection (Raynaud et al. 2002). In addition, a large fraction of bacilli isolated from tuberculosis patient sputum samples have been shown to contain lipid bodies similar to those observed in hypoxic, nonreplicating *M. tuberculosis* (Garton et al. 2002, 2008; Daniel et al. 2004; Deb et al. 2006). The dominant component of these lipid bodies consists of triacylglycerol (TAG), which intracellular bacilli seem to synthesize directly from fatty acids released from host TAG (Daniel et al. 2011). *Mycobacterium tuberculosis*-infected macrophages have similarly been found to undergo foamy maturation followed by migration of *M. tuberculosis*-containing phagosomes toward newly formed intracellular lipid bodies, which ultimately encapsulate the pathogen (Peyron et al. 2008). Such encapsulation provides *M. tuberculosis* with potential access to nutrients in the form of fatty acids from host

TAG that can be stored in its own lipid bodies as potential source of energy, even when not replicating. LipY a TAG hydrolase (Deb et al. 2006; Mishra et al. 2007) has been detected in the *M. tuberculosis* cytosol and on the bacterial surface suggesting that it is involved in hydrolysis of intrabacterial, stored TAGs as well as TAGs from host lipid bodies (Dedieu et al. 2012).

Studies of genetically engineered *M. tuberculosis* mutant strains have provided similar evidence of fatty acids in the host niche. Bacteria grown on fatty acid-derived acetyl-CoA rely on gluconeogenesis and the anaplerotic glyoxylate shunt (Fig. 1). Accordingly, *M. tuberculosis* mutants lacking the glyoxylate shunt enzyme isocitrate lyase (ICL) or phosphoenolpyruvate carboxykinase (PEPCK), catalyzing the first committed of gluconeogenesis, were shown to be incapable of growing on fatty acids in vitro and found to be profoundly attenuated during the acute and chronic phases of mouse infections (McKinney et al. 2000; Muñoz-Elías and McKinney 2005; Marrero et al. 2010; Blumenthal et al. 2010). Together, these mutants implicated fatty acid metabolism as essential for growth and survival of *M. tuberculosis* in vivo.

More detailed interpretations of these phenotypes, however have proven challenging. For example, recent work showed that *M. tuberculosis* ICL functions as a bifunctional isocitrate and methylisocitrate lyase (Muñoz Elías et al. 2006), thus making it unclear which of its bifunctional metabolic roles in the glyoxylate shunt and methylcitrate cycles explains the profound attenuation of ICL-deficient *M. tuberculosis* in vivo. Loss of methylisocitrate lyase activity is predicted to result in the accumulation of toxic propionyl-CoA metabolites (Upton and McKinney 2007), while loss of isocitrate lyase activity is expected to impair carbon assimilation through inactivation of the glyoxylate shunt. While disruption of the methylcitrate cycle via deletion of the downstream enzymes, *prpDC* encoding methylcitrate synthase and methylcitrate dehydratase (Fig. 1), attenuated *M. tuberculosis* growth on propionate and in isolated macrophages, replication and persistence in mice was unimpaired, suggesting the primacy of the glyoxylate shunt (Muñoz Elías et al. 2006). However, absence of methylcitrate cycle activity can also be compensated for by the vitamin B12-dependent methylmalonyl pathway (Savvi et al. 2008; Griffin et al. 2012), while the ability of *M. tuberculosis* to scavenge vitamin B12 from the host or synthesize it during infection remains to be investigated.

The failure of *M. tuberculosis* lacking PEPCK to replicate in resting macrophages or during acute mouse infections further supports the view that *M. tuberculosis* relies on gluconeogenesis for biomass during growth within a host. However, the profound death following PEPCK deletion or silencing in vivo at early and at late stages of the infection remains unexplained. In contrast to the ICL deficient mutant, the growth defect of *M. tuberculosis* lacking PEPCK in media containing fatty acids could be rescued by supplementation with glucose or glycerol (Marrero et al. 2010). Lack of PEPCK may, however, cause metabolite perturbations that might sensitize *M. tuberculosis* to stresses encountered within the host and lead to its death. Together, these examples illustrate that, while powerful, the use of genetically engineered *M. tuberculosis* strains as a bioprobes

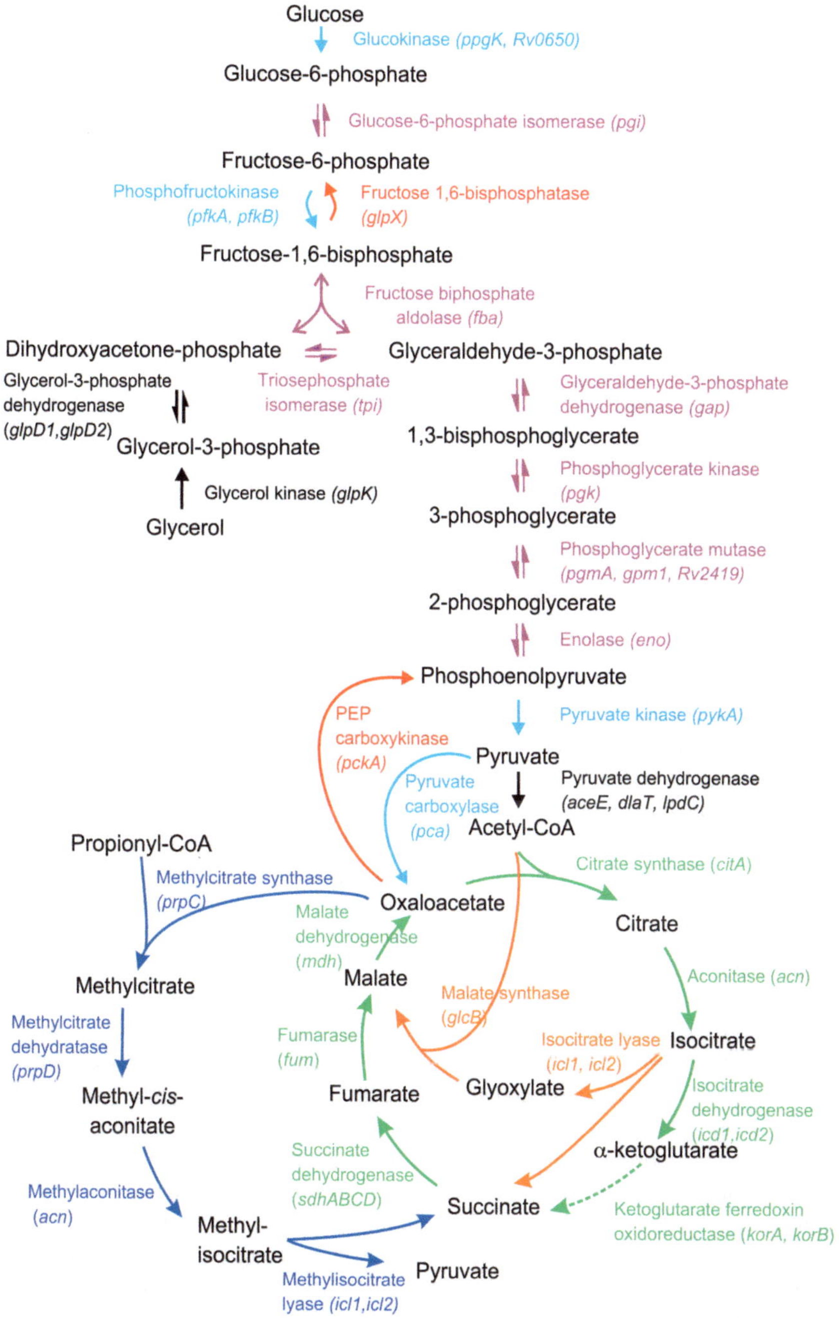

Fig. 1 Central metabolic pathways of *M. tuberculosis*. Enzymes and their encoding genes are color coded to reflect their dedicated pathways: glycolysis (*light blue*), gluconeogenesis (*red*), glycolysis and gluconeogenesis (*purple*), TCA cycle (*green*), glyoxylate shunt (*orange*), and methylcitrate cycle (*dark blue*). Reactions that convert α-ketoglutarate into succinate are detailed in Fig. 2

of the host–pathogen interaction can be complicated by their limited biochemical resolution, as exemplified by their inability to determine if the attenuation of a given enzymatic mutant is due to depletion of its product or an intoxication arising from the accumulation of an upstream or distantly related metabolite. Nonetheless, that fatty acids and lipids are relevant carbon substrates during *M. tuberculosis* growth and persistence in macrophages and during infections remain the prevailing paradigm.

3.2 Good Cholesterol?

Cholesterol is an essential constituent of mammalian cell membranes where it plays important structural and regulatory roles (Munro 2003; Brown and Goldstein 2008; Yuan et al. 2012). It is also a component of lipid droplets, which accumulate in alveolar macrophages of tuberculosis patients and in foamy macrophages of mouse granulomas (Russell et al. 2009). In fact, *M. tuberculosis* can induce the formation of foamy, lipid-loaded macrophages, and electron microscopy showed that the bacilli containing phagosomes are found in close proximity to intracellular nonmembrane bound lipid bodies (Peyron et al. 2008; Melo and Dvorak 2012). These foamy macrophages appear to provide a hospitable niche for *M. tuberculosis* as they protect the bacilli from direct contact with lymphocytes, appear to have lost bactericidal activity and might facilitate the acquisition of lipid and fatty acid nutrients. That *M. tuberculosis* utilizes cholesterol in vivo is supported by a wealth of data. It encodes a dedicated cholesterol uptake system and the complete pathway for its degradation (van der Geize et al. 2007), can catabolize cholesterol in vitro and use it as sole carbon source for replication (Pandey and Sassetti 2008; Griffin et al. 2012). Catabolism of cholesterol is predicted to yield propionyl-CoA, acetyl-CoA, and pyruvate (van der Geize et al. 2007) and cholesterol contributes to *M. tuberculosis* propionyl-CoA pool in vitro and when it resides in macrophages (Yang et al. 2009; Griffin et al. 2012).

Mycobacterium tuberculosis mutants whose ability to degrade cholesterol is impaired via deletion of the *mce4*-encoded cholesterol transporter, or abolished due to lack of enzymes required for cholesterol catabolism, fail to persist normally in the chronic phase of mouse and guinea pig infections (Pandey and Sassetti 2008; Nesbitt et al. 2009; Yam et al. 2009; Hu et al. 2010). This suggested that cholesterol is an important nutrient in vivo, but also indicates access to alternative carbon sources prior to the onset of adaptive immunity. A mutant lacking the "intracellular growth (igr)" locus, that encodes enzymes important for early steps in cholesterol degradation was similarly attenuated during early mouse infections (Chang et al. 2009). This phenotype has been attributed to a presumptive intoxication from cholesterol metabolites as disruption of the Mce4 cholesterol transporter suppressed the in vivo growth defect of the irg mutant. In contrast, *M. tuberculosis* lacking 3β-hyroxysteroid dehydrogenase (HSD), the enzyme catalyzing the first step in cholesterol catabolism, grew normally in guinea pig lungs and persisted normally for up to 7 weeks suggesting that cholesterol is not an essential nutrient source for *M. tuberculosis* during infection of guinea pigs (Yang et al. 2011). The dispensability of HSD could be explained if *M. tuberculosis* has access not only to intact cholesterol, but also the product of HSD, cholest-4-en-3-one; or it expresses a second, regulated HSD that is only induced during infection. However, while these data further highlight the complexity of interpreting the phenotypes of gene deletion mutants, they clearly support a role for cholesterol as a "good" carbon source for *M. tuberculosis* during chronic infection.

3.3 Carb Counting and the Role of Sugar Transporters

While the weight of prevailing evidence has clearly implicated fatty acids as an essential carbon source for *M. tuberculosis* in vivo, roles for other carbon sources remain conspicuously unresolved. Recent work showed that, unlike most bacterial pathogens, *M. tuberculosis* is capable of utilizing multiple carbon substrates simultaneously (de Carvalho et al. 2010a; Rhee et al. 2011). In addition, genetic studies have clearly demonstrated that the LpqY-SugA-SugB-SugC carbohydrate transporter is required for normal growth of *M. tuberculosis* in mouse lungs and spleens (Kalscheuer et al. 2010). This transporter was shown to be highly specific for uptake of the disaccharide trehalose, which is not present in mammals, but can be released by *M. tuberculosis* from trehalose-containing cell wall glycolipids (Kalscheuer et al. 2010). The intracellular fate of recycled trehalose remains to be identified; it might serve as a precursor for α-glucans, for trehalose-containing glycolipids, which are very prevalent in the mycolic acid cell wall, and as a reservoir of glucose. Trehalose biosynthesis is mediated by three biosynthetic pathways (De Smet et al. 2000) and is required for growth of *M. smegmatis*, likely because it serves as an essential precursor for cell wall biosynthesis. *In vitro*, *M. tuberculosis* can grow with trehalose as sole carbon source (Kalscheuer et al. 2010) presumably due to conversion of trehalose to glucose by trehalase (Carroll et al. 2007). Trehalose hydrolysis by trehalase yields two molecules of glucose, while trehalose phosphorylase catalyzes the phosphorolytic cleavage of trehalose into glucose 1-phosphate and glucose (Argüelles 2000). A phosphate dependent trehalase yet, has been purified from *M. smegmatis* and *M. tuberculosis* contains a homolog. Yet, despite the requirement for phosphate, there was no evidence of phosphorolytic cleavage, suggesting that the enzyme produces two molecules of glucose. No matter how it is metabolized, it is intriguing to speculate that trehalose serves as carbon store in *M. tuberculosis* in form of cell wall glycolipids such as trehalose dimycolate. Within the host, *M. tuberculosis* may also have a direct access to glucose and glucose-phosphate. *Salmonella typhimurium* uses primarily glucose when replicating in the phagosome (Bowden et al. 2009), while glucose-6-phosphate serves as a carbon source for intracellular *Listeria monocytogenes* (Chico-Calero et al. 2001). Uptake of glucose-6-phosphate is mediated by a specific hexose-phosphate transporter, which is present in several intracellular pathogens such as *L. monocytogenes*, *Shigella flexneri*, *Salmonella enterica*, and *Chlamydia pneumonia* (Schwoppe et al. 2002; Fuchs et al. 2011), but has not been identified in *M. tuberculosis*. Glucose-6-phosphate is an important precursor for anabolic processes, especially cell wall and nucleotide biosynthesis and serves as a source of reducing power in mycobacteria (Hasan et al. 2010). Whether glucose metabolism is important for the generation of biomass during in vivo growth remains to be shown. *M. tuberculosis* expresses two hexokinases (Cole et al. 1998). Transposon mutant analysis suggested that one of them, encoded by *ppgK*, is important for replication in mouse spleens (Sassetti and Rubin 2003), but this awaits confirmation with gene deletion mutants. Thus, while it is clear that *M. tuberculosis* relies on trehalose recycling during mouse infections, the intracellular fate of trehalose remains to be determined and the importance of host-derived glucose or glucose-6-phosphate at any stage of the infection unresolved.

3.4 The Glycerol Paradox

Early studies showed that *M. tuberculosis* growth was most strongly enhanced with the inclusion of glycerol in the culture medium. Glycerol-fed *M. tuberculosis* grows faster and the bacilli reach a higher density than when metabolizing other carbon sources such as glucose or fatty acids, which has thus led to the use of glycerol in virtually all standard mycobacterial growth media (Dubos 1947; Edson 1951). Recent work, however, demonstrated that *M. tuberculosis* lacking glycerol kinase GlpK, which is essential for the first step in glycerol catabolism (Fig. 1), was unable to utilize glycerol as sole carbon source in vitro, yet replicated and persisted like wild type *M. tuberculosis* in mouse lungs suggesting that glycerol is not a critical carbon source for *M. tuberculosis*, at least in the mouse model (Pethe et al. 2010). After uptake, glycerol is phosphorylated by glycerol kinase and glycerol-3-phosphate dehydrogenase converts the resulting glycerol-3-phosphate (glycerol-3-P) to dihydroxyacetone phosphate, which enters glycolysis and gluconeogenesis (Fig. 1). The possibility of a redundant glycerol kinase induced in the intracellular environment during infection thus remains to be evaluated. An alternative model consistent with the foregoing observations is that *M. tuberculosis* utilizes glycerol-3-P liberated from phospholipids and metabolizes it via glycerol-3-P dehydrogenase into dihydroxyacetone phosphate, at the same time perhaps metabolizing the released free fatty acid chains. *E. coli* expresses glycerophosphodiester phosphodiesterase, which can convert phospholipid derived glycerophosphodiesters into glycerol-3-P (Larson et al. 1983), however this enzyme has not been identified in *M. tuberculosis*. The pathogen might instead utilize host-derived glycerol-phosphate consistent with the observation that intracellular *M. tuberculosis* induced expression of the *ugp* operon encoding a putative ABC transporter for glycerol-3-P (Schnappinger et al. 2003). *M. tuberculosis* contains two genes encoding probable glycerol-3-phosphate dehydrogenases, *glpD1* and *glpD2* (Fig. 1) and the latter is required for normal growth on solid media containing glycerol and glucose (Griffin et al. 2011). Their importance during host infection has not yet been investigated. The phagosome might not provide easy access to glycerol-3-P; however, *M. tuberculosis* that escaped to the cytosol (van der Wel et al. 2007; Houben et al. 2012) or to the extracellular space might face a less restricted nutritional environment. Of note, there is ample evidence that glycerol and glycerol-3-P serve as major carbon sources for several cytosolic pathogens including *Shigella flexneri* and cytosolic *Listeria monocytogenes* (Eisenreich et al. 2010). In summary, while *M. tuberculosis* does not appear to utilize glycerol to replicate or persist during mouse infections, glycerol-3-P potentially derived from host phospholipids might serve as alternative carbon source in vivo.

3.5 Eating and Breathing at the Same Time: Effects of Oxygen and Carbon Dioxide Tensions on **Mycobacterium tuberculosis***'s Central Carbon Metabolism?*

Putting aside the specific nutrients consumed in the host, *M. tuberculosis* resides within granulomas, which are frequently hypoxic (Via et al. 2008) and situated in the CO_2-rich environment of the lung. Not surprisingly, *M. tuberculosis*'s metabolic activity has been found strongly influenced by oxygen tension and CO_2 availability. That *M. tuberculosis* can utilize CO_2 as source of carbon has been known for many years (Nishihara 1954); however, the functional relevance of this was unknown until recently. ^{13}C metabolic flux analysis revealed that in a carbon-limited chemostat *M. tuberculosis* dissimilated pyruvate via the glyoxylate shunt and incorporated CO_2 into central carbon metabolism as shown by CO_2-derived ^{13}C incorporation into several amino acids and produced succinyl-CoA (Beste et al. 2011). Importantly, in conditions of low oxygen tension, at levels that mimic those identified in lung granulomas, *M. tuberculosis* generated fumarate though the reductive TCA cycle, assimilated CO_2, and used fumarate as an electron sink (Watanabe et al. 2011). Reduction of fumarate generates succinate, which *M. tuberculosis* actively secretes into the extracellular environment thereby maintaining an energized membrane (Watanabe et al. 2011). This led to the hypothesis that, in hypoxic granulomas, *M. tuberculosis* might reverse its TCA cycle, incorporate CO_2 and accumulate and secrete succinate to maintain a state of persistence, if provided a glycolytic carbon source. While PEPCK was found to catalyze the conversion of oxaloacetate to PEP in vitro in oxygenated conditions, it is possible that it may also function in the reverse direction, especially during nonreplicative persistence, contributing to CO_2 utilization and energy generation instead of anabolism and biomass production (Zhang et al. 2010; Watanabe et al. 2011).

These studies separately emphasize the inherent dependence of all genetic studies of *M. tuberculosis* metabolism on the limitations of the specific host model used. The mouse strains most commonly used for such studies do not form necrotic and caseating granulomas in response to *M. tuberculosis* infection (McMurray et al. 1996; Rhoades et al. 1997; Flynn 2006). Accordingly granulomas from C57BL/6 mice were not severely hypoxic in contrast to those in guinea pigs, rabbits, and nonhuman primates (Aly et al. 2006; Tsai et al. 2006; Via et al. 2008). This difference in pathology and oxygen tension might be associated with an altered nutritional environment and different metabolic adaptations of the bacilli within the lung of mice compared to humans, nonhuman primates, rabbits, and guinea pigs. Notwithstanding, hypoxia is likely not the only cue triggering CO_2 utilization as this is also associated with slow growth (Beste et al. 2011); even in mouse lesions the oxygen concentration may be reduced and growth of *M. tuberculosis* is significantly slowed with the onset of adaptive immunity so that metabolic adaptations become essential that enable the pathogen to maintain its energy metabolism during phases of persistence.

3.6 Adapting to Nutritionally Diverse Environments

As described above, *M. tuberculosis* encounters a diverse and dynamic array of microenvironments during infection that require the pathogen to adapt its carbon and energy metabolism accordingly. While the intracellular phagosomal environment is clearly one of the chief niches *M. tuberculosis* resides in, the bacilli can actively escape into the cytosol (van der Wel et al. 2007; Houben et al. 2012), are found in the acellular center of caseating granulomas (Kaplan et al. 2003), and must survive at least for some time in air during transmission. In addition, when the host responds to infection, the intracellular niches change. Several examples indicate that *M. tuberculosis* is metabolically flexible and well capable of utilizing and realigning different metabolic pathways to optimally exploit available nutrients or adapt to nutrient deficiency. Many bacteria use available carbons substrates selectively: when presented with a mix of carbon substrates, they first catabolize their preferred carbon source, which in model organisms such as *E. coli* and *Bacillus subtilis* is glucose and only when this preferred substrate is exhausted will they catabolize secondary substrates (Görke and Stülke 2008). This carbon catabolite repression is mediated by different sophisticated regulatory mechanisms and represents one of the most studied global control systems in bacteria. Similar to some highly host-adapted pathogens such as *Chlamydia trachomatis*, *M. tuberculosis* lacks classical carbon catabolite repression and instead regulates its growth through simultaneous co-catabolism of multiple carbon sources (Nicholson and Chiu 2004; de Carvalho et al. 2010a). When fed with ^{13}C-labled glucose and acetate, *M. tuberculosis* metabolized glucose preferentially through glycolysis and the pentose phosphate pathway and converted acetate mostly into intermediates of the TCA cycle. Yet, at the same time, it incorporated low amounts of acetate-derived carbon into glycolytic/gluconeogenic and pentose phosphate intermediates and dextrose-derived carbon into intermediates of the TCA cycle. *M. tuberculosis* thus metabolizes different carbon substrates simultaneously to feed different pathways and is capable to segregate carbon flow through the same metabolic pathway but in opposite directions. This unusual metabolic network topology likely benefits the capacity of *M. tuberculosis* to adapt to different host niches.

Other examples of metabolic plasticity come from *M. tuberculosis*'s TCA cycle enzymes. An *M. tuberculosis* mutant, whose pyruvate dehydrogenase (PDH) complex was inactivated due to the deletion of *dlaT,* coding for dihydrolipoamide dehydrogenase, the E2 of PDH, upregulated a branched-chain keto acid dehydrogenase (BCKADH) complex (Venugopal et al. 2011). BCKADH activity prevented accumulation of pyruvate, branched-chain amino acids, and branched-chain keto acids in DlaT deficient *M. tuberculosis*. The genes encoding BCKADH were also upregulated in response to nutrient starvation (Betts et al. 2002) suggesting that catabolism of branched-chain keto and/or amino acids in nutrient restricted environments might fuel *M. tuberculosis*'s energy metabolism.

Metabolic capabilities aside, the nature of *M. tuberculosis*'s TCA cycle has proven surprisingly difficult to dissect. It was thought to lack a functional α-ketoglutarate dehydrogenase (KDH) complex (Tian et al. 2005b) and the predicted E1 enzyme was found to function as a thiamine diphosphate-dependent α-ketoglutarate decarboxylase (KGD) (Tian et al. 2005a). This enzyme also has carboligase activity and produces 2- hydroxy-3-oxoadipate (HOA) from α-ketoglutarate and glyoxylate (de Carvalho et al. 2010b), thereby perhaps detoxifying the cell from high glyoxylate concentrations or producing biosynthetic precursors. The discovery of an aerotolerant anaerobic-type α-KG ferredoxin oxidoreductase (KOR) that could link the oxidative and reductive branch of the TCA cycle and produce succinyl-CoA from α-KG and CoA-SH further supported the model that *M. tuberculosis* lacks KDH (Baughn et al. 2009). Recent work, however, revealed that *M. tuberculosis*'s KGD is in fact multifunctional and in vitro capable of catalyzing KG dehydrogenase, KG decarboxylase, and HOA synthase activities (Wagner et al. 2011) (Fig. 2). KGD and KOR together could enable *M. tuberculosis* to convert α-ketoglutarate into succinate via three parallel pathways linking the oxidative and reductive branch of the TCA cycle. *M. tuberculosis* lacking KOR grew like wt when provided with either heightened CO_2 concentration, in the absence of fatty acids, or when the glyoxylate cycle was inhibited (Baughn et al. 2009). KOR thus appears to be necessary to provide succinyl-CoA, CO_2 and reducing equivalents during growth on fatty acids, thereby fueling gluconeogenesis. A different not mutually exclusive interpretation of the data is that KOR is important when the reverse TCA cycle, which requires CO_2, is active. Operation of the reverse TCA cycle has only been demonstrated with glycolytic carbon sources and in a hypoxic environment (Watanabe et al. 2011), however it is plausible that additional metabolic cues and environmental conditions

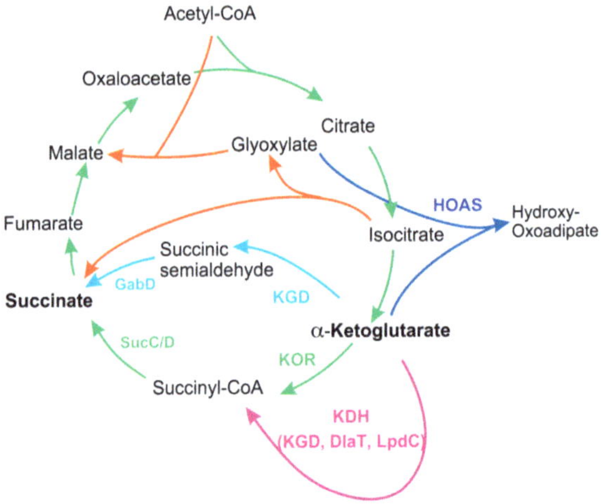

Fig. 2 *M. tuberculosis*'s TCA cycle reactions. α-Ketoglutarate (α-KG) can be converted to succinate via (1) α-KG ferredoxin oxidoreductase (KOR, Rv2254c, Rv2455c), (2) α-KG decarboxylase (KGD, Rv1248c) and (3) α-KG dehydrogenase complex (KDH) consisting of KGD, dihydrolipoamide dehydrogenase (DlaT, Rv2215) and lipoamide dehydrogenase (LpdC, Rv0462). KGD also functions as 2-hydroxy-3-oxoadipate synthase (HOAS). GabD, succinic semialdehyde dehydrogenase (GabD1, Rv0234c; GabD2, Rv1731); SucC/D, succinyl- CoA synthetase (SucC, Rv0951; SucD, Rv0952)

trigger reversal of the cycle. Notwithstanding, these alternative functional pathways of the TCA cycle might provide metabolic flexibility required for growth and survival in nutritionally diverse environments and allow the pathogen to consume various carbon sources.

3.7 How Do Others Do It?

Elegant work has investigated the intracellular lifestyle and metabolic requirements of *S. typhimurium* (reviewed in Eisenreich et al. 2010; Dandekar 2012), a pathogen that remains within a vacuole throughout its life inside the host. This revealed that metabolism of carbohydrates including glucose via glycolysis is essential for intracellular growth of *Salmonella* in macrophages and during mouse infections (Bowden et al. 2009; Paterson et al. 2009; Götz and Goebel 2010). Fatty acid catabolism and gluconeogenesis, by contrast, were dispensable for virulence of *S. typhimurium* in mice (Tchawa Yimga et al. 2006). This differs significantly from the metabolic pathways required by *M. tuberculosis* and suggests that intracellular pathogens, despite sharing the intraphagosomal habitat, have evolved distinctive carbon acquisition strategies. Host-derived amino acids are essential for the intracellular growth of *Legionella pneumophila*, which parasitizes various protozoan species, but can also be transmitted to humans where it infects and replicates in alveolar macrophages causing Legionnaires' disease (Sauer et al. 2005; Wieland et al. 2005; Newton et al. 2010). *L. pneumophila* replicates within a specialized endosome-derived vacuole that maintains a neutral pH and acquires characteristics of the endoplasmic reticulum. While it seems to depend on metabolism of amino acids (George et al. 1980), it can also utilize glucose via the Entner-Doudoroff pathway (Eylert et al. 2010). Recent work revealed that *Legionella* 'hijacks' the host proteasome to generate amino acids from poly-ubiquitinated proteins (Price et al. 2011). Whether this strategy is exploited by other intracellular pathogens remains to be seen; *M. tuberculosis* might rely on its own proteasome for the generation of amino acids as carbon substrates during nutrient starvation, as a proteasome mutant was unable to survive in stationary phase, during chronic mouse infections and in a model of strict carbon starvation (Gandotra et al. 2010).

It is possible that some of the differences in nutrient dependence between intracellular *M. tuberculosis* and *Salmonella* or *Legionella* may be explained by the ability of *M. tuberculosis* to gain access to the cytosol. *L. monocytogenes*, which replicates in the cytosol, relies predominantly on glycerol for intracellular growth and also utilizes sugar phosphates (Eisenreich et al. 2010; Fuchs et al. 2012). In contrast, glucose seems not to play a role (Stoll and Goebel 2010) and there is no evidence that *L. monocytogenes* accesses lipids or catabolizes fatty acids inside a host. In fact, *L. monocytogenes* lacks the genes for β–oxidation and the glyoxylate shunt (Eisenreich et al. 2010); its TCA cycle is interrupted (Eisenreich et al. 2006) and it requires pyruvate carboxylase (PYC) for the synthesis of oxaloacetate from pyruvate bypassing TCA cycle activity (Schar et al. 2010). This is essential for in vivo growth as a mutant lacking PYC was strongly attenuated in a mouse sepsis model (Schar et al. 2010). Major differences thus exist with respect to both the carbon sources that different pathogens scavenge from their host cells and the metabolic pathways they employ for biomass and energy generation.

It is similarly possible that pathogens may manipulate host metabolism to directly or indirectly modifying the metabolic environment they are confronted with. For example, macrophage activation by IFN-γ and proinflammatory cytokines induces NADPH oxidase and iNOS. NADPH oxidase consumes ATP, whose production requires flux through glycolysis; generation of NO requires arginine, is accompanied by recycling of citruline and inhibition of ornithine and polyamine biosynthesis (Gordon 2003; Mori and Gotoh 2004; Naderer and McConville 2007). In contrast, macrophages activated by Th2-type cytokines IL-4 or IL-13 are characterized by increased arginase activity and production of polyamine biosynthesis precursors (Gordon 2003). *Leishmania* parasites induce a Th2 response in susceptible mouse strains (Sacks and Anderson 2004) and benefit from the increased availability of essential amino acids and polyamines supporting their intracellular replication (Naderer and McConville 2007).

Finally, hormone-, growth factor-, and nutrient-regulated signaling pathways control the metabolism of mammalian host cells (Levine and Puzio-Kuter 2010). Pathogens might directly interfere or modulate such signaling pathways in pathogen-specific ways thereby altering host cell metabolism and controlling nutrient accessibility.

4 Concluding Remarks

Multiple lines of evidence have implicated *M. tuberculosis*' metabolic network as a central mediator of its pathogenicity. However, knowledge of how it achieves this remains incomplete. Most textbooks depict metabolism as a housekeeping activity of all cells relegated to bulk restocking of biosynthetic precursors and energy. We believe that such views are both outdated and incomplete. They misequate evolutionary conservation with biological law, neglect the selective pressure of an organism's ecologic and nutritional niche on its physiology, and dissociate metabolism from cell physiology. Moreover, it remains a fact that cells have evolved an extensive array of mechanisms to regulate the activity of individual pathways and that metabolism serves as the chemical arbiter of all cellular reactions.

Existing studies have focused on compiling a molecular inventory and homology-based reconstruction of *M. tuberculosis*' metabolic network. However, facts do not equate into knowledge nor knowledge into understanding. Key challenges thus await. Chief among them is a more detailed understanding of the specific biochemical needs served by *M. tuberculosis*' metabolic network during each phase of its nutrient-limited life cycle. Studies of the *M. tuberculosis* cell surface components have revealed their distinct host regulatory activities, ranging from phagocytosis by alveolar macrophages to phagolysosomal maturation of bone marrow-derived macrophages to secretion of pro-inflammatory cytokines. These are associated with specific cell wall lipids and carbohydrates, and presumably derive from distinct metabolic pathways that await elucidation (Torrelles and Schlesinger 2010; Mishra et al. 2011; Neyrolles and Guilhot 2011). Knowledge of such links, however, is likely to require more integrated views of metabolism that include balanced homeostasis of essential noncarbon metabolites, such as sulfur, nitrogen, and phosphorous, as well as metals including but not limited to Fe, Cu, Zn, Mg, Mn, and Mo.

An equally important area of *M. tuberculosis* metabolism in need of further study is its nutrient uptake and transport mechanisms (Niederweis 2008). Considerable knowledge about lipid transport across the cytoplasmic membrane has been gained (Jackson and Stadthagen 2007), and transporters of nitrogen, phosphate, sulfates, and several metal ions have been identified using bio-informatic methods (Niederweis 2008). Such analyses, for example, helped predict five carbohydrate transporters in *M. tuberculosis,* whereas *M. smegmatis* was predicted to encode 28 such transporters, underscoring important differences in their physiology (Titgemeyer et al. 2007). Direct biochemical and structural understanding of these and most other nutrient transport processes however remains scarce, while transport across the outer membrane of *M. tuberculosis* is even less well understood (Niederweis 2008; Niederweis et al. 2010).

A larger mystery awaiting investigation is how *M. tuberculosis* senses various nutrients and different forms of nutrient limitations. Some bacteria can regulate their cell size in response to nutrient availability (Schaechter et al. 1958; Chien et al. 2012). In *B. subtilis*, the glucosyl transferase UgtP serves as UDP-glucose dependent metabolic sensor of nutrient availability to coordinate cell size with growth rate thus linking nutritional information from the environment with cell division (Weart et al. 2007). Whether *M. tuberculosis* employs similar nutritional regulatory pathways remains to be investigated. Addressing these questions may thus help further elucidate the intracellular lifestyle of *M. tuberculosis* and its interaction with the host.

Acknowledgments We thank Werner Goebel for critical reading of the manuscript, insightful comments, and valuable suggestions.

References

Alonso S, Pethe K, Russell DG, Purdy GE (2007) Lysosomal killing of *Mycobacterium* mediated by ubiquitin-derived peptides is enhanced by autophagy. Proc Natl Acad Sci USA 104:6031–6036. doi:10.1073/pnas.0700036104

Aly S, Wagner K, Keller C et al (2006) Oxygen status of lung granulomas in *Mycobacterium tuberculosis*-infected mice. J Pathol 210:298–305. doi:10.1002/path.2055

Argüelles JC (2000) Physiological roles of trehalose in bacteria and yeasts: a comparative analysis. Arch Microbiol 174:217–224. doi:10.1007/s002030000192

Barry CE, Boshoff HI, Dartois V et al (2009) The spectrum of latent tuberculosis: rethinking the biology and intervention strategies. Nat Rev Micro 7:845–855. doi:10.1038/nrmicro2236

Baughn AD, Garforth SJ, Vilchèze C, Jacobs WR (2009) An anaerobic-type alpha-ketoglutarate ferredoxin oxidoreductase completes the oxidative tricarboxylic acid cycle of *Mycobacterium tuberculosis*. PLoS Pathog 5:e1000662. doi:10.1371/journal.ppat.1000662

Beste DJV, Bonde B, Hawkins N et al (2011) C metabolic flux analysis identifies an unusual route for pyruvate dissimilation in *Mycobacteria* which requires isocitrate lyase and carbon dioxide fixation. PLoS Pathog 7:e1002091. doi:10.1371/journal.ppat.1002091

Betts JC, Lukey PT, Robb LC et al (2002) Evaluation of a nutrient starvation model of *Mycobacterium tuberculosis* persistence by gene and protein expression profiling. Mol Microbiol 43:717–731

Bloch H, Segal W (1956) Biochemical differentiation of *Mycobacterium tuberculosis* grown in vivo and in vitro. J Bacteriol 72:132–141

Blumenthal A, Trujillo C, Ehrt S, Schnappinger D (2010) Simultaneous analysis of multiple *Mycobacterium tuberculosis* knockdown mutants in vitro and in vivo. PLoS One 5:e15667. doi:10.1371/journal.pone.0015667

Bodnar KA, Serbina NV, Flynn JL (2001) Fate of *Mycobacterium tuberculosis* within murine dendritic cells. Infect Immun 69:800–809. doi:10.1128/IAI.69.2.800-809.2001

Boshoff H, Barry C (2005) Tuberculosis—metabolism and respiration in the absence of growth. Nat Rev Micro 3:70–80

Bowden S, Rowley G, Hinton J, Thompson A (2009) Glucose and glycolysis are required for the successful infection of macrophages and mice by *Salmonella enterica* serovar *Typhimurium*. Inf Immun 77:3117

Brown MS, Goldstein JL (2008) Cholesterol feedback: from Schoenheimer's bottle to Scap's MELADL. J Lipid Res 50:S15–S27. doi:10.1194/jlr.R800054-JLR200

Carroll JD, Pastuszak I, Edavana VK et al (2007) A novel trehalase from *Mycobacterium smegmatis*—purification, properties, requirements. FEBS J 274:1701–1714. doi:10.1111/j.1742-4658.2007.05715.x

Chang JC, Miner MD, Pandey AK et al (2009) Igr genes and *Mycobacterium tuberculosis* cholesterol metabolism. J Bacteriol 191:5232–5239. doi:10.1128/JB.00452-09

Chen L, Vitkup D (2007) Distribution of orphan metabolic activities. Trends Biotechnol 25:343–348. doi:10.1016/j.tibtech.2007.06.001

Chico-Calero I, Suárez M, González-Zorn B et al (2001) Hpt, a bacterial homolog of the microsomal glucose- 6-phosphate translocase, mediates rapid intracellular proliferation in *Listeria*. Proc Natl Acad Sci USA 99:431–436. doi:10.1073/pnas.012363899

Chien A-C, Hill NS, Levin PA (2012) Cell size control in bacteria. Curr Biol 22:R340–R349. doi:10.1016/j.cub.2012.02.032

Choi H-S, Rai PR, Chu HW et al (2002) Analysis of nitric oxide synthase and nitrotyrosine expression in human pulmonary tuberculosis. Am J Respir Crit Care Med 166:178–186

Cole ST, Brosch R, Parkhill J et al (1998) Deciphering the biology of *Mycobacterium tuberculosis* from the complete genome sequence. Nature 393:537–544

Côtes K, Bakala N'Goma JC, Dhouib R et al (2008) Lipolytic enzymes in *Mycobacterium tuberculosis*. Appl Microbiol Biotechnol 78:741–749. doi:10.1007/s00253-008-1397-2

Dandekar T, Astrid F, Jasmin P, Hensel M (2012) *Salmonella enterica*: a surprisingly well-adapted intracellular lifestyle. Front Microbiol 3:164. doi:10.3389/fmicb.2012.00164

Daniel J, Deb C, Dubey VS et al (2004) Induction of a novel class of diacylglycerol acyltransferases and triacylglycerol accumulation in *Mycobacterium tuberculosis* as it goes into a dormancy-like state in culture. J Bacteriol 186:5017–5030. doi:10.1128/JB.186.15.5017-5030.2004

Daniel J, Maamar H, Deb C et al (2011) *Mycobacterium tuberculosis* uses host triacylglycerol to accumulate lipid droplets and acquires a dormancy-like phenotype in lipid-loaded macrophages. PLoS Pathog 7:e1002093. doi:10.1371/journal.ppat.1002093.t003

de Carvalho LPS, Fischer SM, Marrero J et al (2010a) Metabolomics of *Mycobacterium tuberculosis* reveals compartmentalized co-catabolism of carbon substrates. Chem Biol 17:1122–1131. doi:10.1016/j.chembiol.2010.08.009

de Carvalho LPS, Zhao H, Dickinson CE et al (2010b) Activity-based metabolomic profiling of enzymatic function: identification of Rv1248c as a mycobacterial 2-Hydroxy-3-oxoadipate synthase. Chem Biol 17:323–332. doi:10.1016/j.chembiol.2010.03.009

De Smet KA, Weston A, Brown IN et al (2000) Three pathways for trehalose biosynthesis in mycobacteria. Microbiology 146(Pt 1):199–208

Deb C, Daniel J, Sirakova TD et al (2006) A novel lipase belonging to the hormone-sensitive lipase family induced under starvation to utilize stored triacylglycerol in *Mycobacterium tuberculosis*. J Biol Chem 281:3866–3875. doi:10.1074/jbc.M505556200

Dedieu L, Serveau-Avesque C, Kremer L, Canaan S (2012) Mycobacterial lipolytic enzymes: a gold mine for tuberculosis research. Biochimie 95(1):66–73. doi:10.1016/j.biochi.2012.07.008 (in press)

Dubos R (1947) Media for tubercle bacilli. Am Rev Tuberc 56:334–345

Edson NL (1951) The intermediary metabolism of the mycobacteria. Bacteriol Rev 15:147–182

Eisenreich W, Dandekar T, Heesemann J, Goebel W (2010) Carbon metabolism of intracellular bacterial pathogens and possible links to virulence. Nat Rev Micro 8:401–412

Eisenreich W, Slaghuis J, Laupitz R et al (2006) ^{13}C isotopologue perturbation studies of Listeria monocytogenes carbon metabolism and its modulation by the virulence regulator PrfA. Proc Natl Acad Sci USA 103:2040–2045. doi:10.1073/pnas.0507580103

Elkington PT, D'Armiento JM, Friedland JS (2011) Tuberculosis immunopathology: the neglected role of extracellular matrix destruction. Sci Transl Med 3:71ps6. doi:10.1126/scitranslmed.3001847

Eum S-Y, Kong J-H, Hong M-S et al (2010) Neutrophils are the predominant infected phagocytic cells in the airways of patients with active pulmonary TB. Chest 137:122–128. doi:10.1378/chest.09-0903

Eylert E, Herrmann V, Jules M et al (2010) Isotopologue Profiling of *Legionella pneumophila*: role of serine and glucose as carbon substrates. J Biol Chem 285:22232–22243. doi:10.1074/jbc.M110.128678

Flynn JL (2006) Lessons from experimental *Mycobacterium tuberculosis* infections. Microbes Inf 8:1179–1188. doi:10.1016/j.micinf.2005.10.033

Fratazzi C, Arbeit RD, Carini C et al (1999) Macrophage apoptosis in mycobacterial infections. J Leukoc Biol 66:763–764

Fuchs T, Eisenreich W, Heesemann J, Goebel W (2011) Metabolic adaptation of human pathogenic and related nonpathogenic bacteria to extra-and intracellular habitats. FEMS Microbiol Rev 36:435–462. doi:10.1111/j.1574-6976.2011.00301.x

Fuchs TM, Eisenreich W, Kern T, Dandekar T (2012) Toward a systemic understanding of *Listeria monocytogenes* metabolism during infection. Front Microbiol. doi:10.1186/1471-2105-11-77

Gandotra S, Lebron MB, Ehrt S (2010) The *Mycobacterium tuberculosis* proteasome active site threonine is essential for persistence yet dispensable for replication and resistance to nitric oxide. PLoS Pathog 6:e1001040. doi:10.1371/journal.ppat.1001040

Garton NJ, Christensen H, Minnikin DE et al (2002) Intracellular lipophilic inclusions of mycobacteria in vitro and in sputum. Microbiology 148:2951–2958

Garton NJ, Waddell SJ, Sherratt AL et al (2008) Cytological and transcript analyses reveal fat and lazy persister-like bacilli in tuberculous sputum. PLoS Med 5:e75. doi:10.1371/journal.pmed.0050075

George J, Pine L, Reeves M, Harrell WK (1980) Amino acid requirements of *Legionella pneumophila*. J Clin Microbiol 11:286–291

Gordon S (2003) Alternative activation of macrophages. Nat Rev Immunol 3:23–35. doi:10.1038/nri978

Görke B, Stülke J (2008) Carbon catabolite repression in bacteria: many ways to make the most out of nutrients. Nat Rev Micro 6:613–624. doi:10.1038/nrmicro1932

Götz A, Goebel W (2010) Glucose and glucose 6-phosphate as carbon sources in extra-and intracellular growth of enteroinvasive *Escherichia coli* and *Salmonella enterica*. Microbiology 156:1176

Griffin JE, Gawronski JD, Dejesus MA et al (2011) High-resolution phenotypic profiling defines genes essential for mycobacterial growth and cholesterol catabolism. PLoS Pathog 7:e1002251. doi:10.1371/journal.ppat.1002251

Griffin JE, Pandey AK, Gilmore SA et al (2012) Cholesterol catabolism by *Mycobacterium tuberculosis* requires transcriptional and metabolic adaptations. Chem Biol 19:218–227. doi:10.1016/j.chembiol.2011.12.016

Gutierrez M, Master S, Singh S et al (2004) Autophagy is a defense mechanism inhibiting BCG and *Mycobacterium tuberculosis* survival in infected macrophages. Cell 119:753–766

Hasan MR, Rahman M, Jaques S et al (2010) Glucose 6-phosphate accumulation in mycobacteria: implications for a novel F420-dependent anti-oxidant defense system. J Biol Chem 285:19135–19144. doi:10.1074/jbc.M109.074310

Houben D, Demangel C, van Ingen J et al (2012) ESX-1-mediated translocation to the cytosol controls virulence of mycobacteria. Cell Microbiol 14:1287–1298. doi:10.1111/j.1462-5822.2012.01799.x

Hu Y, van der Geize R, Besra GS et al (2010) 3-Ketosteroid 9Î ± -hydroxylase is an essential factor in the pathogenesis of *Mycobacterium tuberculosis*. Mol Microbiol 75:107–121. doi:10.1111/j.1365-2958.2009.06957.x

Jackson M, Stadthagen G, Gicquel B (2007) Long-chain multiple methyl-branched fatty acid-containing lipids of Mycobacterium tuberculosis: biosynthesis, transport, regulation and biological activities. Tuberculosis 87:78–86. doi:10.1016/j.tube.2006.05.003

James PE, Grinberg OY, Michaels G, Swartz HM (1995) Intraphagosomal oxygen in stimulated macrophages. J Cell Physiol 163:241–247. doi:10.1002/jcp.1041630204
Kalscheuer R, Weinrick B, Veeraraghavan U et al (2010) Trehalose-recycling ABC transporter LpqY-SugA-SugB-SugC is essential for virulence of *Mycobacterium tuberculosis*. Proc Natl Acad Sci 107:21761–21766. doi:10.1073/pnas.1014642108
Kaplan G, Post FA, Moreira AL et al (2003) *Mycobacterium tuberculosis* growth at the cavity surface: a microenvironment with failed immunity. Inf Immun 71:7099–7108. doi:10.1128/IAI.71.12.7099-7108.2003
Kim BH, Shenoy AR, Kumar P et al (2011) A family of IFN- -Inducible 65-kD GTPases protects against bacterial infection. Science 332:717–721. doi:10.1126/science.1201711
Kim M-J, Wainwright HC, Locketz M et al (2010) Caseation of human tuberculosis granulomas correlates with elevated host lipid metabolism. EMBO Mol Med 2:258–274. doi:10.1002/emmm.201000079
Larson TJ, Ehrmann M, Boos W (1983) Periplasmic glycerophosphodiester phosphodiesterase of *Escherichia coli*, a new enzyme of the glp regulon. J Biol Chem 258:5428–5432
Levine AJ, Puzio-Kuter AM (2010) The control of the metabolic switch in cancers by oncogenes and tumor suppressor genes. Science 330:1340–1344. doi:10.1126/science.1193494
MacMicking JD (2003) Immune control of tuberculosis by IFN-γ-Inducible LRG-47. Science 302:654–659. doi:10.1126/science.1088063
MacMicking JD, North RJ, LaCourse R et al (1997) Identification of nitric oxide synthase as a protective locus against tuberculosis. Proc Natl Acad Sci USA 94:5243–5248
Marrero J, Rhee KY, Schnappinger D et al (2010) Gluconeogenic carbon flow of tricarboxylic acid cycle intermediates is critical for *Mycobacterium tuberculosis* to establish and maintain infection. Proc Natl Acad Sci 107:9819–9824. doi:10.1073/pnas.1000715107
McKinney J, zu Bentrup K, Muñoz-Elías E et al (2000) Persistence of *Mycobacterium tuberculosis* in macrophages and mice requires the glyoxylate shunt enzyme isocitrate lyase. Nature 406:735–738
Melo RCN, Dvorak AM (2012) Lipid body–phagosome interaction in macrophages during infectious diseases: host defense or pathogen survival strategy? PLoS Pathog 8:e1002729. doi:10.1371/journal.ppat.1002729.t001
McMurray DN, Collins FM, Dannenberg AMJ, Smith DW (1996) Pathogenesis of experimental tuberculosis in animal models. Curr Top Microbiol Immunol 215:157–179
Mishra AK, Driessen NN, Appelmelk BJ, Besra GS (2011) Lipoarabinomannan and related glycoconjugates: structure, biogenesis and role in *Mycobacterium tuberculosis* physiology and host-pathogen interaction. FEMS Microbiol Rev 35:1126–1157. doi:10.1111/j.1574-6976.2011.00276.x
Mishra KC, De Chastellier C, Narayana Y et al (2007) Functional role of the PE domain and immunogenicity of the Mycobacterium tuberculosis triacylglycerol hydrolase LipY. Inf Immun 76:127–140. doi:10.1128/IAI.00410-07
Mori M, Gotoh T (2004) Arginine metabolic enzymes, nitric oxide and infection. J Nutr 134:2820S–2825S
Mukamolova GV, Turapov O, Malkin J et al (2010) Resuscitation-promoting factors reveal an occult population of tubercle bacilli in sputum. Am J Respir Crit Care Med 181:174–180. doi:10.1164/rccm.200905-0661OC
Munro S (2003) Lipid rafts: elusive or illusive? Cell 115:377–388
Muñoz Elías E, Upton A, Cherian J, McKinney J (2006) Role of the methylcitrate cycle in *Mycobacterium tuberculosis* metabolism, intracellular growth, and virulence. Mol Microbiol 60:1109–1122
Muñoz-Elías E, McKinney J (2005) *Mycobacterium tuberculosis* isocitrate lyases 1 and 2 are jointly required for in vivo growth and virulence. Nat Med 11:638–644
Murthy PS, Sirsi M, Ramakrishnan T (1962) Tricarboxylic acid cycle and related enzymes in cell-free extracts of *Mycobacterium tuberculosis* H37Rv. Biochem J 84:263
Naderer T, McConville MJ (2007) The Leishmania-macrophage interaction: a metabolic perspective. Cell Microbiol 10:301–308. doi:10.1111/j.1462-5822.2007.01096.x

Nathan C (2009) Taming tuberculosis: a challenge for science and society. Cell Host Microbe 5:220–224. doi:10.1016/j.chom.2009.02.004

Nathan C, Shiloh MU (2000) Reactive oxygen and nitrogen intermediates in the relationship between mammalian hosts and microbial pathogens. Proc Natl Acad Sci USA 97:8841–8848

Nesbitt NM, Yang X, Fontan P et al (2009) A thiolase of *Mycobacterium tuberculosis* is required for virulence and production of androstenedione and androstadienedione from cholesterol. Inf Immun 78:275–282. doi:10.1128/IAI.00893-09

Newton HJ, Ang DKY, van Driel IR, Hartland EL (2010) Molecular pathogenesis of infections caused by legionella pneumophila. Clin Microbiol Rev 23:274–298. doi:10.1128/CMR.00052-09

Neyrolles O, Guilhot C (2011) Recent advances in deciphering the contribution of *Mycobacterium tuberculosis* lipids to pathogenesis. Tuberculosis 91:187–195. doi:10.1016/j.tube.2011.01.002

Neyrolles O, Hernández-Pando R, Pietri-Rouxel F et al (2006) Is adipose tissue a place for *Mycobacterium tuberculosis* persistence? PLoS One 1:e43. doi:10.1371/journal.pone.0000043.t002

Nicholson TL, Chiu K, Stephens RS (2004) *Chlamydia trachomatis* lacks an adaptive response to changes in carbon source availability. Inf Immun 72:4286–4289. doi:10.1128/IAI.72.7.4286-4289.2004

Niederweis M (2008) Nutrient acquisition by mycobacteria. Microbiology 154:679–692. doi:10.1099/mic.0.2007/012872-0

Niederweis M, Danilchanka O, Huff J et al (2010) Mycobacterial outer membranes: in search of proteins. Trends Microbiol 18:109–116. doi:10.1016/j.tim.2009.12.005

Nishihara H (1954) Studies on the metabolism of the tubercle bacillus with the use of radioactive substrates in the presence and absence of streptomycin. J Biochem 41:167–181

Pandey A, Sassetti C (2008) Mycobacterial persistence requires the utilization of host cholesterol. Proc Natl Acad Sci 105:4376

Paterson GK, Cone DB, Northen H et al (2009) Deletion of the gene encoding the glycolytic enzyme triosephosphate isomerase (tpi) alters morphology of Salmonella entericaserovar Typhimurium and decreases fitness in mice. FEMS Microbiol Lett 294:45–51. doi:10.1111/j.1574-6968.2009.01553.x

Pethe K, Sequeira PC, Agarwalla S et al (2010) A chemical genetic screen in *Mycobacterium tuberculosis* identifies carbon-source-dependent growth inhibitors devoid of in vivo efficacy. Nature Commun 1:1–8. doi:10.1038/ncomms1060

Peyron P, Vaubourgeix J, Poquet Y et al (2008) Foamy macrophages from tuberculous patients' granulomas constitute a nutrient-rich reservoir for *M. tuberculosis* persistence. PLoS Pathog 4:e1000204. doi:10.1371/journal.ppat.1000204.t003

Philips JA, Ernst JD (2012) Tuberculosis pathogenesis and immunity. Annu Rev Pathol Mech Dis 7:353–384. doi:10.1146/annurev-pathol-011811-132458

Price CTD, Al-Quadan T, Santic M et al (2011) Host proteasomal degradation generates amino acids essential for intracellular bacterial growth. Science 334:1553–1557. doi:10.1126/science.1212868

Purdy GE, Russell DG (2007) Lysosomal ubiquitin and the demise of *Mycobacterium tuberculosis*. Cell Microbiol 9:2768–2774. doi:10.1111/j.1462-5822.2007.01039.x

Raynaud C, Guilhot C, Rauzier J et al (2002) Phospholipases C are involved in the virulence of *Mycobacterium tuberculosis*. Mol Microbiol 45:203–217

Rhee KY, de Carvalho LPS, Bryk R et al (2011) Central carbon metabolism in *Mycobacterium tuberculosis*: an unexpected frontier. Trends Microbiol 19:307–314. doi:10.1016/j.tim.2011.03.008

Rhoades ER, Frank AA, Orme IM (1997) Progression of chronic pulmonary tuberculosis in mice aerogenically infected with virulent *Mycobacterium tuberculosis*. Tuber Lung Dis 78:57–66

Rich AR (1946) The pathogenesis of tuberculosis, 2nd ed. Charles C Thomas, Publisher, Springfied, Ill, Illinois

Rieder HL (1999) Epidemiologic basis of tuberculosis control 1–162 In: international union against tuberculosis and lung disease (IUATLD)

Rohde K, Yates RM, Purdy GE, Russell DG (2007) *Mycobacterium tuberculosis* and the environment within the phagosome. Immunol Rev 219:37–54. doi:10.1111/j.1600-065X.2007.00547.x

Roy S, Sharma S, Sharma M et al (2004) Induction of nitric oxide release from the human alveolar epithelial cell line A549: an in vitro correlate of innate immune response to *Mycobacterium tuberculosis*. Immunology 112:471–480. doi:10.1046/j.1365-2567.2004.01905.x

Russell D, Barry C, Flynn J (2010) Tuberculosis: what we don't know can, and does, hurt us. Science 328:852

Russell DG (2001) *Mycobacterium tuberculosis*: here today, and here tomorrow. Nat Rev Mol Cell Biol 2:569–577. doi:10.1038/35085034

Russell DG (2006) Who puts the tubercle in tuberculosis? Nat Rev Micro 5:39–47. doi:10.1038/nrmicro1538

Russell DG, Cardona P-J, Kim M-J et al (2009) Foamy macrophages and the progression of the human tuberculosis granuloma. Nat Immunol 10:943–948. doi:10.1038/ni.1781

Sacks D, Anderson C (2004) Re-examination of the immunosuppressive mechanisms mediating non-cure of *Leishmania* infection in mice. Immunol Rev 201:225–238. doi:10.1111/j.0105-2896.2004.00185.x

Sakula A (1983) Robert Koch: Centenary of the discovery of the tubercle bacillus, 1882. Can Vet J 24:127–131

Sambandamurthy VK, Wang X, Chen B et al (2002) A pantothenate auxotroph of *Mycobacterium tuberculosis* is highly attenuated and protects mice against tuberculosis. Nat Med 8:1171–1174. doi:10.1038/nm765

Sassetti CM, Boyd DH, Rubin EJ (2001) Comprehensive identification of conditionally essential genes in mycobacteria. Proc Natl Acad Sci USA 98:12712–12717. doi:10.1073/pnas.231275498

Sassetti CM, Rubin EJ (2003) Genetic requirements for mycobacterial survival during infection. Proc Natl Acad Sci USA 100:12989–12994. doi:10.1073/pnas.2134250100

Sauer J-D, Bachman MA, Swanson MS (2005) The phagosomal transporter A couples threonine acquisition to differentiation and replication of Legionella pneumophila in macrophages. Proc Natl Acad Sci USA 102:9924–9929. doi:10.1073/pnas.0502767102

Savvi S, Warner DF, Kana BD et al (2008) Functional characterization of a vitamin B12-dependent methylmalonyl pathway in *Mycobacterium tuberculosis*: implications for propionate metabolism during growth on fatty acids. J Bacteriol 190:3886–3895. doi:10.1128/JB.01767-07

Schaechter M, Maaløe O, Kjeldgaard NO (1958) Dependency on medium and temperature of cell size and chemical composition during balanced growth of *Salmonella typhimurium*. J Gen Microbiol 19:592–606. doi:10.1099/00221287-19-3-592

Schaible UE, Sturgill-Koszycki S, Schlesinger PH, Russell DG (1998) Cytokine activation leads to acidification and increases maturation of *Mycobacterium avium*-containing phagosomes in murine macrophages. J Immunol 160:1290–1296

Schar J, Stoll R, Schauer K et al (2010) Pyruvate carboxylase plays a crucial role in carbon metabolism of extra- and intracellularly replicating *Listeria monocytogenes*. J Bacteriol 192:1774–1784. doi:10.1128/JB.01132-09

Schnappinger D, Ehrt S, Voskuil MI et al (2003) Transcriptional adaptation of *Mycobacterium tuberculosis* within macrophages: insights into the phagosomal environment. J Exp Med 198:693–704. doi:10.1084/jem.20030846

Schwoppe C, Winkler HH, Neuhaus HE (2002) Properties of the Glucose-6-Phosphate transporter from Chlamydia pneumoniae (HPTcp) and the glucose-6-phosphate sensor from *Escherichia coli* (UhpC). J Bacteriol 184:2108–2115. doi:10.1128/JB.184.8.2108-2115.2002

Shenoy AR, Wellington DA, Kumar P et al (2012) GBP5 promotes NLRP3 inflammasome assembly and immunity in mammals. Science 336:481–485. doi:10.1126/science.1217141

Singh G, Singh G, Jadeja D, Kaur J (2010) Lipid hydrolizing enzymes in virulence: *Mycobacterium tuberculosis*as a model system. Crit Rev Microbiol 36:259–269. doi:10.3109/1040841X.2010.482923

Stoll R, Goebel W (2010) The major PEP-phosphotransferase systems (PTSs) for glucose, mannose and cellobiose of *Listeria monocytogenes*, and their significance for extra- and intracellular growth. Microbiology 156:1069–1083. doi:10.1099/mic.0.034934-0

Sturgill-Koszycki S, Schlesinger PH, Chakraborty P et al (1994) Lack of acidification in *Mycobacterium* phagosomes produced by exclusion of the vesicular proton-ATPase. Science 263:678–681

Tchawa Yimga M, Leatham MP, Allen JH et al (2006) Role of Gluconeogenesis and the Tricarboxylic Acid Cycle in the Virulence of *Salmonella enterica* Serovar *Typhimurium* in BALB/c Mice. Inf Immun 74:1130–1140. doi:10.1128/IAI.74.2.1130-1140.2006

Tian J, Bryk R, Itoh M et al (2005a) Variant tricarboxylic acid cycle in *Mycobacterium tuberculosis*: Identification of α-ketoglutarate decarboxylase. Proc Natl Acad Sci USA 102:10670

Tian J, Bryk R, Shi S et al (2005b) *Mycobacterium tuberculosis* appears to lack α-ketoglutarate dehydrogenase and encodes pyruvate dehydrogenase in widely separated genes. Mol Microbiol 57:859–868. doi:10.1111/j.1365-2958.2005.04741.x

Timm J, Post FA, Bekker L-G et al (2003) Differential expression of iron-, carbon-, and oxygen-responsive mycobacterial genes in the lungs of chronically infected mice and tuberculosis patients. Proc Natl Acad Sci USA 100:14321–14326. doi:10.1073/pnas.2436197100

Titgemeyer F, Amon J, Parche S et al (2007) A genomic view of sugar transport in *Mycobacterium smegmatis* and *Mycobacterium tuberculosis*. J Bacteriol 189:5903

Torrelles JB, Schlesinger LS (2010) Diversity in *Mycobacterium tuberculosis* mannosylated cell wall determinants impacts adaptation to the host. Tuberculosis 90:84–93. doi:10.1016/j.tube.2010.02.003

Tsai MC, Chakravarty S, Zhu G et al (2006) Characterization of the tuberculous granuloma in murine and human lungs: cellular composition and relative tissue oxygen tension. Cell Microbiol 8:218–232. doi:10.1111/j.1462-5822.2005.00612.x

Upton A, McKinney J (2007) Role of the methylcitrate cycle in propionate metabolism and detoxification in Mycobacterium smegmatis. Microbiology 153:3973

van der Geize R, Yam K, Heuser T et al (2007) A gene cluster encoding cholesterol catabolism in a soil actinomycete provides insight into *Mycobacterium tuberculosis* survival in macrophages. Proc Natl Acad Sci USA 104:1947–1952. doi:10.1073/pnas.0605728104

van der Wel N, Hava D, Houben D et al (2007) *M. tuberculosis* and *M. leprae* translocate from the phagolysosome to the cytosol in myeloid cells. Cell 129:1287–1298. doi:10.1016/j.cell.2007.05.059

Vandal OH, Pierini LM, Schnappinger D et al (2008) A membrane protein preserves intrabacterial pH in intraphagosomal *Mycobacterium tuberculosis*. Nat Med 14:849–854. doi:10.1038/nm.1795

Venugopal A, Bryk R, Shi S et al (2011) Virulence of *Mycobacterium tuberculosis* depends on lipoamide dehydrogenase, a member of three multienzyme complexes. Cell Host Microbe 9:21–31. doi:10.1016/j.chom.2010.12.004

Via LE, Lin PL, Ray SM et al (2008) Tuberculous granulomas are hypoxic in guinea pigs, rabbits, and nonhuman primates. Inf Immun 76:2333–2340. doi:10.1128/IAI.01515-07

Vieira OV, Botelho RJ, Grinstein S (2002) Phagosome maturation: aging gracefully. Biochem J 366:689–704. doi:10.1042/BJ20020691

Vogt G, Nathan C (2011) In vitro differentiation of human macrophages with enhanced antimycobacterial activity. J Clin Investig 121:3889–3901. doi:10.1172/JCI57235

Voskuil MI, Schnappinger D, Visconti KC et al (2003) Inhibition of respiration by nitric oxide induces a *Mycobacterium tuberculosis* dormancy program. J Exp Med 198:705–713. doi:10.1084/jem.20030205

Wagner T, Bellinzoni M, Wehenkel A et al (2011) Functional plasticity and allosteric regulation of α-ketoglutarate decarboxylase in central mycobacterial metabolism. Chem Biol 18:1011–1020. doi:10.1016/j.chembiol.2011.06.004

Watanabe S, Zimmermann M, Goodwin MB et al (2011) Fumarate reductase activity maintains an energized membrane in anaerobic *Mycobacterium tuberculosis*. PLoS Pathog 7:e1002287. doi:10.1371/journal.ppat.1002287

Weart RB, Lee AH, Chien A-C et al (2007) A metabolic sensor governing cell size in bacteria. Cell 130:335–347. doi:10.1016/j.cell.2007.05.043

Wheeler P, Blanchard J (2005) General Metabolism and biochemical pathways of tubercle bacilli. In: Cole ST, Eisenach KD, McMurray DN, Jacobs WR Jr (ed) Tuberculosis and the tubercle bacillus. ASM Press, Washington DC

Wieland H, Ullrich S, Lang F, Neumeister B (2005) Intracellular multiplication of *Legionella pneumophila* depends on host cell amino acid transporter SLC1A5. Mol Microbiol 55:1528–1537. doi:10.1111/j.1365-2958.2005.04490.x

Wolf AJ, Linas B, Trevejo-Nuñez GJ et al (2007) *Mycobacterium tuberculosis* infects dendritic cells with high frequency and impairs their function in vivo. J Immunol 179:2509–2519

Yam KC, D'angelo I, Kalscheuer R et al (2009) Studies of a ring-cleaving dioxygenase illuminate the role of cholesterol metabolism in the pathogenesis of *Mycobacterium tuberculosis*. PLoS Pathog 5:e1000344. doi:10.1371/journal.ppat.1000344.t003

Yang X, Gao J, Smith I et al (2011) Cholesterol is not an essential source of nutrition for *Mycobacterium tuberculosis* during infection. J Bacteriol 193:1473–1476. doi:10.1128/JB.01210-10

Yang X, Nesbitt NM, Dubnau E et al (2009) Cholesterol metabolism increases the metabolic pool of propionate in *Mycobacterium tuberculosis*. Biochemistry 48:3819–3821. doi:10.1021/bi9005418

Youmans GP (1979) Tuberculosis. W.B. Saunders Company, Philadelphia

Yuan Y, Li P, Ye J (2012) Lipid homeostasis and the formation of macrophage-derived foam cells in atherosclerosis. Protein Cell 3:173–181. doi:10.1007/s13238-012-2025-6

Zhang X, Shanmugam KT, Ingram LO (2010) Fermentation of glycerol to succinate by metabolically engineered strains of *Escherichia coli*. Appl Environ Microbiol 76:2397–2401. doi:10.1128/AEM.02902-09

Surviving the Macrophage: Tools and Tricks Employed by *Mycobacterium tuberculosis*

Rajesh Jayachandran, Somdeb BoseDasgupta and Jean Pieters

Abstract *Mycobacterium tuberculosis* has evolved to withstand one of the most inhospitable cells within the human body, namely the macrophage, a cell that is normally geared toward the destruction of any invading microbe. How *M. tuberculosis* achieves this is still incompletely understood; however, a number of mechanisms are now known that provide advantages to *M. tuberculosis* for its survival and proliferation inside the macrophage. While some of these mechanisms are mediated by factors released by *M. tuberculosis*, others rely on host components that are being hijacked to benefit survival of *M. tuberculosis* within the macrophage as well to avoid the generation of an effective immune response. Here, we describe several of these mechanisms, also pointing out the potential usage of this knowledge toward the development of novel strategies to treat tuberculosis. Furthermore, we attempt to put the 'macrophage niche' into context with other intracellular pathogens and discuss some of the generalities as well as specializations that *M. tuberculosis* employs to survive.

Contents

R. Jayachandran · S. BoseDasgupta · J. Pieters (✉)
Biozentrum, University of Basel, Klingelbergstrasse 50/70, 4056 Basel, Switzerland
e-mail: jean.pieters@unibas.ch

Current Topics in Microbiology and Immunology (2012) 374: 189–209

DOI: 10.1007/82_2012_273

Published Online: 16 November 2012

1 Introduction

Mycobacterium tuberculosis, probably one of the most successful pathogens on Earth, has coexisted with the human population for centuries and in the course of doing so managed to evade the human immune system at multiple levels. Normally the human immune system is quite efficient in mounting an effective immune response against both extracellular and intracellular pathogens, resulting in the rapid clearance of an acute infectious state. Although microbes are usually cleared by the immune system, several pathogens, not only *M. tuberculosis,* but also other bacterial pathogens such as *Salmonella enterica, Helicobacter pylori* as well as parasitic pathogens including *Leishmania*, *Plasmodium*, and *Toxoplasma* can escape immune recognition, thereby resulting in a chronic infectious condition that can either remain unnoticed or turn debilitating resulting in severe morbidity and mortality of the affected individual (Davies et al. 1993; Kaye et al. 1994; Hassett 1996; Kima et al. 1996; Restrepo et al. 1999; Allen et al. 2000; Iniesta et al. 2001; Parkhill et al. 2001; Townsend et al. 2001; Wain et al. 2001; Hughes and Galan 2002; Mahdavi et al. 2002; Tateda et al. 2003; Zheng and Jones 2003; Franchi et al. 2006; Ling et al. 2006; Pieters 2008; Raoust et al. 2009; Lang et al. 2012).

The tremendous success of *M. tuberculosis* as a pathogen is due to the fact that these bacilli have developed a plethora of strategies to counteract the bactericidal activities of the host macrophages, thereby successfully establishing a niche for long-term survival inside the affected individual (Pieters 2008;Gengenbacher and Kaufmann 2012; Goldberg et al. 2012; Soldati and Neyrolles 2012).

M. tuberculosis is transmitted through aerosols that are released by individuals harboring an active tuberculosis infection. These aerosols, containing living tubercle bacilli, reach the alveoli, where the bacteria can be efficiently phagocytosed by alveolar macrophages. Such internalization can be mediated by means of various receptors that ultimately result in the bacterial engulfment into the phagosome (Aderem and Underhill 1999). It is at this early stage, namely immediately following alveolar internalization, that *M. tuberculosis* initiates a response to avoid immune destruction: instead of allowing the macrophage to rapidly shuttle the bacteria to the destructive environment of the lysosome,

M. tuberculosis manages to interfere with the normal phagosomal maturation pathway by blocking its transfer to lysosomes (Armstrong and Hart 1971). As a result, *M. tuberculosis* remains in the relative hospitable environment of the phagosomes, where it can survive for decades.

The default route for phagocytosed cargo includes a series of maturation steps carried out by fusion and fission, which gradually result in the delivery of cargo to lysosomes, in which efficient degradation is ensured by the activity of various destructive mechanisms (Vieira et al. 2002; Rohde et al. 2007). These destructive mechanisms include vacuolar acidification, oxidative burst via production of reactive oxygen species and nitrogen species, changes in ion fluxes, and the activation of lytic enzymes (Aderem and Underhill 1999; Nathan and Shiloh 2000; Shiloh and Nathan 2000; Akira et al. 2006; Uematsu and Akira 2006; Rohde et al. 2007; Pieters 2008; Soldati and Neyrolles 2012).

As mentioned, *M. tuberculosis* actively avoids lysosomal delivery, and as a result, mycobacterial phagosomes are characterized by the absence of the characteristic set of late endosomal and lysosomal markers such as Rab7, v-ATPase, LAMP1, and the two PI(3)P binding proteins EEA1 (early endosome antigen 1) and Hrs (hepatocyte growth factor-regulated tyrosine kinase substrate) (Chakraborty et al. 1994; Vergne et al. 2004). In addition, *M. tuberculosis* utilizes several proteins as well as lipid molecules to arrest phagosome maturation at an early stage (Philips 2008; Pieters 2008; Steinberg and Grinstein 2008). In the following section, the various routes employed by mycobacterial to enter into its host cells, the macrophages, as well as the different host and mycobacterial factors which they employ to modulate phagosomal trafficking for their prolonged survival are discussed.

2 The Process of Mycobacterial Entry

Entry of *M. tuberculosis* can be mediated by a variety of receptors expressed on the surface of the phagocytic cells, including C-type lectin receptors, scavenger receptors, and complement receptors. The various C-type lectin receptors involved in mycobacterial uptake include the mannose receptor and the dendritic cell-specific intracellular adhesion molecule 3 grabbing nonintegrin (DC-SIGN) (Schafer et al. 2009; Ehlers 2010). These receptors recognize mannosylated molecules such as lipoarabinomannan (ManLAM) or phosphatidylinositol mannan (PIM) of the mycobacterial cell wall (Torrelles et al. 2006). Dectin 1 is another phagocyte receptor whose mycobacterial ligand is unknown (Yadav and Schorey 2006; Reid et al. 2009). However, the precise nature of the various *M. tuberculosis* cell wall components interacting with these receptors can influence the outcome of the mycobacterium–phagocyte interaction: for example, pathogenic mycobacteria posses ManLAM, which elicits lower innate immune responses compared to the arabinosylated lipoarabinomannan (AraLAM) of nonpathogenic mycobacteria (Briken et al. 2004). Also, mannosylated mycobacterial surface molecules can enhance mycobacterial opsonization thereby influencing their mode of uptake (Cywes et al. 1997). Furthermore, macrophage inducible C-type lectin (Mincle) recognizes the mycobacterial glycolipid trehalose dimycolate (TDM) and plays an important role in cytokine signaling in response to TDM (Ishikawa et al. 2009).

Mycobacteria can be internalized by complement receptors both in opsonized and nonopsonized forms. Interestingly, the macrophage integrin CR3 plays an important role in the entry of mycobacteria where the complement cleavage products were found to bind directly to the mycobacterial cell wall, resulting in mycobacterial opsonization and recognition via the complement receptor type 3 (Schorey et al. 1997). Nonopsonic entry of mycobacteria can occur as well via complement receptor type 3 but utilizes a distinct binding site, results in a different signaling pathway and involves RacI and Cdc42 (Le Cabec et al. 2000, 2002). In order to eradicate infections in opsonin-poor sites such as lungs, other receptors including the mannose receptor, scavenger receptor, CD14, or β-glucan may help in the ingestion process (Le Cabec et al. 2002).

Another key component that regulates mycobacterial entry is cholesterol (Gatfield and Pieters 2000). Cholesterol is the major sterol constituent of eukaryotic organisms, involved in stabilization of membranes as well as being a factor in hormonal and other signaling pathways (Simons and Ikonen 1997; Brown and London 2000; Lingwood et al. 2009). Nonpathogenic mycobacteria, especially the fast growing bacilli, degrade cholesterol from their culture media and are able to grow on cholesterol as the sole carbon source (Av-Gay and Sobouti 2000). In contrast, slow-growing mycobacteria, including *M. tuberculosis* and *M. bovis* BCG, cannot degrade or use cholesterol as a carbon source though they internalize, modify, and accumulate it from liquid growth media (Av-Gay and Sobouti 2000). Interestingly, cholesterol was found to be an essential component for entry of mycobacteria into macrophages; depletion of macrophage plasma membrane cholesterol drastically reduced mycobacterial uptake (Gatfield and Pieters 2000). The precise role for cholesterol in mycobacterial entry remains to be established. Cholesterol may be required as a plasma membrane component to provide direct entry via the plasma membrane (Simons and Ikonen 1997); alternatively, the presence of cholesterol allows the stable expression of signaling molecules at the plasma membrane that assist mycobacterial entry (Peyron et al. 2000; Brown and London 2000).

3 Host Factors and *M. tuberculosis* Survival Within Macrophages

3.1 Modulation of Lysosomal Delivery

Macrophages are probably among the most hostile cells of the immune system, geared toward destruction of anything that can be phagocytosed. Possibly as a result of their long-term interaction with macrophages, pathogenic mycobacteria have developed a plethora of strategies to combat the bactericidal milieu that is encountered upon ingestion, most prominently by blocking phagosome maturation via a number of different strategies. For instance, mycobacterial phagosomes fail to acidify due to mycobacterial interference with recruitment of the vesicular proton ATPase pump, fail to acquire late markers of the endocytic pathway such as Rab7, and reduce the levels of phosphatidylinositol 3-phosphate (PI3P) on their phagosomal membranes (Chakraborty et al. 1994; Sturgill-Koszycki et al. 1994; Fratti et al. 2003; Vergne et al. 2004).

Multiple host factors play an active role in the modulation of the endocytic trafficking pathway of mycobacteria. Importantly, one of the immune cell-activating cytokines that are released following an infection is interferon γ (IFNγ), a macrophage-activating cytokine that induces several genes involved in the modulation of intracellular trafficking of mycobacteria (Boehm et al. 1997). For example, IFNγ upregulates the transcriptional machinery to produce immune related GTPases (Feng et al. 2004) and guanylate-binding proteins (Gbps) (Kim et al. 2011), and at the same time IFNγ modulates a plethora of macrophage activities involved at all levels of the host defense (see also below).

3.2 *Modulation of Autophagy*

Autophagy is a potent immune effector mechanism against intracellular pathogens, which is the process wherein the intracellular pathogen along with bacterial products and damaged organelles are sequestered into an autophagosome for degradation in autolysosomes (Singh et al. 2006). Pathogenic mycobacteria have been known to secrete factors, which apart from preventing phagosome maturation can also prevent autophagy. For example, the secreted phosphatase SapM is known to reduce the levels of PI3P at the phagosome membrane, which on the one hand modulates phagosome maturation, and on the other hand results in the activation of STAT6 to inhibit autophagy (Deretic et al. 2006; Harris et al. 2007). An important molecule is the below described IFNγ-induced protein LRG47 has been shown to promote autophagy; in the absence of LRG-47, the mycobacterial phagosomes were found to carry lower levels of the v-ATPase that additionally hindered the process of phagosomal maturation by preventing acidification of the mycobacterial phagosome (see below and Taylor et al. 2004).

3.3 *Phosphoinositides*

Phosphoinositides play critical roles in regulating a diverse array of signal transduction processes. Class I phosphoinositol 3 kinase (PI3K) generates PI3P in a spatially restricted manner on the nascent phagosome, while class III PI3K which generates PI3P, function during phagosome maturation by recruiting the early endosomal antigen-1 (EEA1) (Fratti et al. 2003; Roth 2004). Thus, PI3P, a membrane trafficking regulatory lipid essential for phagosomal acquisition of lysosomal constituents (Vergne et al. 2005), is retained on phagosomes harboring dead mycobacteria but is continuously eliminated from phagosomes containing live bacilli (Fratti et al. 2003; Roth 2004). Interestingly, several mycobacterial products contribute to inhibition of phagosomal maturation, including the lipid phosphatase SapM that hydrolyzes PI3P (Saleh and Belisle 2000; Vergne et al. 2005), as well as the tyrosine phosphatase PtpA which, upon secretion, dephosphorylates and inactivates the host protein Vacuolar Protein Sorting (VPS33B), a regulator of membrane fusion (Grundner et al. 2005; Bach et al. 2008). Inactive VPS33B cannot generate GTP-activated Rab7 thereby resulting in a block in phagosome maturation (Bach et al. 2008). The different mechanisms used by pathogenic mycobacteria to interfere with phosphoinositide signaling may reflect the importance of this pathway in the regulation of phagosomal trafficking (Grinstein 2010).

3.4 Coronin 1

One host factor important for intracellular survival of *M. tuberculosis* is coronin 1. Coronin 1, also known as Tryptophan Aspartate containing COat protein (TACO, or P57), is highly expressed in macrophages and other leukocytes, and was found to be exclusively associated with phagosomes harboring living mycobacteria (Hasan et al. 1997; Ferrari et al. 1999). Coronin 1 is a member of a large family of proteins characterized by the presence of WD repeat domains and is widely distributed in the animal kingdom, from *Dictyostelium discoideum* to primates including humans (Rybakin and Clemen 2005). Coronin 1 is distributed both in the subcortical and the cytosolic fractions of the macrophages, and following phagocytosis of cargo, coronin 1 becomes recruited at the cytosolic surface of the phagosome (Gatfield et al. 2005). In the course of the normal process of phagocytosis, coronin 1 is slowly released from the cytosolic surface of the maturing phagosome, ultimately resulting in fusion of this vesicle with lysosomes transferring the internal contents for degradation. When macrophages are infected with live pathogenic mycobacteria, the release of coronin 1 from the mycobacterial phagosome is prevented and coronin 1 is retained for a prolonged duration of time resulting in a block in fusion with or maturation into degradative lysosomes (Ferrari et al. 1999; Gatfield and Pieters 2000, 2003).

Infection of coronin 1-deficient macrophages, either isolated from mice harboring a targeted gene disruption of coronin 1 or through siRNA-mediated knockdown revealed that mycobacteria are rapidly delivered to lysosomes resulting in an efficient degradation and killing of the mycobacteria, revealing the importance of coronin 1 for mycobacterial survival (Jayachandran et al. 2007, 2008).

How mycobacterial phagosomes manage to retain the coronin 1 coat for prolonged times remains unknown. One report demonstrated that pathogenic mycobacteria secrete a factor, called lipoamide dehydrogenase, which potentially interacts by means of cholesterol-dependent interactions with coronin 1, thereby preventing the release of coronin 1 from the mycobacterial phagosome (Deghmane et al. 2007), a scenario consistent with a role for cholesterol in mycobacterial pathogenesis (Gatfield and Pieters 2000).

3.5 Mycobacterial Survival, Coronin 1, and Calcium Signaling

The activity via which coronin 1 excerpts its protective effect on intracellularly residing mycobacteria was somewhat serendipitously found to occur via the modulation of calcium mobilization. In fact, analysis of mice lacking coronin 1 revealed that these mice were effectively depleted of peripheral T lymphocytes (Foger et al. 2006; Jayachandran et al. 2007; Mueller et al. 2008). Notably, T cell survival is absolutely dependent on calcium mobilization, that in turn induces calcineurin activation, thereby activating the T cell essential transcription factor NFAT (Winslow et al. 2003); Indeed, direct analysis of Ca^{2+} mobilization revealed that coronin 1 is essential for activating Ca^{2+}-dependent signaling following T-cell receptor engagement (Mueller et al. 2008) as well as in B cells following B-cell receptor triggering (Combaluzier et al. 2009). Interestingly, mycobacterial infection of macrophages similarly induces a coronin 1-dependent calcineurin activation, which is essential for preventing mycobacterial destruction in lysosomes. Importantly, the calcineurin inhibitors cyclosporine and FK506 block mycobacterial survival and allow the efficient lysosomal transfer and clearance of *M. tuberculosis* in lysosomes. Thus, in leukocytes, coronin 1 is an essential molecule in the activation of Ca^{2+}-dependent signal transduction (Jayachandran et al. 2007; Pieters 2008).

Coronin 1-dependent activation of calcium mobilization was observed for nonopsonized live mycobacterial entry (Jayachandran et al. 2007), which is in apparent contradiction of previous work showing a suppression of calcium mobilization upon mycobacterial entry (Malik et al. 2000, 2003). How can these results be reconciled? In fact, in the studies by Malik et al. (2000, 2003), all mycobacteria were fully opsonized with complement prior to macrophage infection, which indeed results in a suppression rather than an elevation of cytosolic calcium ((Malik et al. 2000, 2003) and RJ and JP, unpublished results). One reason why mycobacterial entry may result in either suppression or enhancement of Ca^{2+} mobilization, thereby activating different downstream signaling cascades, could be due to the fact that there are distinct recognition domains on the complement receptor type 3, which is involved in recognition and uptake of mycobacteria depending on whether the bacilli are opsonized or nonopsonized at the time of entry (Le Cabec et al. 2000, 2002). Indeed, it is known that opsonized versus nonopsonized entry of mycobacteria does affects recruitment of downstream signaling components, subcellular localization of mycobacteria, and the composition of mycobacterial phagosome (Le Cabec et al. 2000, 2002).

4 Bacterial Factors and *M. tuberculosis* Survival Within Macrophages

Most human pathogens have a repertoire of different virulent factors that are shed or secreted in the host cell milieu at defined stages of the infection. In pathogenic mycobacteria, several of such virulent factors have been characterized that can interfere with various host factors at different stages of infection, finally deciding the outcome of the infection.

4.1 Exported Repetitive Protein

One of the first mycobacterial virulence factors characterized was the exported repetitive protein (Erp). This 36 kDa secreted protein is required for mycobacterial survival within cultured macrophages as well as in mice (Berthet et al. 1998). Erp is also known to maintain cell wall integrity. Erp interacts with two membrane proteins Rv1417 and Rv2617c of which the former is linked to the riboflavin metabolism pathway. Further studies need to be done to determine the exact mode of action of this virulence factor (Cosma et al. 2006; Klepp et al. 2009).

4.2 Eukaryotic-Like Protein Kinases and Phosphatases

Interestingly, mycobacteria encode 11 different eukaryotic-like protein kinases that either have been retained during evolution or may have arrived in the *M. tuberculosis* genome through horizontal gene transfer (Ponting et al. 1999; Han and Zhang 2001). Most of these kinases, apart from protein kinase G and K, are transmembrane proteins, exposing the kinase domain within the bacterial cytosol, and therefore are likely to regulate mycobacterial signal transduction processes and morphology (Kang et al. 2005; Thakur et al. 2008; Thakur and Chakraborti 2008; Wehenkel et al. 2008). Protein kinase A and B (PknA and PknB) are the best characterized in this group of mycobacterial kinases and have been shown to play multiple roles in cell division and morphology (Thakur et al. 2008; Thakur and Chakraborti 2008). Protein kinase D (PknD), on the other hand, has been demonstrated to play an important role in regulating mycobacterial gene expression by phosphorylating alternative sigma-factor regulators (Greenstein et al. 2007). More recently, PknD has also been shown to play an active role in invasion of the central nervous system via vascular endothelium of the brain (Be et al. 2012).

As mentioned, two of the mycobacterial kinases, protein kinase G and protein kinase K, lack a transmembrane domain, thus potentially being able to be released by mycobacteria into the host cell environment (Av-Gay and Everett 2000; Houben et al. 2009). PknG is specifically expressed in virulent mycobacteria such as *M. bovis* BCG or *M. tuberculosis* but is absent in nonpathogenic mycobacteria such as *M. smegmatis*. Importantly, the human pathogen *M. leprae* that is considered to have retained the most important genes required for pathogenicity within its downsized genome, expresses PknG (Cole et al. 2001; Houben et al. 2009). Analysis of the phenotype of pathogenic mycobacteria lacking PknG revealed that this kinase is essential for promoting survival within host macrophages (Cowley et al. 2004; Walburger et al. 2004). Upon infection, PknG from pathogenic mycobacteria is released into the cytosol of macrophages where it prevents lysosomal delivery and degradation (Walburger et al. 2004; Houben et al. 2006; Scherr et al. 2009b). In fast growing mycobacteria, expression of PknG is suppressed, which is possibly due to the presence of regulatory elements through inefficient ribosome binding in the *pknG* upstream region or by interactions with other regulatory factors such as regulatory RNA or proteins (Houben et al. 2009). Wfen mycobacteria are engineered to be deficient for PknG or upon chemical inhibition of PknG from pathogenic species, this results in rapid mycobacterial transfer to lysosomes (Scherr et al. 2007). Interestingly, a small molecular weight inhibitor of PknG has been identified, named AX20017, which effectively blocks proliferation of *M. tuberculosis* within macrophages (Walburger et al. 2004). The availability of the crystal structure of PknG in complex with AX20017 may contribute to the development of a novel class of drugs to block the survival of *M. tuberculosis* within macrophages (Scherr et al. 2007, 2009a; Warner and Mizrahi 2007).

4.3 Mycobacterial Proteases

A putative Zn^{2+}-dependent metalloprotease (Zmp1) released by mycobacteria was proposed to interfere with phagolysosome biogenesis by interfering with caspase 1-dependent activation and secretion of interleukin-1β. In the absence of this metalloprotease, this activation is suppressed and mycobacteria are eliminated within lysosomes, suggesting that the production of Zmp1 is required for virulence and survival of *M. tuberculosis* (Master et al. 2008).

4.4 Enhanced Intracellular Survival Protein

The ‘enhanced intracellular survival’ (Eis) protein of virulent mycobacteria inhibits the c-Jun NH2-terminal protein kinase (JNK)-dependent reactive oxygen species production via its aminoglycoside acetyltransferase activity (Wei et al. 2000; Shin et al. 2010; Kim et al. 2012). Eis is thermostable and active in its hexameric form, where it dampens TNFα and IL-10 production, and thereby prevents macrophage activation, inflammation, and autophagy (Ganaie et al. 2011). Eis rapidly acetylates lysine-55 of dual-specificity protein phosphatase 16 (DUSP16)/mitogen-activated protein kinase phosphatase-7 (MKP-7), a JNK-specific phosphatase. As a result, *M. tuberculosis* Eis initiates the inhibition of JNK-dependent autophagy, phagosome maturation, and ROS generation by acetylating DUSP16/MKP-7.

4.5 ESX-1 Secretion System/ESAT 6

Many of the virulence factors employed by pathogenic mycobacteria have to be translocated to the host cytosol in order to effectively subvert macrophage function. While several secretion systems have been characterized in mycobacteria (see also Chapter by van der Woude, Luirink and Bitter), recent work has focused on the early secretory antigenic target of 6 kDa (ESAT-6) system 1 (ESX-1) (Abdallah et al. 2007; DiGiuseppe Champion and Cox 2007). Esx-1 is encoded by the RD1 (region of difference 1) genomic region and has been demonstrated to be a major virulence determinant related to the modulation of phagosome maturation (Gao et al. 2004; Majlessi et al. 2005; Brodin et al. 2006; DiGiuseppe Champion and Cox 2007). Infection of macrophages with *M. tuberculosis* displaying defects in Esx-1-dependent secretion results in increased lysosomal delivery and killing of the bacteria (Tracy Tan 2006). On the other hand, *M. bovis* BCG, which lacks the RD1 region, can still block phagosome maturation, and in addition, the known secreted products ESAT-6 (early secretory antigenic target 6, EsxA), CFP-10 (culture filtrate protein 10, EsxB) (Welin and Lerm 2011), and EspA (Fortune et al. 2005) have been found to be dispensable with regard to the arrest of phagosome maturation, indicating the existence of other effector proteins that could as well play similar roles in modulating mycobacterial trafficking that promotes their survival in host macrophages (Gao et al. 2004; MacGurn and Cox 2007).

The ESX-1 system has also been implicated in the escape of mycobacteria to the cytosol. Cryoelectron microscopic studies detected a certain percentage of mycobacteria residing outside the phagosome in the cytoplasm of macrophages, and this percentage of nonphagosomal mycobacteria increased with time post infection (van der Wel et al. 2007). Similarly, mycobacteria lacking the RD1 region fail to escape from the phagosome (Smith et al. 2008). Interestingly, a reduced pH present within phagolysosomes causes an increased expression of the ESX-1 gene cluster (Chen et al. 2012). ESX-1 genes have also been suggested to be involved in establishing pores within the phagolysosome membrane, through which mycobacteria may escape into the cytosol (de Jonge et al. 2007). The exact contribution of the cytosolic escape pathway to mycobacterial virulence remains, however, to be established; it could well be that depending on the precise nature of the macrophage–mycobacterium interaction, different strategies are being used by the pathogen, in some cases blocking phagosome maturation, and in others escape from the phagosome to the cytosol (Welin and Lerm 2011).

5 Macrophage Activation as a Response to Mycobacterial Infection

Many of the strategies that *M. tuberculosis* employs to resist destruction within macrophages operate in nonactivated macrophages; however, one of the consequences of an infection is the release of macrophage-activating cytokines by other immune cells. Activated macrophages, which are particularly efficient in clearing microbial load (North 1974) are known to induce a plethora of genes including the aforementioned LRG-47 and Gbp1 which help to promote phagolysosome formation resulting in mycobacterial destruction. Also, macrophage activation helps to overcome the reduced levels of phosphoinositides at the phagosomal membrane thereby mediating the exchange of Rab5 on the phagosome membrane by Rab7 leading to phagosome–lysosome fusion (Vergne et al. 2005). Macrophage activation can also initiate autophagy, resulting in the elimination of pathogenic mycobacteria by transferring these to lysosomes. Furthermore, macrophage activation results in an upregulation of reactive oxide and nitric oxide products that are particularly effective in inducing mycobacterial killing (Nathan and Shiloh 2000; Nathan 2006).

Macrophage activation is primarily mediated by the host cytokine IFNγ, that has a broad spectrum of activities. IFNγ is critical for resistance to infection, exerting its effects through the induction of broad transcriptional programs involving $\sim$2,000 genes, many of which remain uncharacterized (Waddell et al. 2010). For example, LRG-47 (LPS-stimulated RAW 264.7 macrophage protein 47), which is a member of the IFNγ regulated family of p47 GTPases, that plays an important role in containment of trypanosomiasis and salmonellosis infection, is needed for controlling mycobacterial infection (MacMicking et al. 2003; Feng et al. 2004; Henry et al. 2007; Shenoy et al. 2012), see also the chapter by MacMicking et al. LRG47 may act by associating with the mycobacterial phagosomal membrane and accelerating lysosomal fusion (Hunn and Howard 2010). Along with LRG-47, other prominent members of the families of guanosine triphosphatases are the 65–73-kD Gbps and 285-kD very large inducible GTPases (Vligs/Gvins) (Kim et al. 2011). The immunity-related GTPase family M protein 1 (Irgm1) can enhance the formation of autophagosomes that in turn control mycobacterial pathogenicity (Hunn and Howard 2010), see also below). Recent analysis shows that at least four guanylate-binding proteins, namely Gbp1, Gbp6, Gbp7, and Gbp10, confer cell-autonomous immunity to mycobacterial infection within macrophages by inducing the expression of hundreds of genes that include phagocyte oxidase, antimicrobial peptides, and autophagy effectors which thereby help to combat infection. These guanylate-binding proteins are involved in regulating the activity of several host defense proteins to kill intracellular mycobacteria (Kim et al. 2011). Thus, specific 65-kD Gbps coordinates a potent oxidative and vesicular trafficking program to protect the host from infection.

IFNγ also limits the inflammation response by halting the translation of the inflammatory genes through the activation of the IFNγ-activated inhibitor of translation (GAIT) complex genes, comprising glutamyl-prolyl tRNA synthetase (EPRS), NS1-associated protein 1 (NSAP1), ribosomal protein L13a, and glyceraldehyde-3-phosphate dehydrogenase (GAPDH). This complex binds defined 3' untranslated region elements within a family of inflammatory mRNAs and suppresses their translation. IFNγ-dependent phosphorylation, and consequent release of EPRS and L13a from the tRNA multisynthetase complex and 60S ribosomal subunit, respectively, regulates GAIT complex assembly (Mukhopadhyay et al. 2009). Whether or not pathogenic mycobacteria are able to modulate this process to prevent macrophage activation remains to be analyzed.

Several *M. tuberculosis* products are known to interfere with macrophage activation. In particular, LAM is known to potently prevent IFNγ-mediated macrophage activation (Sibley et al. 1988, 1990) that may play a role in the ability of *M. tuberculosis* to sustain itself within macrophages, e.g., within granuloma's, for prolonged times, despite the presence of activated lymphocytes. Mycobacteria-generated Eis (see above), by inhibiting host redox-dependent signaling is also known to prevent macrophage activation (Shin et al. 2010). Finally, *M. tuberculosis* utilizes the proteasome to interfere with the damaging activity of nitric oxide that is generated within activated macrophages (Darwin et al. 2003; Pieters and Ploegh 2003), a process that is a potential target for antimycobacterial drugs (Lin et al. 2009; Cheng and Pieters 2010). Thus, *M. tuberculosis* has evolved several ways to circumvent the macrophage activation response in order to cope with host responses.

Besides IFNγ toll-like receptor ligands can also potently activate infected macrophages, resulting in the generation of nitric oxide intermediates as well as modulating phagosome–lysosome fusion (Janeway and Medzhitov 2002; Kleinnijenhuis et al. 2011; Saiga et al. 2011). For example, mycobacterial DNA in the form of palindromic sequences including the 5'-CG-3'motif, now called CpG motif, are known to activate TLR9 (Hemmi et al. 2000). Also, the mycobacterial cell wall components LAM, lipomannan (LM), and PIM are ligands for TLR2 (Ryffel et al. 2006). The in vivo importance of the different toll-like receptor-mediated signaling pathways in host defense to mycobacteria was highlighted in studies using mice lacking MyD88, which is a critical and common component of TLR signaling. MyD88-deficient mice are highly susceptible to airborne infection with mycobacteria (Drennan et al. 2004).

It has long been realized that individuals having reduced levels of vitamin D have an increased susceptibility to tuberculosis, but the molecular basis of this increased susceptibility was unknown. Interestingly, it turns out that vitamin D, upon interaction with the vitamin D receptor, promotes the production of the antimicrobial peptide cathelicidin, an effective antimycobacterial peptide (Liu et al. 2006, 2007). The role of vitamin D in controlling mycobacterial defense responses is highlighted by the fact that people predisposed to lower vitamin D levels in serum have a higher incidence of active tuberculosis (Liu et al. 2006; Adams et al. 2009). Furthermore, vitamin D is required for several of the antimicrobial effects induced by IFNγ (Fabri et al. 2011), suggesting that the vitamin D pathway appears to be an important host defense mechanism to overcome the ability of *M. tuberculosis* to evade host innate immune mechanisms at multiple levels.

6 How Unique is *M. tuberculosis* with Regard to Its Survival Strategy?

As is clear from the above description, pathogenic mycobacteria employ a plethora of survival strategies. But how unique are these strategies, in relation to other intracellular pathogens? Although initially it was suspected that diverse intracellular pathogens employ common strategies to circumvent immune responses, it is becoming increasingly clear that each pathogen is modulating the immune response in a rather unique way. This may on the one hand have to do with the distinct niches in which different pathogens reside, and on the other hand with the uniqueness of the separate pathogens in terms of their biology and growth requirements.

For example, there are several other pathogenic microbes that avoid lysosomal delivery: *Listeria* spp ensure escape from the phagosome at an early stage by secreting a lysogenic factor, listeriolysin, that allows their escape into the cytosol (Lemeland et al. 1974; Tilney and Portnoy 1989; Hurme and Cossart 1999); Also, *Brucella* spp as well as *Legionella* are phagocytosed but then home to the endoplasmic reticulum for survival and replication (Meresse et al. 1999). *Chlamydia* is another example of an intracellular pathogen that avoids lysosomal transfer, but in this case by residing in a compartment that is believed to be completely segregated from the endocytic pathway (Hackstadt et al. 1997). The intracellular fate of *Salmonella* is also distinct from that of mycobacteria; whereas in some cases, *Salmonella*-containing vacuoles have characteristics of Golgi-like organelles (Lahiri et al. 2010), in other studies *Salmonella* spp. were found to thrive and replicate within lysosomal organelles (Rosales-Reyes et al. 2005). One pathogen that probably comes the closest to employing strategies similar to pathogenic mycobacteria is *Leishmania*. This parasite is normally phagocytosed by macrophages, but once inside the macrophage it utilizes cell surface components to efficiently block phagosome–lysosome fusion (Mauel 1990; Desjardins and Descoteaux 1997). Taken together, although some overlap exist in the strategies employed, most pathogens seem to have evolved their preferred ways to interfere with and circumvent host innate immune responses.

7 Conclusion

Research over the past few decades is beginning to provide the details of how *M. tuberculosis* manages to withstand the hostile environment of the macrophages. The strategies employed range from hijacking several host proteins for their own survival to the intracellular release of a plethora of molecules that are either shed or secreted into the host cells at different stages of infection. Furthermore, the exact route via which *M. tuberculosis* enters the macrophage can also dictate the outcome of the pathogen–host balance. Together, the knowledge that is accumulating on the different survival strategies employed by *M. tuberculosis* is helping the development of compounds that interfere with these pathways, holding promises for the availability of novel drugs to combat tuberculosis.

Acknowledgments Research in the laboratory of Jean Pieters is funded by the Swiss National Science Foundation and the Optimus Foundation. Somdeb BoseDasgupta is a recipient of an EMBO Long Term Fellowship and Rajesh Jayachandran is a Cloetta Medical Research fellow.

References

Abdallah AM, Gey van Pittius NC, Champion PA et al (2007) Type VII secretion–mycobacteria show the way. Nat Rev Microbiol 5:883–891

Adams JS, Ren S, Liu PT et al (2009) Vitamin d-directed rheostatic regulation of monocyte antibacterial responses. J Immunol 182:4289–4295

Aderem A, Underhill DM (1999) Mechanisms of phagocytosis in macrophages. Annu Rev Immunol 17:593–623

Akira S, Uematsu S, Takeuchi O (2006) Pathogen recognition and innate immunity. Cell 124:783–801

Allen LA, Schlesinger LS, Kang B (2000) Virulent strains of *Helicobacter pylori* demonstrate delayed phagocytosis and stimulate homotypic phagosome fusion in macrophages. J Exp Med 191:115–128

Armstrong JA, Hart PD (1971) Response of cultured macrophages to *Mycobacterium tuberculosis*, with observations on fusion of lysosomes with phagosomes. J Exp Med 134:713–740

Av-Gay Y, Everett M (2000) The eukaryotic-like Ser/Thr protein kinases of *M. tuberculosis*. Trends Microbiol 8:238–244

Av-Gay Y, Sobouti R (2000) Cholesterol is accumulated by mycobacteria but its degradation is limited to non-pathogenic fast-growing mycobacteria. Can J Microbiol 46:826–831

Bach H, Papavinasasundaram KG, Wong D, Hmama Z, Av-Gay Y (2008) *M. tuberculosis* virulence is mediated by PtpA dephosphorylation of human vacuolar protein sorting 33B. Cell Host Microbe 3:316–322

Be NA, Bishai WR, Jain SK (2012) Role of *M. tuberculosis* pknD in the pathogenesis of central nervous system tuberculosis. BMC Microbiol 12:7

Berthet FX, Lagranderie M, Gounon P et al (1998) Attenuation of virulence by disruption of the *M. tuberculosis* erp gene. Science 282:759–762

Boehm U, Klamp T, Groot M, Howard JC (1997) Cellular responses to interferon-gamma. Annu Rev Immunol 15:749–795

Briken V, Porcelli SA, Besra GS, Kremer L (2004) Mycobacterial lipoarabinomannan and related lipoglycans: from biogenesis to modulation of the immune response. Mol Microbiol 53:391–403

Brodin P, Majlessi L, Marsollier L et al (2006) Dissection of ESAT-6 system 1 of *M. tuberculosis* and impact on immunogenicity and virulence. Infect Immun 74:88–98
Brown DA, London E (2000) Structure and function of sphingolipid- and cholesterol-rich membrane rafts. J Biol Chem 275:17221–17224
Chakraborty P, Sturgill-Koszycki S, Russell DG (1994) Isolation and characterization of pathogen-containing phagosomes. Methods Cell Biol 45:261–276
Chen JM, Boy-Rottger S, Dhar N et al (2012) EspD is critical for the virulence-mediating ESX-1 secretion system in *M. tuberculosis*. J Bacteriol 194:884–893
Cheng Y, Pieters J (2010) Novel proteasome inhibitors as potential drugs to combat tuberculosis. J Mol Cell Biol 2:173–175
Cole ST, Eiglmeier K, Parkhill J et al (2001) Massive gene decay in the leprosy bacillus. Nature 409:1007–1011
Combaluzier B, Mueller P, Massner J, Finke D, Pieters J (2009) Coronin 1 is essential for IgM-mediated Ca^{2+} mobilization in B cells but dispensable for the generation of immune responses in vivo. J Immunol 182:1954–1961
Cosma CL, Klein K, Kim R, Beery D, Ramakrishnan L (2006) Mycobacterium marinum Erp is a virulence determinant required for cell wall integrity and intracellular survival. Infect Immun 74:3125–3133
Cowley S, Ko M, Pick N et al (2004) The *M. tuberculosis* protein serine/threonine kinase PknG is linked to cellular glutamate/glutamine levels and is important for growth in vivo. Mol Microbiol 52:1691–1702
Cywes C, Hoppe HC, Daffe M, Ehlers MR (1997) Nonopsonic binding of *M. tuberculosis* to complement receptor type 3 is mediated by capsular polysaccharides and is strain dependent. Infect Immun 65:4258–4266
Darwin KH, Ehrt S, Gutierrez-Ramos JC, Weich N, Nathan CF (2003) The proteasome of *M. tuberculosis* is required for resistance to nitric oxide. Science 302:1963–1966
Davies DG, Chakrabarty AM, Geesey GG (1993) Exopolysaccharide production in biofilms: substratum activation of alginate gene expression by *Pseudomonas aeruginosa*. Appl Environ Microbiol 59:1181–1186
de Jonge MI, Pehau-Arnaudet G, Fretz MM et al (2007) ESAT-6 from *M. tuberculosis* dissociates from its putative chaperone CFP-10 under acidic conditions and exhibits membrane-lysing activity. J Bacteriol 189:6028–6034
Deghmane AE, Soualhine H, Bach H et al (2007) Lipoamide dehydrogenase mediates retention of coronin-1 on BCG vacuoles, leading to arrest in phagosome maturation. J Cell Sci 120:2796–2806
Deretic V, Singh S, Master S et al (2006) *Mycobacterium tuberculosis* inhibition of phagolysosome biogenesis and autophagy as a host defence mechanism. Cell Microbiol 8:719–727
Desjardins M, Descoteaux A (1997) Inhibition of phagolysosomal biogenesis by the Leishmania lipophosphoglycan. J Exp Med 185:2061–2068
DiGiuseppe Champion PA, Cox JS (2007) Protein secretion systems in Mycobacteria. Cell Microbiol 9:1376–1384
Drennan MB, Nicolle D, Quesniaux VJ et al (2004) Toll-like receptor 2-deficient mice succumb to *M. tuberculosis* infection. Am J Pathol 164:49–57
Ehlers S (2010) DC-SIGN and mannosylated surface structures of *M. tuberculosis*: a deceptive liaison. Eur J Cell Biol 89:95–101
Fabri M, Stenger S, Shin DM et al (2011) Vitamin D is required for IFN-gamma-mediated antimicrobial activity of human macrophages. Sci Transl Med 3:104ra102
Feng CG, Collazo-Custodio CM, Eckhaus M et al (2004) Mice deficient in LRG-47 display increased susceptibility to mycobacterial infection associated with the induction of lymphopenia. J Immunol 172:1163–1168
Ferrari G, Langen H, Naito M, Pieters J (1999) A coat protein on phagosomes involved in the intracellular survival of mycobacteria. Cell 97:435–447

Foger N, Rangell L, Danilenko DM, Chan AC (2006) Requirement for coronin 1 in T lymphocyte trafficking and cellular homeostasis. Science 313:839–842.
Fortune SM, Jaeger A, Sarracino DA et al (2005) Mutually dependent secretion of proteins required for mycobacterial virulence. Proc Natl Acad Sci USA 102:10676–10681
Franchi L, Amer A, Body-Malapel M et al (2006) Cytosolic flagellin requires Ipaf for activation of caspase-1 and interleukin 1beta in *salmonella*-infected macrophages. Nat Immunol 7:576–582
Fratti RA, Chua J, Vergne I, Deretic V (2003) *Mycobacterium tuberculosis* glycosylated phosphatidylinositol causes phagosome maturation arrest. Proc Natl Acad Sci USA 100:5437–5442
Ganaie AA, Lella RK, Solanki R, Sharma C (2011) Thermostable hexameric form of Eis (Rv2416c) protein of *M. tuberculosis* plays an important role for enhanced intracellular survival within macrophages. PLoS One 6:e27590
Gao LY, Guo S, McLaughlin B, Morisaki H, Engel JN, Brown EJ (2004) A mycobacterial virulence gene cluster extending RD1 is required for cytolysis, bacterial spreading and ESAT-6 secretion. Mol Microbiol 53:1677–1693
Gatfield J, Pieters J (2000) Essential role for cholesterol in entry of mycobacteria into macrophages. Science 288:1647–1650
Gatfield J, Pieters J (2003) Molecular mechanisms of host-pathogen interaction: Entry and survival of mycobacteria in macrophages. Adv Immunol 82:45–96
Gatfield J, Albrecht I, Zanolari B, Steinmetz MO, Pieters J (2005) Association of the Leukocyte Plasma Membrane with the Actin Cytoskeleton through Coiled Coil-mediated Trimeric Coronin 1 Molecules. Mol Biol Cell 16:2786–2798
Gengenbacher M, Kaufmann SH (2012) *Mycobacterium tuberculosis*: success through dormancy. FEMS Microbiol Rev 36:514–532
Goldberg DE, Siliciano RF, Jacobs WR, Jr (2012) Outwitting evolution: fighting drug-resistant TB, malaria, and HIV. Cell 148:1271–1283
Greenstein AE, MacGurn JA, Baer CE, Falick AM, Cox JS, Alber T (2007) M. tuberculosis Ser/Thr protein kinase D phosphorylates an anti–anti-sigma factor homolog. PLoS Pathog 3:e49
Grinstein S (2010) Imaging signal transduction during phagocytosis: phospholipids, surface charge, and electrostatic interactions. Am J Physiol. Cell Physiol 299:C876–881
Grundner C, Gay LM, Alber T (2005) *Mycobacterium tuberculosis* serine/threonine kinases PknB, PknD, PknE, and PknF phosphorylate multiple FHA domains. Protein Sci 14:1918–1921
Hackstadt T, Fischer ER, Scidmore MA, Rockey DD, Heinzen RA (1997) Origins and functions of the chlamydial inclusion. Trends Microbiol 5:288–293
Han G, Zhang C–C (2001) On the origin of Ser/Thr kinases in a prokaryote. FEMS Microbiol Lett 200:79–84
Harris J, De Haro SA, Master SS, Keane J, Roberts EA, Delgado M, Deretic V (2007) T helper 2 cytokines inhibit autophagic control of intracellular *M. tuberculosis*. Immunity 27:505–517
Hasan Z, Schlax C, Kuhn L, Lefkovits I, Young D, Thole J, Pieters J (1997) Isolation and characterization of the mycobacterial phagosome: segregation from the endosomal/lysosomal pathway. Mol Microbiol 24:545–553
Hassett DJ (1996) Anaerobic production of alginate by *Pseudomonas aeruginosa*: alginate restricts diffusion of oxygen. J Bacteriol 178:7322–7325
Hemmi H, Takeuchi O, Kawai T et al (2000) A Toll-like receptor recognizes bacterial DNA. Nature 408:740–745
Henry SC, Daniell X, Indaram M et al (2007) Impaired macrophage function underscores susceptibility to *Salmonella* in mice lacking Irgm1 (LRG-47). J Immunol 179:6963–6972
Houben EN, Nguyen L, Pieters J (2006) Interaction of pathogenic mycobacteria with the host immune system. Curr Opin Microbiol 9:76–85
Houben EN, Walburger A, Ferrari G et al (2009) Differential expression of a virulence factor in pathogenic and non-pathogenic mycobacteria. Mol Microbiol 72:41–52
Hughes EA, Galan JE (2002) Immune response to *Salmonella*: location, location, location? Immunity 16:325–328

Hunn JP, Howard JC (2010) The mouse resistance protein Irgm1 (LRG-47): a regulator or an effector of pathogen defense? PLoS Pathog 6:e1001008
Hurme R, Cossart P (1999) Invasion of mammalian cells by Listeria monocytogenes. In: Gordon S (ed) Phagocytosis: Microbial invasion. JAI Press Inc., Stamford, pp 59–79
Iniesta V, Gomez-Nieto LC, Corraliza I (2001) The inhibition of arginase by N(omega)-hydroxy-l-arginine controls the growth of Leishmania inside macrophages. J Exp Med 193:777–784
Ishikawa E, Ishikawa T, Morita YS et al (2009) Direct recognition of the mycobacterial glycolipid, trehalose dimycolate, by C-type lectin Mincle. J Exp Med 206:2879–2888
Janeway CA, Jr., Medzhitov R (2002) Innate immune recognition. Annu Rev Immunol 20:197–216
Jayachandran R, Sundaramurthy V, Combaluzier B et al (2007) Survival of mycobacteria in macrophages is mediated by coronin 1-dependent activation of calcineurin. Cell 130:37–50
Jayachandran R, Gatfield J, Massner J, Albrecht I, Zanolari B, Pieters J (2008) RNA interference in J774 macrophages reveals a role for Coronin 1 in mycobacterial trafficking but not in actin-dependent processes. Mol Biol Cell 19:1241–1251
Kang CM, Abbott DW, Park ST, Dascher CC, Cantley LC, Husson RN (2005) The *M. tuberculosis* serine/threonine kinases PknA and PknB: substrate identification and regulation of cell shape. Genes Dev 19:1692–1704
Kaye PM, Rogers NJ, Curry AJ, Scott JC (1994) Deficient expression of co-stimulatory molecules on Leishmania-infected macrophages. Eur J Immunol 24:2850–2854
Kim BH, Shenoy AR, Kumar P, Das R, Tiwari S, MacMicking JD (2011) A family of IFN-gamma-inducible 65-kD GTPases protects against bacterial infection. Science 332:717–721
Kim KH, An DR, Song J et al (2012) *Mycobacterium tuberculosis* Eis protein initiates suppression of host immune responses by acetylation of DUSP16/MKP-7. Proc Natl Acad Sci USA 109:7729–7734
Kima PE, Soong L, Chicharro C, Ruddle NH, McMahon-Pratt D (1996) Leishmania-infected macrophages sequester endogenously synthesized parasite antigens from presentation to $CDCD^{4+}$ T cells. Eur J Immunol 26:3163–3169
Kleinnijenhuis J, Oosting M, Plantinga TS, van der Meer JW, Joosten LA, Crevel RV, Netea MG (2011) Autophagy modulates the *M. tuberculosis*-induced cytokine response. Immunology 134:341–348
Klepp LI, Soria M, Blanco FC, Bianco MV, Santangelo MP, Cataldi AA, Bigi F (2009) Identification of two proteins that interact with the Erp virulence factor from *M. tuberculosis* by using the bacterial two-hybrid system. BMC Mol Biol 10:3
Lahiri A, Eswarappa SM, Das P, Chakravortty D (2010) Altering the balance between pathogen containing vacuoles and lysosomes: a lesson from *Salmonella*. Virulence 1:325–329
Lang C, Hildebrandt A, Brand F, Opitz L, Dihazi H, Luder CG (2012) Impaired chromatin remodelling at STAT1-regulated promoters leads to global unresponsiveness of Toxoplasma gondii-Infected macrophages to IFN-gamma. PLoS Pathog 8:e1002483
Le Cabec V, Cols C, Maridonneau-Parini I (2000) Nonopsonic phagocytosis of zymosan and *M. kansasii* by CR3 (CD11b/CD18) involves distinct molecular determinants and is or is not coupled with NADPH oxidase activation. Infect Immun 68:4736–4745
Le Cabec V, Carreno S, Moisand A, Bordier C, Maridonneau-Parini I (2002) Complement receptor 3 (CD11b/CD18) mediates type I and type II phagocytosis during nonopsonic and opsonic phagocytosis, respectively. J Immunol 169:2003–2009
Lemeland JF, Allaire R, Boiron H (1974) [Hemolysin of Listeria monocytogenes (listeriolysin)]. Pathol Biol (Paris) 22:763–770
Lin G, Li D, de Carvalho LP et al (2009) Inhibitors selective for mycobacterial versus human proteasomes. Nature 461:621–626
Ling YM, Shaw MH, Ayala C, Coppens I, Taylor GA, Ferguson DJ, Yap GS (2006) Vacuolar and plasma membrane stripping and autophagic elimination of Toxoplasma gondii in primed effector macrophages. J Exp Med 203:2063–2071
Lingwood D, Kaiser HJ, Levental I, Simons K (2009) Lipid rafts as functional heterogeneity in cell membranes. Biochem Soc Trans 37:955–960

Liu PT, Stenger S, Li H et al (2006) Toll-like receptor triggering of a vitamin D-mediated human antimicrobial response. Science 311:1770–1773

Liu PT, Stenger S, Tang DH, Modlin RL (2007) Cutting edge: vitamin D-mediated human antimicrobial activity against *M. tuberculosis* is dependent on the induction of cathelicidin. J Immunol 179:2060–2063

MacGurn JA, Cox JS (2007) A genetic screen for *M. tuberculosis* mutants defective for phagosome maturation arrest identifies components of the ESX-1 secretion system. Infect Immun 75:2668–2678

MacMicking JD, Taylor GA, McKinney JD (2003) Immune control of tuberculosis by IFN-gamma-inducible LRG-47. Science 302:654–659

Mahdavi J, Sonden B, Hurtig M et al (2002) *Helicobacter pylori* SabA adhesin in persistent infection and chronic inflammation. Science 297:573–578

Majlessi L, Brodin P, Brosch R et al (2005) Influence of ESAT-6 secretion system 1 (RD1) of *M. tuberculosis* on the interaction between mycobacteria and the host immune system. J Immunol 174:3570-3579.

Malik ZA, Denning GM, Kusner DJ (2000) Inhibition of Ca(2 +) signaling by *M. tuberculosis* is associated with reduced phagosome-lysosome fusion and increased survival within human macrophages. J Exp Med 191:287–302

Malik ZA, Thompson CR, Hashimi S, Porter B, Iyer SS, Kusner DJ (2003) Cutting edge: *M. tuberculosis* blocks Ca2+ signaling and phagosome maturation in human macrophages via specific inhibition of sphingosine kinase. J Immunol 170:2811–2815

Master SS, Rampini SK, Davis AS et al (2008) *Mycobacterium tuberculosis* prevents inflammasome activation. Cell Host Microbe 3:224–232

Mauel J (1990) Macrophage-parasite interactions in Leishmania infections. J Leukoc Biol 47:187–193

Meresse S, Steele-Mortimer O, Moreno E, Desjardins M, Finlay B, Gorvel JP (1999) Controlling the maturation of pathogen-containing vacuoles: a matter of life and death. Nat Cell Biol 1:E183–188

Mueller P, Massner J, Jayachandran R et al (2008) Regulation of T cell survival through coronin-1-mediated generation of inositol-1,4,5-trisphosphate and calcium mobilization after T cell receptor triggering. Nat Immunol 9:424–431

Mukhopadhyay R, Jia J, Arif A, Ray PS, Fox PL (2009) The GAIT system: a gatekeeper of inflammatory gene expression. Trends Biochem Sci 34:324–331

Nathan C (2006) Role of iNOS in human host defense. Science 312:1874–1875; author reply 1874–1875

Nathan C, Shiloh MU (2000) Reactive oxygen and nitrogen intermediates in the relationship between mammalian hosts and microbial pathogens. Proc Natl Acad Sci USA 97:8841–8848

North RJ (1974) T cell dependence of macrophage activation and mobilization during infection with *M. tuberculosis*. Infect Immun 10:66–71

Parkhill J, Dougan G, James KD et al (2001) Complete genome sequence of a multiple drug resistant *Salmonella enterica* serovar Typhi CT18. Nature 413:848–852

Peyron P, Bordier C, N'Diaye EN, Maridonneau-Parini I (2000) Nonopsonic phagocytosis of *M. kansasii* by human neutrophils depends on cholesterol and is mediated by CR3 associated with glycosylphosphatidylinositol-anchored proteins. J Immunol 165:5186–5191

Philips JA (2008) Mycobacterial manipulation of vacuolar sorting. Cell Microbiol 10:2408–2415

Pieters J (2008) *Mycobacterium tuberculosis* and the macrophage: maintaining a balance. Cell Host Microbe 3:399–407

Pieters J, Ploegh H (2003) Microbiology. Chemical warfare and mycobacterial defense. Science 302:1900–1902

Ponting CP, Aravind L, Schultz J, Bork P, Koonin EV (1999) Eukaryotic signalling domain homologues in archaea and bacteria. Ancient ancestry and horizontal gene transfer. J Mol Biol 289:729–745

Raoust E, Balloy V, Garcia-Verdugo I, Touqui L, Ramphal R, Chignard M (2009) *Pseudomonas aeruginosa* LPS or flagellin are sufficient to activate TLR-dependent signaling in murine alveolar macrophages and airway epithelial cells. PLoS One 4:e7259

Reid DM, Gow NA, Brown GD (2009) Pattern recognition: recent insights from Dectin-1. Curr Opin Immunol 21:30–37

Restrepo CI, Dong Q, Savov J, Mariencheck WI, Wright JR (1999) Surfactant protein D stimulates phagocytosis of *Pseudomonas aeruginosa* by alveolar macrophages. Am J Respir Cell Mol Biol 21:576–585

Rohde K, Yates RM, Purdy GE, Russell DG (2007) *Mycobacterium tuberculosis* and the environment within the phagosome. Immunol Rev 219:37–54

Rosales-Reyes R, Alpuche-Aranda C, Ramirez-Aguilar Mde L, Castro-Eguiluz AD, Ortiz-Navarrete V (2005) Survival of *Salmonella enterica* serovar Typhimurium within late endosomal-lysosomal compartments of B lymphocytes is associated with the inability to use the vacuolar alternative major histocompatibility complex class I antigen-processing pathway. Infect Immun 73:3937–3944

Roth MG (2004) Phosphoinositides in constitutive membrane traffic. Physiol Rev 84:699–730

Rybakin V, Clemen CS (2005) Coronin proteins as multifunctional regulators of the cytoskeleton and membrane trafficking. Bioessays 27:625–632

Ryffel B, Jacobs M, Parida S, Botha T, Togbe D, Quesniaux V (2006) Toll-like receptors and control of mycobacterial infection in mice. Novartis Found Symp 279:127–139; discussion 139–141, 216–129

Saiga H, Shimada Y, Takeda K (2011) Innate immune effectors in mycobacterial infection. Clin Dev Immunol 2011:347594

Saleh MT, Belisle JT (2000) Secretion of an acid phosphatase (SapM) by *M. tuberculosis* that is similar to eukaryotic acid phosphatases. J Bacteriol 182:6850–6853

Schafer G, Jacobs M, Wilkinson RJ, Brown GD (2009) Non-opsonic recognition of *M. tuberculosis* by phagocytes. J Innate Immun 1:231–243

Scherr N, Honnappa S, Kunz G, et al. (2007) Structural basis for the specific inhibition of protein kinase G, a virulence factor of M. tuberculosis. Proc Natl Acad Sci USA 104:12151–12156

Scherr N, Jayachandran R, Mueller P, Pieters J (2009a) Interference of *M. tuberculosis* with macrophage responses. Indian J Exp Biol 47:401–406

Scherr N, Muller P, Perisa D, Combaluzier B, Jeno P, Pieters J (2009b) Survival of pathogenic mycobacteria in macrophages is mediated through autophosphorylation of protein kinase G. J Bacteriol 191:4546–4554

Schorey JS, Carroll MC, Brown EJ (1997) A macrophage invasion mechanism of pathogenic mycobacteria. Science 277:1091–1093

Shenoy AR, Wellington DA, Kumar P, Kassa H, Booth CJ, Cresswell P, MacMicking JD (2012) GBP5 promotes NLRP3 inflammasome assembly and immunity in mammals. Science 336:481–485

Shiloh MU, Nathan CF (2000) Reactive nitrogen intermediates and the pathogenesis of *Salmonella* and mycobacteria. Curr Opin Microbiol 3:35–42

Shin DM, Jeon BY, Lee HM et al (2010) *Mycobacterium tuberculosis* eis regulates autophagy, inflammation, and cell death through redox-dependent signaling. PLoS Pathog 6:e1001230

Sibley LD, Hunter SW, Brennan PJ, Krahenbuhl JL (1988) Mycobacterial lipoarabinomannan inhibits gamma interferon-mediated activation of macrophages. Infect Immun 56:1232–1236

Sibley LD, Adams LB, Krahenbuhl JL (1990) Inhibition of interferon-gamma-mediated activation in mouse macrophages treated with lipoarabinomannan. Clin Exp Immunol 80:141–148

Simons K, Ikonen E (1997) Functional rafts in cell membranes. Nature 387:569–572

Singh SB, Davis AS, Taylor GA, Deretic V (2006) Human IRGM induces autophagy to eliminate intracellular mycobacteria. Science 313:1438–1441

Smith J, Manoranjan J, Pan M et al (2008) Evidence for pore formation in host cell membranes by ESX-1-secreted ESAT-6 and its role in *M. marinum* escape from the vacuole. Infect Immun 76:5478–5487

Soldati T, Neyrolles O (2012) Mycobacteria and the intraphagosomal environment: take it with a pinch of salt(s)! Traffic 13:1042–1052
Steinberg BE, Grinstein S (2008) Pathogen destruction versus intracellular survival: the role of lipids as phagosomal fate determinants. J Clin Invest 118:2002–2011
Sturgill-Koszycki S, Schlesinger PH, Chakraborty P et al (1994) Lack of acidification in Mycobacterium phagosomes produced by exclusion of the vesicular proton-ATPase. Science 263:678–681
Tateda K, Ishii Y, Horikawa M et al (2003) The *Pseudomonas aeruginosa* autoinducer N-3-oxododecanoyl homoserine lactone accelerates apoptosis in macrophages and neutrophils. Infect Immun 71:5785–5793
Taylor GA, Feng CG, Sher A (2004) p47 GTPases: regulators of immunity to intracellular pathogens. Nat Rev Immunol 4:100–109
Thakur M, Chakraborti PK (2008) Ability of PknA, a mycobacterial eukaryotic-type serine/threonine kinase, to transphosphorylate MurD, a ligase involved in the process of peptidoglycan biosynthesis. Biochem J 415:27–33
Thakur M, Chaba R, Mondal AK, Chakraborti PK (2008) Interdomain interaction reconstitutes the functionality of PknA, a eukaryotic type Ser/Thr kinase from *M. tuberculosis*. J Biol Chem 283:8023–8033
Tilney LG, Portnoy DA (1989) Actin filaments and the growth, movement, and spread of the intracellular bacterial parasite, Listeria monocytogenes. J Cell Biol 109:1597–1608
Torrelles JB, Azad AK, Schlesinger LS (2006) Fine discrimination in the recognition of individual species of phosphatidyl-myo-inositol mannosides from *M. tuberculosis* by C-type lectin pattern recognition receptors. J Immunol 177:1805–1816
Townsend SM, Kramer NE, Edwards R et al (2001) *Salmonella enterica* serovar Typhi possesses a unique repertoire of fimbrial gene sequences. Infect Immun 69:2894–2901
Tracy Tan WLLDCASGJL (2006) The ESAT-6/CFP-10 secretion system of <i> *M. marinum* </i> modulates phagosome maturation. Cellular Microbiology 8:1417–1429
Uematsu S, Akira S (2006) Innate immune recognition of viral infection. Uirusu 56:1–8
van der Wel N, Hava D, Houben D et al (2007) *M. tuberculosis* and *M. leprae* translocate from the phagolysosome to the cytosol in myeloid cells. Cell 129:1287–1298
Vergne I, Fratti RA, Hill PJ, Chua J, Belisle J, Deretic V (2004) *Mycobacterium tuberculosis* phagosome maturation arrest: mycobacterial phosphatidylinositol analog phosphatidylinositol mannoside stimulates early endosomal fusion. Mol Biol Cell 15:751–760
Vergne I, Chua J, Lee HH, Lucas M, Belisle J, Deretic V (2005) Mechanism of phagolysosome biogenesis block by viable *M. tuberculosis*. Proc Natl Acad Sci USA 102:4033–4038
Vieira OV, Botelho RJ, Grinstein S (2002) Phagosome maturation: aging gracefully. Biochem J 366:689–704
Waddell SJ, Popper SJ, Rubins KH, Griffiths MJ, Brown PO, Levin M, Relman DA (2010) Dissecting interferon-induced transcriptional programs in human peripheral blood cells. PLoS One 5:e9753
Wain J, Pham VB, Ha V, et al. (2001) Quantitation of bacteria in bone marrow from patients with typhoid fever: relationship between counts and clinical features. J Clin Microbiol 39:1571–1576
Walburger A, Koul A, Ferrari G et al (2004) Protein kinase G from pathogenic mycobacteria promotes survival within macrophages. Science 304:1800–1804
Warner DF, Mizrahi V (2007) The survival kit of *M. tuberculosis*. Nat Med 13:282–284
Wehenkel A, Bellinzoni M, Grana M et al (2008) Mycobacterial Ser/Thr protein kinases and phosphatases: physiological roles and therapeutic potential. Biochim Biophys Acta 1784: 193–202
Wei J, Dahl JL, Moulder JW, Roberts EA, O'Gaora P, Young DB, Friedman RL (2000) Identification of a *M. tuberculosis* gene that enhances mycobacterial survival in macrophages. J Bacteriol 182:377–384
Welin A, Lerm M (2011) Inside or outside the phagosome? The controversy of the intracellular localization of *M. tuberculosis*. Tuberculosis (Edinb)

Winslow MM, Neilson JR, Crabtree GR (2003) Calcium signalling in lymphocytes. Curr Opin Immunol 15:299–307

Yadav M, Schorey JS (2006) The beta-glucan receptor dectin-1 functions together with TLR2 to mediate macrophage activation by mycobacteria. Blood 108:3168–3175

Zheng PY, Jones NL (2003) *Helicobacter pylori* strains expressing the vacuolating cytotoxin interrupt phagosome maturation in macrophages by recruiting and retaining TACO (coronin 1) protein. Cell Microbiol 5:25–40

Host–Pathogen Interactions During *Mycobacterium tuberculosis* infections

Sarah A. Stanley and Jeffery S. Cox

Abstract The intimate and persistent connection between *Mycobacterium tuberculosis* and its human host suggests that the pathogen has evolved extensive mechanisms to evade eradication by the immune system. In particular, the organism has adapted to replicate within phagocytic cells, especially macrophages, which are specialized to kill microbes. Over the past decade of *M. tuberculosis* research, the means to manipulate both the organism and the host has ushered in an exciting time that has uncovered some of the mechanisms of the innate macrophage-pathogen interactions that lie at the heart of *M. tuberculosis* pathogenesis, though many interactions likely still await discovery. In this chapter, we will delve into some of these advances, with an emphasis on the interactions that occur on the cellular level when *M. tuberculosis* cells encounter macrophages. In particular, we focus on two major aspects of *M. tuberculosis* biology regarding the proximal physical interface between the bacterium and host, namely the interactions with the phagosomal membrane as well as the distinctive mycobacterial cell wall. Importantly, some of the emerging paradigms in *M. tuberculosis* pathogenesis and host response represent common themes in bacterial pathogenesis, such as the role of host cell membrane perforation in intracellular survival and host response. However, the array of unique bacterial lipid mediators and their interaction with host cells highlights the unique biology of this persistent pathogen.

S. A. Stanley
School of Public Health, University of California, 500C Li Ka Shing Hall, Berkeley, CA 94720-3370, USA

J. S. Cox (✉)
Department of Microbiology and Immunology, University of California, 600 16th St. Box 2200, MBGH Room N372B, San Francisco, CA 94143-2200, USA
e-mail: jeffery.cox@ucsf.edu

Current Topics in Microbiology and Immunology (2013) 374: 211–241
DOI: 10.1007/82_2013_332

Published Online: 24 July 2013

Contents

1 *Mycobacterium tuberculosis*–Macrophage Interactions

How *M. tuberculosis* adapts to and manipulates its intracellular environment in order to replicate and persist is a central question in the field that continues to both fascinate and perplex researchers. Over four decades of experimental work supports the conventional notion that the in vivo niche of *M. tuberculosis* is primarily the membrane-bound phagosome of macrophages, though growth in other cell types and even extracellular spaces is also likely important (Wolf et al. 2007). Certainly, the profound ability of *M. tuberculosis* to avoid killing and to replicate within macrophage cells is important for the establishment of infection and for virulence, but the question of how bacilli can interact with and manipulate macrophage functions while confined inside the phagosome lumen is an unsolved conundrum. Indeed, the notion that *M. tuberculosis* can alter membrane traffic, acquire nutrients, etc., while sealed within a phagosome that is impermeant to even protons seems implausible. Accumulating evidence points to the importance of the interactions that occur between *M. tuberculosis* and its surrounding phagosomal membrane, and specifically cytosolic access mediated by pore formation, indicating that not all *M. tuberculosis* cells during infection remain entirely lumenal. In light of this evidence, we argue that the standard model of the intracellular location of *M. tuberculosis* should be revised. In particular, recent data demonstrate that *M. tuberculosis* actively punctures the surrounding phagosomal membrane allowing for cytosolic access, though this activity is less robust when compared to cytosolic bacterial pathogens such as *Listeria monocytogenes*. In view of this work, and in combination with a careful examination of the classic literature, it is also becoming clear that *M. tuberculosis* cells quickly become non-uniform in their intracellular location after infection of naïve macrophage, with some bacteria localized to endosomal compartments and others targeted to lysosomes via the autophagy pathway. This heterogeneity is augmented by the fact that the intracellular environment of bacilli changes over time (Rohde et al. 2012) and that the behavior of individual *M. tuberculosis* cells within a population is not uniform (Armstrong and Hart 1975). Thus, the notion that *M. tuberculosis* adopts one particular intracellular location does not encapsulate the diversity of cell's fate that arise during infection. These two key concepts, membrane-perforation and temporal changes in intracellular location, have been generally underappreciated in reviews of the literature but these must be incorporated to accurately understand how *M. tuberculosis* interacts with its host cell. We propose a model that seeks to more thoroughly encompass these dynamic interactions between *M. tuberculosis* and macrophage that embrace the heterogeneous response phenomenon of phagosome rupture and concomitant cytosolic access.

1.1 Evidence Supporting the "Classic" View

Numerous studies of *M. tuberculosis*-infected macrophages over the past 40 years support the predominant view that bacilli reside in a compartment that is bounded by phagosomal membrane and that these organelles fail to fuse with lysosomes. Classic electron microscopy (EM) studies showed that after 1–4 days of infection, live *M. tuberculosis* exists inside membrane-bound phagosomes and that for the majority of phagosomes, do not fuse with lysosomes (Armstrong and Hart 1971, 1975). These were important and seminal studies showing that phagosomal membranes continuously surround *M. tuberculosis* after infection and strongly suggested that *M. tuberculosis* avoids the lysosome in order to replicate. In fact, these authors noted that bacteria without surrounding host membranes were "never seen" (Armstrong and Hart 1971; Wolf et al. 2007). A number of subsequent immuno-EM studies corroborated these results showing that membranes were uniformly present surrounding the bacilli and that the *M. tuberculosis*-containing vacuole included markers of endosomes (Clemens and Horwitz 1995, 1996; Rohde et al. 2012; Xu et al. 1994). Microinjection of anti-TB antibodies into the cytosol of infected cells failed to bind to the bacterial surface, and although these results were negative, they further helped to instill the idea that *M. tuberculosis* does not access the cytosol (Armstrong and Hart 1975; Clemens et al. 2002). Microscopy studies using endosomal membrane markers showed that *M. tuberculosis* exists in the lumen of an early endosome-like compartment that is accessible to luminal flow and that *M. tuberculosis* actively prevents acquisition of lysosomal markers (Armstrong and Hart 1971, 1975; Clemens and Horwitz 1996; Crowle et al. 1991; Sturgill-Koszycki et al. 1994). The mechanism of phagosome maturation arrest has been extensively studied and involves bacterial manipulation of a number of host molecules, including the phosphoinositide composition of the cytosolic face of the phagosomal membrane (Vergne et al. 2005), sphingosine kinase (Malik et al. 2003), and Coronin-1 (Jayachandran et al. 2007). These results are consistent with the notion that *M. tuberculosis* actively inhibits the maturation of its phagosome along the endolysosomal pathway and resides in a compartment that has not matured into a phagolysosome. It is these data, as well as many other corroborating studies, that have established a dominant paradigm in the field that *M. tuberculosis* resides within the confines of the phagosome (Russell 2011). How *M. tuberculosis* bacilli would be able to control membrane trafficking from within the phagosome lumen, however, has remained mysterious.

More recent EM studies also corroborate the classic view. Rohde et al. (2012) carefully followed *M. tuberculosis*' intracellular localization and bacterial replication dynamics during long-term macrophage infections and found that most of the replicating bacteria were present inside membrane bound compartments of the cell. Likewise, Lee et al. used an inducible GFP reporter system to show that the *M. tuberculosis* cells that reside in "less mature" phagosomes are more metabolically active than bacteria in more acidified phagosomes. Both of these reports show that *M. tuberculosis* can replicate within membrane bound compartments of the cell and suggest that the subset of bacterial cells that are delivered to the lysosome are killed, consistent with the initial EM studies (Armstrong and Hart 1975). However, while these studies support the classical view, both groups noted that by 3–4 days post-infection, a significant proportion of bacteria (12–25 %) did not have "definitive" host membranes surrounding them and observed evidence of phagosomal "membrane damage." Whether this was due to actual dissolution of membranes, or simply a limitation of membrane visualization inherent to EM analysis, was unclear.

Finally, direct pH measurements of the local environment surrounding bacilli indicate the pH reaches approximately 6.0 in naïve macrophages (Rohde et al. 2007), suggesting that most of the bacteria is sealed inside a compartment that can support the generation of a proton gradient and thus separate from the cytosol. Likewise, the demonstration that bacterial resistance to acidification is required for intracellular survival also suggests that survival in a membrane-bound, acidified compartment is an important part of the *M. tuberculosis* lifecycle inside naïve macrophages (Vandal et al. 2008).

1.2 Evidence Supporting an Alternative View: Phagosomal Escape

A number of localization studies, both older and more recent, challenge the classic paradigm by providing evidence for phagosomal escape during *M. tuberculosis* infection (Houben et al. 2012; Leake et al. 1984; McDonough et al. 1993; Myrvik et al. 1984; van der Wel et al. 2007). In particular, these EM studies reported that a significant number of *M. tuberculosis* bacilli were observed *without* visible host membranes and, in most cases, this occurred only after several days of infection and not during the early stages after infection. The percentage of bacteria identified without membranes increased during the course of infection, with 12–40 % of bacilli localized to the cytosol depending on the *M. tuberculosis* strain (Houben et al. 2012). While the interpretation of these results that *M. tuberculosis* exists outside the confines of the phagosome is antithetical to the classic view, these findings are relatively consistent with the phagosomal "damage" observed by Hohde et al. and Lee et al. at later time points of infection. It should be noted that in all of these cases, the sole means of experimentation was EM, which can be subject to artifacts and the key issue of being able definitively determining the *absence* of a membrane can be limited by experimental staining of membranes. Regardless, these studies provide direct support for the idea that a sub-population of *M. tuberculosis* escape the phagosomal membrane during infection.

Importantly, the reports by van der Wel et al. (2007) and Houben et al. (2012) demonstrated that the ESX-1 specialized secretion system is required for phagosomal escape, indicating that this is a specific activity of *M. tuberculosis*. ESX-1 mutant bacteria localized exclusively inside membranous compartments, induced less host cell death, and were attenuated for growth. While these studies draw a correlation between phagosomal escape and virulence, the question of whether the localization of *M. tuberculosis* in the cytosol is important for growth and virulence is still not entirely clear as this secretion system likely mediates multiple interactions with host cells.

A number of recent reports provide increasing evidence supporting the notion that the ESX-1 pathway allows *M. tuberculosis* to gain "access" to the cytosol. Use of a beta-lactamase/fluorescence substrate reporter system demonstrated that the enzyme secreted by *M. tuberculosis* has access to cytosolic substrate even during the earliest stages of infection (Manzanillo et al. 2012; Siméone et al. 2012). Thus, lumenal contents were able to mix with the cytoplasm, a phenomenon that, like phagosomal escape, was dependent on ESX-1. Mycobacterial-derived products, including peptidoglycan fragments and DNA, interact with and are recognized by host cytosolic receptors, including Nod proteins and the IFI201 DNA receptor (Divangahi et al. 2008; Manzanillo et al. 2012; Pandey et al. 2009; Stanley et al. 2007). Likewise, Wong and Jacobs showed that phagosomal damage mediated by ESX-1 also led to activation of NLRP3, a cytosolic inflammasome component that activates caspase-1 (Wong and Jacobs 2011). Importantly, most of these studies showed that cytosolic access occurs extremely early, only hours after macrophage infection, at which time all bacilli are surrounded by phagosomal membranes. Thus, there is ample evidence that ESX-1 functions early during infection to permeabilize the membrane, which precedes overt escape of the bacterium from the phagosome.

Although the two sets of data supporting the classic and alternative views may seem diametrically opposed, there are a number of parallels between these studies suggesting that a synthesis and re-evaluation of the complexity of the intracellular fate of *M. tuberculosis* may reconcile the paradox. First, it is important to note that these different studies used different *M. tuberculosis* strains and phagocytic cells, and variation in the robustness of membrane permeabilization, perhaps due to differences in ESX-1 secretion, could give rise to disparities in the experimental results. Indeed, many studies using models of *M. tuberculosis*, including the BCG vaccine strain and *M. avium,* must be analyzed with caution as these strains naturally lack ESX-1 (Gey Van Pittius et al. 2001; Mahairas et al. 1996). Likewise, as mentioned above, even different strains of *M. tuberculosis* give rise to considerable differences in phagosome escape (Houben et al. 2012). Most importantly, it is clear that there are significant differences in the timing of events after infection. In any of these studies, there is ample evidence that the events occurring during the first few hours versus later stages of infection are different, and thus a temporal understanding of these events will be critical.

1.3 Role of ESX-1 in Phagosomal Membrane Permeabilization

Importantly, the phenomena of cytosolic access (early) and phagosomal escape (late) require the ESX-1 specialized secretion system, a key mediator of *M. tuberculosis* pathogenesis (Houben et al. 2012; van der Wel et al. 2007). ESX-1 modulates host-cell functions, presumably by translocating bacterial effectors into the host (Abdallah et al. 2007). *M. tuberculosis* mutants lacking ESX-1 are defective for replication within macrophages and are severely attenuated in animal models of infection, but the mechanism by which this system functions to promote infection remains unclear (Guinn et al. 2004; Hsu et al. 2003; Stanley et al. 2003). There is growing evidence that ESAT-6, a major substrate of the ESX-1 pathway, directly perturbs membranes (De Jonge et al. 2007; Hsu et al. 2003). Although in vitro, this activity is relatively weak compared to well-known pore forming toxins, it has been proposed that *M. tuberculosis* utilizes the ESX-1 system to deliver ESAT-6 to create conduits in phagosomal membranes, allowing limited perforation of the lipid bilayer early after infection. The mechanism of ESAT-6 pore forming activity, however, remains a critical area of investigation.

An important report by Stamm et al. (2003) demonstrated that *Mycobacterium marinum*, a poikilothermic pathogen of fish and amphibians and closely related to *M. tuberculosis*, patently escapes the phagosome early after infection and recruits host actin to move within the cytoplasm. Because this phenomenon has not been observed with *M. tuberculosis*, it is likely that motility represents a major difference between *M. marinum* and *M. tuberculosis*. Importantly, Smith et al. (2008) showed that ESX-1 is required for phagosomal escape, consistent with the evidence that ESAT-6 has membrane-lytic activity. We envision that pore formation is a common activity between these two pathogens, but it is *M. marinum*'s unique ability to recruit host actin that propels the bacteria away from permeabilized membranes, whereas *M. tuberculosis* remains relatively static. The ability of some pathogens to move within the cell is likely important for cell–cell spread (Smith et al. 2008) as well as autophagy evasion (see below) (Yoshikawa et al. 2009). Indeed, we envision that this capacity was lost in the evolution of *M. tuberculosis*, perhaps tampering virulence to its host and promoting persistence. The mechanism by which *M. marinum* recruits and utilizes actin remains unknown.

The notion that *M. tuberculosis* would permeabilize host membranes is not without precedence as other "vacuolar pathogens", such as *Salmonella enterica* serovar Typhimurium (*S.* Typhimurium), have also been observed to perturb host membranes and, at low frequency, found in the cytoplasm (Thurston et al. 2009). Because this membrane damage requires the type III secretion apparatus, it has been suggested that cytosolic access is due to membrane damage as a result of exuberant pore formation via this needle-like structure. In this way, *S.* Typhimurium must puncture membranes in order to manipulate the host cell by delivering host-interacting factors, but that perforation leads to phagosomal escape by some of the bacteria. In the same way, ESX-1-mediated phagosomal permeabilization may serve to deliver virulence factors into, or acquire nutrients from, the host cell cytosol. Perhaps, then, it is not that *M. tuberculosis* necessarily escapes phagosomal membranes to grow in the cytosol, but that it must simply gain access to the cytosol. We hypothesize that most (if not all) intracellular bacteria must, at some point, require interactions with the cytosol. Thus, the balance of pore formation mediated by *M. tuberculosis* and host repair mechanisms may ultimately dictate the abundance of membranes surrounding the bacterium (Welin and Lerm 2012).

1.4 Consequences of Membrane Permeabilization

A growing concept in the field is that macrophages innately discriminate pathogens from non-pathogens early after phagocytosis, and tailor their responses to match the level of the threat (Vance et al. 2009). One way this is achieved is through sensing membrane permeabilization, either directly or via the recognition of bacterial molecules in the cytosol (Thurston et al. 2012; Vance et al. 2009). Here, we consider the role of *M. tuberculosis* pore formation in the elicitation of two important host responses—induction of type I interferon and activation of autophagy.

1.4.1 Type I Interferon

The perforation of the phagosomal membrane by ESX-1 opens up a myriad of possibilities for interactions between *M. tuberculosis* and its host (Fig. 1). Although ESX-1 has a profound influence on the host–pathogen interaction during *M. tuberculosis* infection, a growing body of work indicates that ESX-1 pore formation contributes to *M. tuberculosis* virulence by modulating host innate immune responses of macrophages. For example, elicitation of type I interferon (IFN) by *M. tuberculosis* infection of both murine and human macrophages requires the ESX-1 secretion system (Novikov et al. 2011; Stanley et al. 2003). Although type II IFN (IFN-γ) is critical for activating host defenses, type I IFNs (IFN-α and IFN-β) can negatively regulate host resistance to *M. tuberculosis* in mouse models of infection (Manca et al. 2001). Bacterial pathogens that replicate

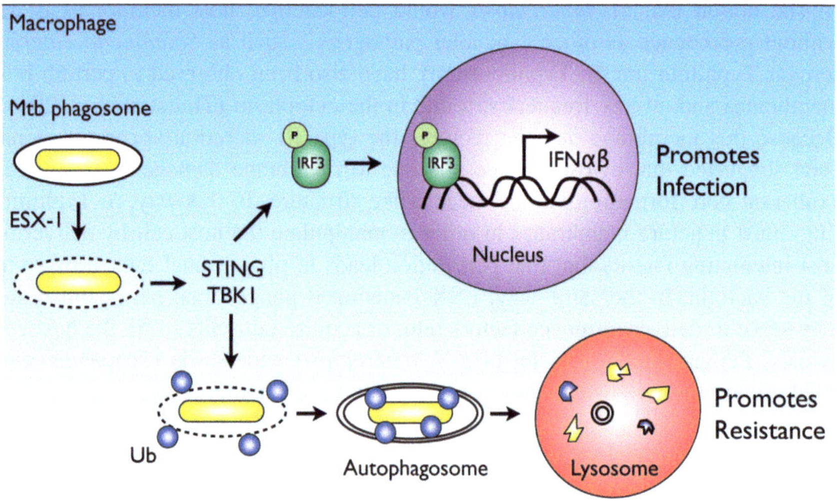

Fig. 1 ESX-1-mediated permeabilization of the phagosomal membrane leads to specific, pathogen-associated innate responses early after macrophage infection. After phagocytosis, permeabilization of the phagosomal membranes surrounding *M. tuberculosis* occurs, likely mediated by the ESX-1 substrate ESAT-6. Although the phagosome lumen and cytosolic compartments can mix, the level of membrane damage appears to be limited. Permeabilization leads to recognition of bacterial products, including *M. tuberculosis* DNA, by cytosolic receptors. Activation of the STING/TBK1/IRF3 signaling pathway leads to induction of a transcriptional response that includes Type I interferons, which appear to promote bacterial infection. Conversely, STING and TBK1 also activate targeting of a sub-population of the bacteria to the ubiquitin-mediated autophagy pathway. Via both cell-intrinsic killing and cell-extrinsic effects on the immune system, this pathway plays a critical role in host resistance to infection. Patent localization of *M. tuberculosis* in the cytosol at later time points may be the result of exuberant ESX-1 activity, tipping the balance between permeabilization and repair

in the cytoplasm of macrophages, such as *L. monocytogenes* and *Franciscella tularensis*, induce IFN-β transcription as part of a large transcriptional response controlled by the "cytosolic surveillance pathway" (CSP) (Henry and Monack 2007; McCaffrey et al. 2004). CSP activation by these species occurs early after infection specifically by cytoplasmic bacteria, whereas mutants unable to breach phagosomal membranes fail to induce transcription (Henry and Monack 2007; Leber et al. 2008). The CSP is controlled by the host transcription factor, interferon regulatory factor 3 (IRF3), which is activated via phosphorylation by the TBK1 kinase. It is thought that upon phagosomal membrane rupture, bacterial products are granted cytosolic access and recognized by cytosolic receptors that lead to IRF3 activation (Vance et al. 2009).

Importantly, *M. tuberculosis* elicits the same innate immune transcriptional response induced by cytoplasmic pathogens. *M. tuberculosis* induces IFN production during macrophage infection via activation of the CSP from within the phagosome. This response requires the IFI204 DNA receptor which stimulates the STING/TBK1/IRF3 signaling axis, the same host components as required for the interferon stimulatory DNA (ISD) pathway (Ishikawa et al. 2009b; Stetson et al. 2008). Extracellular mycobacterial DNA is the critical ligand for CSP activation and the ESX-1 system allows extracellular mycobacterial DNA access to cytoplasmic DNA receptors via limited permeabilization of the phagosomal membrane (Manzanillo et al. 2012). Surprisingly, CSP activation is required for *M. tuberculosis* pathogenesis as *Irf3*$^{-/-}$ mice are profoundly resistant to infection. Importantly, artificially breaking down the membrane via the secretion of LLO restores Nod and IRF3 signaling (Manzanillo et al. 2012; Pandey et al. 2009).

Because virulence-associated protein secretion systems activate cytosolic responses in other pathogens (Monroe et al. 2010), we envision that the primary role of ESAT-6-mediated membrane damage is to provide access of secreted virulence factors to enter the host cytosol, and extracellular DNA exposure to cytosolic receptors is an indirect consequence of membrane perforation. Moreover, sensing of *M. tuberculosis* DNA likely takes place during human tuberculosis, as ESX-1 mediated activation of the CSP is operative in human macrophages and is likely responsible for the robust type I IFN signature associated specifically with active disease (Berry et al. 2010; Novikov et al. 2011).

Although Type I IFN is absolutely critical for resistance to viruses, there is growing literature about the role of IFN-α/β in bacterial infections (Barber 2011; Monroe et al. 2010). Importantly, recent work has shown that type I IFN suppresses type II IFN responses in human macrophages during infection (Teles et al. 2013), as well as inhibits IL-1β, a cytokine that promotes *M. tuberculosis* clearance (Novikov et al. 2011). Thus, it is likely that differential cytokine responses mediated by IFN-α/β contribute to the phenotype of the *Irf3*$^{-/-}$ mouse. It is important to note that while IFN-α/β likely plays a role in this phenotype, the *Irf3*$^{-/-}$ mouse is much more resistant to *M. tuberculosis* infection than the *Ifnar*$^{-/-}$ mouse that is only deficient for type I IFN signaling (Stanley et al. 2007). These results indicate that other immune modulators regulated by ESX-1 activation of the CSP contribute to pro-*M. tuberculosis* inflammatory immune responses.

Since bacterial infection correlates with type I IFN production, it is suggestive that one fitness advantage for pore formation is that it represents a strategy by which *M. tuberculosis* promotes infection by triggering cytosolic responses (Novikov et al. 2011). The phenotype of the *Irf3*$^{-/-}$ mouse is certainly consistent with this notion. However, if cytosolic access by *M. tuberculosis* provides a fitness advantage in vivo, an alternative view is that exposure to the cytosol is an inevitable consequence of membrane perforation that allows virulence proteins a passageway into the cytosol. In this way, while CSP activation initiates profound inflammatory responses, *M. tuberculosis* may have evolved to require these changes in host tissues to activate virulence mechanisms or produce an environment conducive to growth. Consistent with this idea, recent studies have shown that *S.* Typhimurium promotes inflammatory immune responses which, in turn, enhances persistent infection (Arpaia et al. 2011; Winter et al. 2010). Thus, *M. tuberculosis* may have evolved a dependency that requires the effects of robust innate immune signaling triggered by ESX-1-mediated cytosolic access. In either case, it is likely that excessive host immunopathology triggered by ESX-1 activation of the CSP contributes to the susceptibility of wild-type mice to M. tuberculosis (Philips and Ernst 2012).

1.4.2 Autophagy

Cytosolic access mediated by *M. tuberculosis* ESX-1 also triggers autophagy, an evolutionarily conserved process in eukaryotes whereby cytoplasmic components are enveloped and sequestered by double-membrane structures, termed phagophores that subsequently fuse to lysosomes for degradation (Deretic and Levine 2009). The functions of autophagy are multifaceted and have profound effects on eukaryotic cell biology. In response to nutrient limiting conditions, general autophagy serves a catabolic function by mediating non-selective consumption of organelles and other cellular components to generate substrates for energy metabolism and protein synthesis (Deretic and Levine 2009). In contrast, selective autophagy functions to specifically renovate the cell by targeting cytosolic protein aggregates and specific organelles for removal through the use of ubiquitin-mediated targeting. In this way, "cargo" destined to be degraded in the lysosome is first ubiquitinated and subsequently recognized by ubiquitin-autophagy adaptors. Recognition leads to sequestering them into autophagosomes, delivering them to the lysosome. Importantly, cytoplasmic pathogens can also be ubiquitinated and targeted to autophagy, which plays an important role in innate defense against invading intracellular pathogens (Zhao et al. 2008). The prevailing view is that autophagy functions to eliminate intracellular microbes that enter into the cytosol. Indeed, the work of the Deretic's group was the first to show that autophagy is a cell-intrinsic pathway by which macrophages can destroy bacteria (Gutierrez et al. 2004). Subsequent studies using *S.* Typhimurium and *L. monocytogenes* infection of cultured epithelial cells have shown that bacteria which exit the endosomal pathway and enter into the cytosol are ubiquitinated and delivered to autophagosomes via recognition by the cytosolic autophagy receptors p62 and NDP52 (Thurston et al. 2009; Zheng et al. 2009).

Much of the early work on the role of autophagy in mycobacterial clearance was performed using BCG (Gutierrez et al. 2004; Singh et al. 2006; Zhao et al. 2008). Curiously, BCG does not undergo selective autophagy and recruitment of LC3 to the phagosomal membrane unless autophagy is experimentally induced. Although inducing autophagy exogenously via starvation, treatment with rapamycin, interferon gamma (IFN-γ), vitamin D3, or genetic depletion of autophagy inhibitors can lead to decreased bacterial replication in macrophages (Gutierrez et al. 2004; Kumar et al. 2010; Singh et al. 2006; Yuk et al. 2009), recent work shows that wild-type *M. tuberculosis* engages the selective autophagy pathway early after phagosome biogenesis (Watson et al. 2012). Wild-type *M. tuberculosis* cells elicit ubiquitin-mediated autophagy targeting in resting macrophages, resulting in the delivery of bacilli to lysosomes (Watson et al. 2012). As with *M. tuberculosis*, ESX-1 secretion is also required for localization of ubiquitin to *M. marinum* during macrophage infection (Collins et al. 2009; Lerena and Colombo 2011).

There are at least four critical steps in the pathway by which *M. tuberculosis* cells are recognized by host macrophages and targeted to autophagosomes (Fig. 1). First, the bacterial ESX-1 secretion system initiates the interaction by permeabilizing the phagosomal membrane early after phagocytosis. Second, permeabilization exposes the surface of the bacteria and/or the phagosomal membrane that is recognized by components of the cytosolic DNA pathway, including STING, to initiate autophagy targeting (Watson et al. 2012). Third, a population of the engulfed bacteria become associated with, and surrounded by, host ubiquitin chains. Most of the labeled bacteria are associated with K63-linked chains but some K48 linkages are also detectable on a lower percentage of bacteria. Fourth, the ubiquitin LC3 binding autophagy adaptors, p62 and NDP52, are required to recruit autophagy components to create a phagophore surrounding the bacilli, a process that also requires the TBK1 kinase and ATG5 (Pilli et al. 2012; Watson et al. 2012). Once targeted to the ubiquitin-mediated autophagy pathway, bacilli-containing autophagosomes are matured via fusion with lysosomes to create autophagolysosomes. Given the timing of LC3-recruitment and the development of the subpopulation of killed bacteria described by Rohde et al. (Rohde et al. 2012) and Armstrong and Hart (Armstrong and Hart 1971), it is likely that autophagy is responsible for this phenomenon. Although delivery of this population of bacteria to the lysosome is responsible for limited bacterial killing by macrophages ex vivo, the entire autophagy pathway in macrophages is a major determinant of host resistance to *M. tuberculosis* infection in vivo as mice lacking *Atg5* in monocytes are extremely susceptible to infection (Watson et al. 2012).

1.5 Role of Autophagy During M. tuberculosis *Infection*

Given the modest effect of ATG5 on *M. tuberculosis* survival in macrophages, it may be surprising that the *Atg5*$^-$ mice are so profoundly susceptible to infection. Indeed, *Atg5*$^-$ mice succumb to infection with kinetics nearly identical to that of mice that completely lack an adaptive immune system or are missing key activators of macrophage activation, such as IFN-γ (MacMicking et al. 2003). However, it is becoming clear that autophagy plays a broader role in innate and adaptive immune responses than simply leading to direct killing of microbes in the lysosome. For example, autophagy enhances antigen presentation in dendritic cells (Jagannath et al. 2009) and negatively regulates inflammasome activation (Nakahira et al. 2011; Saitoh et al. 2008; Zhao et al. 2008). Indeed, the pronounced increase in inflammatory cytokines during infection of *Atg5*$^-$ mice, such as IL-1β, suggests that inflammasome signaling is augmented in *Atg5*$^-$ mice during *M. tuberculosis* infection (Shi et al. 2012). Therefore, while delivery of bacteria to the lysosome plays a direct role in acute bacterial restriction, autophagy may play a more pronounced role in overall control via non-cell autonomous effects on innate and specific immune responses.

The findings that DNA is a bacterial-derived molecule recognized by host cells to target *M. tuberculosis* to autophagy, and that STING and TBK1 are required for this process, reveals a surprisingly broad connection between cytosolic DNA detection and innate immune responses to pathogens. Indeed, introduction of exogenous DNA into the cytosol of cells triggers three distinct yet inter-related responses: autophagy, cytokine signaling, and inflammasome activation (Hornung et al. 2009; Roberts et al. 2009). STING is a required factor for both autophagy (Saitoh et al. 2009; Watson et al. 2012) and cytokine signaling (Ishikawa et al. 2009b), and functions by sensing cyclic di-nucleotides either directly from bacteria or endogenously catalyzed in response to the DNA-activated cyclic di-nucleotide synthase, cGAS (Sun et al. 2013). Upon recognition of DNA, STING activation leads to ubiquitination and recruitment of both TBK1 and NDP52. It will be interesting to determine if other ubiquitin binding adaptors implicated in autophagy, including NBR1 and optineurin, are also important for this pathway (Kirkin et al. 2009; Wild et al. 2011). Although the molecular connections between DNA and ubiquitination remain unknown, STING itself is a plausible ubiquitinated target as it colocalizes to bacteria and is ubiquitinated during DNA stimulation (Tsuchida et al. 2010). It is likely that cytoplasmic DNA detection also targets other pathogens that access the cytosol to the autophagy pathway. Although it remains to be seen if autophagy targeting of other bacterial pathogens depends on DNA sensing and STING, it is becoming clear that this same pathway is also operational to limit dsDNA virus infection which expose their genome during virion assembly in the cytoplasm (McFarlane et al. 2011; Rasmussen et al. 2011).

Taken together, ESX-1 and ESAT-6 are active early during infection, allowing pores to be formed that provide a conduit between the phagosomal lumen and cytosol, but at this time point the level of permeabilization does not allow escape from the phagosome. During this early time, within the first few hours post-infection, these pores allow for cytosolic sensing that initiates transcriptional, inflammatory, and autophagic responses. We propose that the observed phagosomal escape is due to prolonged permeabilization, which can eventually lead to membrane dissolution. Moreover, it is becoming clear that a common requirement of intracellular pathogens is the ability to penetrate host cell membranes in order to gain access to the cytosol. The mechanisms described here illustrate that it is the combination of pathogen molecule and spatial distribution of pattern recognition receptors that allow host cells to detect membrane puncture and discriminate pathogens from non-pathogens. It is thus a combination of ESX-1 secretion and exposure of bacterial molecules to the cytosol that constitutes the "pattern of pathogenesis" recognized by host cells to mount innate responses against *M. tuberculosis* (Vance et al. 2009).

1.6 Heterogeneity of Bacterial Cell Fate During Macrophage Infection

It is curious that only one-third of intracellular *M. tuberculosis* are targeted by the ubiquitin pathway and colocalize with LC3. This observation is consistent, however, with many of the EM studies mentioned in Sect. 1.1 regarding

M. tuberculosis trafficking in macrophages which showed that although live *M. tuberculosis* bacilli were able to profoundly inhibit phagosome maturation to lysosomes; approximately a third of bacilli still trafficked to lysosomes (Armstrong and Hart 1971; Houben et al. 2012; Rohde et al. 2012). It is likely that the fate of the lysosomal bacteria in these studies was mediated by autophagic targeting. Likewise, the results of Rohde et al. are consistent with this notion, in which it was found that during the very early events after infection only approximately one-third of the bacteria are targeted to lysosomal compartments, while the other bacilli are not. This work also indicated that this phenomenon is absent at later time points after infection, which is consistent with a reduction in autophagic targeting after an initial burst of activity (Rohde et al. 2012; Watson et al. 2012). Moreover, the vast weight of the evidence supports the idea that while most *M. tuberculosis* reside in early phagosomal compartments, *M. tuberculosis* phagosomes within a host cell exhibit significant variation in their maturation status within the endosomal–lysosomal pathway (Lee et al. 2008). Although we do not yet understand the mechanistic basis for the heterogeneous response, one possibility is that *M. tuberculosis* may employ an autophagy evasion strategy that prevents ubiquitination but is only partially effective. Similarly, though perhaps more appealing, is the possibility that phenotypic variation within the bacterial population drives the heterogenous response. For example, stochastic fluctuations in ESX-1 activity may give rise to cell-to-cell variations in flux of ESAT-6 and thus create variations in membrane permeability from phagosome-to-phagosome (Ohol et al. 2010). In either case, heterogeneity of autophagic targeting may be important for successful infection by *M. tuberculosis*. For example, although bacilli that become targeted to lysosomes may be killed, the resultant increase in pro-inflammatory responses and elicitation of adaptive immune responses may promote and favor the persistence strategy of the rest of the bacterial population. Alternatively, it is possible that the heterogeneity is due to host processes, for example, there may be an exclusive interplay between bacterial targeting and membrane repair. In either case, the ability of *M. tuberculosis* to perforate the membrane without its dissolution is likely under a delicate balance of factors, and fine control of ESX-1 secretion may serve to limit membrane perforation and cytosolic signaling (Ohol et al. 2010).

The other source of heterogeneity is temporal. The state of the infection changes during later stages of infection, leading to morphological and transcriptional changes during these later stages. Although it is not exactly known whether bacilli that reach the more acidified compartments are viable, the evidence strongly suggests that most of the bacterial growth occurs in the non-acidic phagosomal compartment. Using an inducible GFP reporter system, Lee et al. (2008) provide evidence that the *M. tuberculosis* cells that reside in less mature phagosomes are more metabolically active than bacteria in more acidified compartments. Likewise, Rohde et al. (2012) directly showed that most of the bacterial killing occurred in the first day after infection at which time bacterial colocalize with lysosomes. This correlated with bacterial gene expression analysis, suggesting that the intracellular environment changes during the course of macrophage infection. Taken together, this data indicate that during the first 2 days of infection, naïve macrophages are relatively adept at killing at least some bacilli, an activity that likely requires autophagic targeting. Importantly, the authors go on to show that after 2 days of infection, this "bottleneck" that constricts *M. tuberculosis* dissipates, leading to increased viability. An effect that was also observed in vivo (Gill et al. 2009).

We propose a model in which ESX-1 plays an important and multifaceted role during initial infection of macrophages (Fig. 1). We posit that cytosolic access is a fundamental requirement of *M. tuberculosis* (and other intracellular bacterial pathogens), and ESAT-6 serves to perforate the phagosomal membrane early after infection. During this early time inside the phagocyte, bacterial-derived molecular cues are recognized in the cytosol, initiating both the IRF3-dependent transcriptional response and autophagy targeting. We suspect that a subpopulation of bacteria, perhaps due to cell-to-cell fluctuations of ESX-1 secretion, are subject to autophagy targeting and are killed during the early stages of infection. As infection proceeds, cues from the intracellular environment are received and processed, allowing the bacterium to adapt and promote cell survival, perhaps tampering pore formation. Although the causal relationships have not been established, it is clear that there is a great deal of dynamic changes occurring very early after *M. tuberculosis* infection of macrophages.

2 Lipid Virulence Factors of *M. tuberculosis*

The cell wall of *M. tuberculosis* contains numerous complex lipids. These lipids have a dual role as both important structural components and as virulence factors that mediate intimate interactions with host cells. *M. tuberculosis* cell wall lipids confer a unique chemical composition and architecture that distinguishes this bacterium from typical gram-negative and gram-positive bacteria. Surrounding a standard inner membrane is a peptidoglycan–arabinogalactan polymer that is linked to an outer membrane-like structure known as the mycomembrane (Hoffmann et al. 2008). The inner layer of the mycomembrane consists of very long chain branched beta-hydroxy acids known as mycolic acids that are unique to mycobacteria and related actinobacteria and are found covalently linked to arabinogalactan. The outer mycomembrane is comprised of a large variety of non-covalently attached lipids and glycolipids, including additional mycolic acids in the form of the glycolipid trehalose-6,6′-dimycolate (TDM), and a family of structurally related phthiocerol dimycoserosates (DIMs) (Fig. 2). Finally, the surface of *M. tuberculosis* is coated with a capsular layer that contains additional extractable glycans, lipids, and proteins (Sani et al. 2010). Thus, the outer surface is extremely hydrophobic and a reservoir for a myriad of bacterial products that are poised for delivery to host cells.

From the earliest investigations into the virulence of *M. tuberculosis*, it was recognized that the unique cell envelope of this organism confers the fundamental

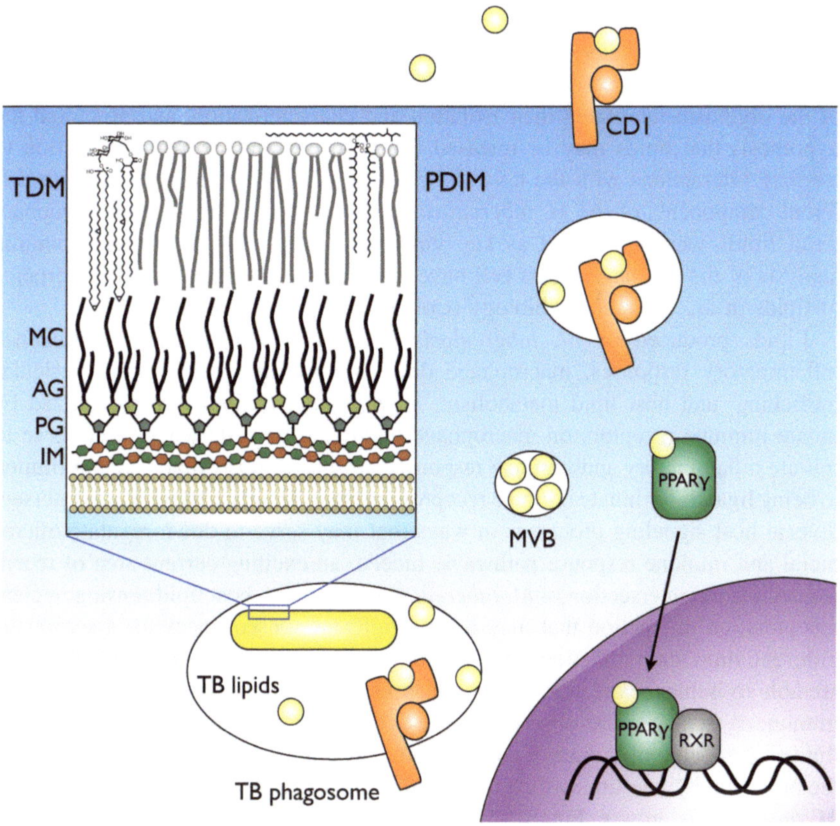

Fig. 2 Lipids from the *M. tuberculosis* cell wall and interactions with host signaling and immune pathways. Non-covalently attached lipids found in the outer leaflet of the *M. tuberculosis* cell wall, including trehalose dimycolate (TDM) and phthiocerol dimycoserosate (PDIM), are poised for release inside of host cells. Lipids released from *M. tuberculosis* traffic outside the phagosome and accumulate in host membranes, including multivesicular bodies (MVBs). *M. tuberculosis* lipid virulence factors may have evolved to mimic host lipids and thereby directly influence innate immune responses of macrophages via interactions with specific signaling pathways. For example, activation of the lipid sensing nuclear receptor PPAR-γ results in a gene expression pathway that is anti-inflammatory and promotes *M. tuberculosis* growth in macrophages. In addition, *M. tuberculosis* lipids released in the phagosome are loaded into CD1 molecules for presentation to CD1 reactive T cells that participate in the adaptive response to infection. Finally, *M. tuberculosis* lipids are exocytosed from infected macrophages where they can exert influence beyond the infected cell. Uptake by neighboring dendritic cells would promote the priming of CD1 reactive cells and may influence the characteristics of the adaptive immune response. Uptake by uninfected macrophages would serve to down-regulate effective innate immune responses in the environment of the granuloma

basis of its ability to modulate host immunity. The effectiveness of *M. tuberculosis* as an immunological adjuvant was found to result from properties of components of the cell envelope, and this observation motivated the earliest investigations into the chemical composition of this structure. Early in the twentieth century it was recognized that *M. tuberculosis* produces an astounding diversity of complex lipids. The fact that many of these lipids are non-covalently attached to the surface of the organism facilitated their isolation and characterization, and promoted the hypothesis that lipids may be released from the cell surface during infection to mediate interactions with the host. Association of specific bioactivities with different components of the *M. tuberculosis* cell wall soon followed, and mycobacterial lipids were confirmed as key virulence factors. Lipidomic and genomic analysis of the *M. tuberculosis* cell have affirmed the complexity and importance of lipids in *M. tuberculosis* biology (Cole et al. 1998; Layre et al. 2011).

Lipids produced by *M. tuberculosis* have been demonstrated to influence inflammatory responses, macrophage differentiation and polarization, vesicular trafficking, and host lipid metabolism. *M. tuberculosis* lipids are recognized by innate immune receptors on macrophage and dendritic cells where they serve to initiate inflammatory and immune responses. *M. tuberculosis* lipids are not limited to being ligands for innate immune receptors as many *M. tuberculosis* lipids intersect diverse host signaling processes in ways that may serve to down-regulate microbicial and immune response pathways. Indeed, an exciting current area of recent research is the intersection of *M. tuberculosis* lipids with host lipid sensing nuclear receptors, an interaction that may have significant consequences for macrophage differentiation upon infection. By regulating the innate response to infection, lipids are able to influence the adaptive immune response thereby exerting effects on the immune system that are observable at a systemic level. Purified *M. tuberculosis* lipids are capable of initiating granulomatous responses, and regulate virulence of the organism by fine-tuning the inflammatory response to a pitch that is optimal for *M. tuberculosis* growth. Finally, *M. tuberculosis* lipids directly influence adaptive immunity through activation of CD1 responsive T-cells and NK T-cells. Clearly, lipids are key mediators of the host–pathogen interaction, with complex effects on host cells and tissues. A full understanding of this complex biology as it relates to human infection is an exciting area of current and future research.

2.1 Phthiocerol Dimycoserosate

Like many of the complex lipids found in the *M. tuberculosis* cell wall, lipids of the phthiocerol family, including a family of structurally related phthiocerol dimycoserosates are found non-covalently attached to the cell surface where they are important for cell wall integrity and intrinsic defense of *M. tuberculosis* against environmental insults. In addition, phthiocerol dimycoserosates of *M. tuberculosis* have long been associated with virulence. Decades ago, spontaneous *M. tuberculosis* mutants were isolated that were found to be attenuated in the guinea pig model of infection, and were soon recognized as deficient for production of phthiocerol dimycoserosate (PDIM) (Goren et al. 1974). The very first genetic screens that sought to identify virulence factors in *M. tuberculosis* provided the first rigorous genetic proof that lipids in general, and PDIM in particular, play a role in *M. tuberculosis* pathogenesis (Camacho et al. 1999; Cox et al. 1999). The first published genetic screen conducted in *M. tuberculosis*, a signature tagged mutagenesis screen for mutants with defective growth in the lungs of mice, identified genes required for the biosynthesis of phthiocerol dimycoserosate (PDIM). Mutants unable to synthesize or secrete PDIM to the cell surface were found to be defective for replication in lung tissue during the acute phase of infection (Cox et al. 1999), establishing PDIM as one of the first virulence factors described in *M. tuberculosis*. In addition, mutants lacking PDIM have an altered colony morphology suggesting that PDIM may play a structural role in the mycobacterial cell wall (Cox et al. 1999). Despite the considerable evidence that PDIM is required for full virulence of *M. tuberculosis*, the mechanism by which PDIM contributes to pathogenesis is unclear. PDIM mutants arrest phagosome maturation to the same degree as wild-type *M. tuberculosis*, and are not defective for growth in inactivated macrophages (MacGurn and Cox 2007; Rousseau et al. 2004), suggesting that the interaction of PDIM with the host may occur at the level of a complex interaction with the immune system. PDIM deficient mutants are defective for growth in interferon-γ (IFN-γ) activated macrophages, pointing to a role for PDIM in resisting the effects of IFN-γ. However, PDIM mutants do not show increased susceptibility to reactive nitrogen intermediates in vitro (Camacho et al. 1999), and are highly attenuated in iNOS deficient mice (Kirksey et al. 2011; Murry et al. 2009). Taken together, the data suggest that PDIM is required for resistance to an IFN-γ dependent mechanism that is iNOS independent. Whether PDIM contributes to resistance through its role in cell wall integrity, or through a more specific host–pathogen interaction is an area of future investigation.

2.2 Mycolic Acids in Inflammatory Responses

Although much attention has been focused on the ability of *M. tuberculosis* to resist or inhibit immune responses, it is clear that inducing an inflammatory response can be beneficial to the bacterium in some contexts. The inflammatory response to infection leads to the formation of granulomas, immune structures that are the hallmark of *M. tuberculosis* infection. The role of the granuloma in the pathogenesis of tuberculosis is complex. Although a robust granulomatous response has been associated with local control of bacterial replication and resolution of disease (Capuano et al. 2003; Lin et al. 2009), granulomas have also been demonstrated to promote bacterial replication and dissemination in *M. marinum* infected fish (Bold and Ernst 2009; Lesley and Ramakrishnan 2008). In addition, necrotic granulomatous lesions with liquid caseated centers are responsible for transmission of disease in the human population. Intriguingly, a single lipid found on the cell surface of virulent mycobacteria, Trehalose-6,6′-dimycolate (TDM), is capable of eliciting a granulomatous response that recapitulates the hallmarks of the granulomatous response induced by infection with the fully virulent organism.

TDM is a glycolipid consisting of two mycolic acids linked by a trehalose sugar, and is the most abundantly produced lipid on the surface of virulent strains of *M. tuberculosis*. TDM has a long and rich history, as it was one of the first lipids associated with the virulence of *M. tuberculosis.* In 1884, Robert Koch observed that *M. tuberculosis* forms microscopically visible serpentine cords in broth culture. Decades later, Middlebrook noted that this cording morphology is observed only in virulent strains of *M. tuberculosis*, initiating the search for a "cord factor" that might confer virulence on the organism (Middlebrook et al. 1947). Subsequent investigations resulted in the isolation of "cord factor", later identified as TDM (Bloch 1950; Noll et al. 1956). Although more recent research has determined that TDM is not required for mycobacterial cording, the contribution of TDM to virulence of *M. tuberculosis* is an area of ongoing investigation. Numerous studies have demonstrated that purified TDM is highly bioactive, with potent effects on mammalian cells (reviewed in Daffe and Draper 1997; Hunter et al. 2006), including a proposed role in the modulation of phagosome maturation (Axelrod et al. 2008; Indrigo et al. 2003), the ability to elicit production of the inflammatory cytokines such as IL-1 and TNF-α (Geisel et al. 2005), and the capacity to stimulate lung granuloma formation (Bekierkunst et al. 1969). Injection of purified beads coated with mycobacterial lipids (including TDM) into mice induces a potent local granulomatous response characterized by the recruitment of immune cells, neovascularization, and the differentiation of macrophages into epithelioid cells and lipid droplet laden foamy macrophages (Kim et al. 2010).

In recent years, the ultrastructure of mycolic acids has become an area of considerable interest, as specific modifications found on mycolic acids have been found to profoundly affect the bioactivity of these molecules. Three distinct subclasses of mycolic acids are found on virulent *M. tuberculosis*, the α-, methoxy-, and keto-mycolic acid forms. These forms of mycolic acids are classified based on specific modifications of the meromycolate chain; α-mycolic acids have two cyclopropane rings, methoxymycolates have a single cyclopropane ring and a methoxy group, and ketomycolates have a single cyclopropane ring and a ketone (Takayama et al. 2005). These modifications are instilled by a specific family of S-adenosyl methionine dependent methyltransferases (George et al. 1995; Yuan and Barry 1996; Yuan et al. 1995). In an elegant series of experiments, genetic manipulation of these mycolic acid-modifying enzymes in *M. tuberculosis* has defined the role that the modifications play in virulence. Inactivation of individual mycolic acid modifying enzymes results in mutants with defined and specific changes in pattern of modifications found on the mycolic acids. Surprisingly, TDM obtained from these *M. tuberculosis* mutants is either hypo-inflammatory or hyper-inflammatory, depending upon the specific modifications of the mycolic acid fine structure resultant in each mutant (Dao et al. 2008; Glickman et al. 2000; Rao et al. 2005, 2006). The in vivo phenotypes of these mutants correlate with the inflammatory properties of the TDM; mutants with hypo-inflammatory TDM are hypervirulent, whereas mutants with hyper-inflammatory TDM are attenuated (Rao et al. 2005, 2006). Likewise, the cyclopropanation modifications have been clearly established to suppress immune responses (Barkan et al. 2012). Taken together, there is strong evidence that in addition to their important structural role in the cell wall, mycolic acids are also crucial virulence determinants that specifically influence the nature of the immune response to *M. tuberculosis* infection.

2.3 M. tuberculosis *Lipids and Interactions with the Host*

Despite abundant evidence that lipids are important virulence factors for *M. tuberculosis*, relatively little is known about how *M. tuberculosis* lipids interact with the host at the molecular level. This is in part because of the difficulty of studying lipids relative to proteins, which are biochemically more tractable and for which more tools are available. Identification of lipid interaction partners is difficult, and although a few receptors for *M. tuberculosis* lipids have been identified (Ishikawa et al. 2009a; Underhill et al. 1999), very little is known about how *M. tuberculosis* lipids specifically interact with the host to achieve effector function. Indeed, although it is possible that lipids exert their effects by interacting with host proteins, it is also possible that interaction with host membranes is a key to understand the role that *M. tuberculosis* lipids play in pathogenesis.

In contrast to protein virulence factors that require active transport and secretion mechanisms to traverse host membranes, the hydrophobic nature of *M. tuberculosis* lipids means that these molecules can readily insert into virtually any biological membrane, including those of the host. Accordingly, *M. tuberculosis* lipids have been observed intercalated into numerous host membranes, including membranes of the endosomal–lysosomal network, and mitochondria. TDM inserts into model phospholipid monolayers where it decreases membrane fluidity and increases passive permeability (Rhoades and Ullrich 2000). Similarly, biochemical studies have suggested that the biological effects of PDIM on host cells may be a consequence of influencing host cell membrane structure (Astarie-Dequeker et al. 2009). Purified PDIM inserts into model membranes, increasing the rigidity of the bilayer, and the ability of PDIM to alter host membranes has been observed in macrophage infections. *Mycobacterium bovis* BCG was found to interact with human macrophages in a manner that led to a change in global membrane fluidity of the host cell, and this was found to be dependent on PDIM. The ability of PDIM to alter host membrane fluidity may influence phagocytosis and subsequent trafficking events in a manner that promotes bacterial survival in the macrophages. Given the evidence for pore formation in phagosomal membranes mediated by *M. tuberculosis*, it is an intriguing possibility that PDIM could participate in this process.

Studies have suggested that *M. tuberculosis* lipids are abundantly produced during macrophage infection, and that these lipids are actively trafficked out of the mycobacterial phagosome (Beatty et al. 2000, 2001; Beatty and Russell 2000; van den Elzen et al. 2005). These lipids include many known to be associated with the outer cell envelope, including TDM and DIMs. Once released from the phagosome, *M. tuberculosis* lipids are observed intercalated into the membranes of host organelles and endocytic compartments including MHC class II rich multivesicular bodies. Ultimately, *M. tuberculosis* lipids are exocytosed from infected macrophages, where they traffic into neighboring uninfected macrophages. As much of the bioactivity of *M. tuberculosis* lipid virulence factors has been ascribed to the ability to modulate inflammatory responses and macrophage differentiation, spread of *M. tuberculosis* lipids from infected cells to uninfected "by-stander" cells is a potential mechanism for expanding the influence of *M. tuberculosis* virulence factors beyond individual infected macrophages (Russell 2011; Russell et al. 2009) (Fig. 2). In addition, the exocytosis of *M. tuberculosis* lipids from infected macrophages may be important for acquisition of lipid antigens by dendritic cells which express class I CD1 molecules and can therefore present these lipid antigens to CD1 reactive T-cells (van den Elzen et al. 2005).

2.4 Complex Crosstalk Between Host and Pathogen Lipid Metabolism

Intriguingly, *M. tuberculosis* lipids may have a major role in influencing host-lipid metabolism and in the development of host immune cells. Foam cells are a type of differentiated macrophage that is filled with host-derived lipids stored in organelles known as lipid bodies. These cells are thought to be relevant to the pathophysiology of human disease as they are observed in granulomas isolated from human patients, most often in association with granulomatous lesions that are necrotic with caseous centers (Peyron et al. 2008). The role that foamy macrophages play in the pathogenesis of tuberculosis has become an area of interesting research. Infection of macrophages with *M. tuberculosis* in vitro has been shown to lead to differentiation into foamy macrophages. This has been attributed to mycolic acids found in the cell wall of *M. tuberculosis*, and purified mycolic acids, including TDM, isolated from virulent mycobacteria can also induce foamy macrophage differentiation in vitro (Korf et al. 2005; Peyron et al. 2008; Puissegur et al. 2004). This effect is likely dictated by the bacteria as the ability to drive foamy macrophage differentiation seems to be specific to oxygenated keto-and hydroxyl mycolic acids (Peyron et al. 2008).

In EM studies of *M. tuberculosis* infected foamy macrophages, lipid bodies formed upon infection quickly associated with and fused with phagosomes, presumably delivering a rich source of lipids to the bacteria. It is well documented that host lipids are an important nutrient source for *M. tuberculosis* in vivo, leading to the hypothesis that lipid bodies found in foamy macrophages are important nutrient sources for persistent infection (Kim et al. 2010; Peyron et al. 2008). The biology of lipid bodies and the role that they may play in infection is complex, as these organelles are also sites for the production of eicosanoids, lipid mediators of inflammation that are important for determining the outcome of *M. tuberculosis* infection (D'Avila et al. 2006). In addition, the association of foamy macrophages with necrotic lesions, and the similarity of lipids found in lipid bodies with those comprising the caseum of necrotic lesions, has led to the hypothesis that foamy macrophages play a key role in the formation of caseous lesions, and therefore the tissue damage and destruction are required for transmission of the disease (Russell et al. 2009). It is interesting to speculate one way that mycolic acids on the surface of *M. tuberculosis* contribute to virulence is by driving the differentiation of foamy macrophages, thereby creating both a nutrient reservoir for bacterial growth and promoting immune responses that favor transmission of the disease.

Mammalian lipids are important regulators of inflammatory signaling pathways, suggesting that some of the complex lipids of *M. tuberculosis* may have evolved to mimic host lipids to activate specific host signaling pathways. The recent finding that *M. tuberculosis* lipids may activate signaling via host lipid sensing nuclear receptors in the macrophage provide insight into the mechanisms by which *M. tuberculosis* lipids may influence host lipid metabolism, macrophage differentiation, and inflammatory responses. Peroxisome proliferator-activated receptors (PPARs) are type II nuclear receptors with well-established roles in cellular differentiation, lipid metabolism, and inflammation (Mahajan et al. 2012; Varga et al. 2011). Of particular relevance to immunity is the ability of PPARs to respond to fatty acid derived eicosanoids. Eicosanoids are potent host signaling molecules that have been proposed to regulate may act as a key aspect of *M. tuberculosis* infection, whether infected macrophages die by apoptosis or necrosis (Behar et al. 2011; Divangahi et al. 2010; Lu et al. 2005; Mahajan et al. 2012; Rajaram et al. 2010), and the balance of inflammation that is critical for determining the outcome of infection at a systemic level (Beckman et al. 1994; Tobin et al. 2010). Responses mediated by activated PPAR-γ include a role as a negative regulator of macrophage activation that inhibits the expression of inflammatory genes, enhances IL-10 expression, and inhibits the macrophage respiratory burst. It is possible that mycobacterial lipid virulence factors have evolved to take advantage of PPAR-γ signaling to alter responses to infection and promote an environment that is conducive to bacterial growth and persistence. In this way, it is possible that some *M. tuberculosis* lipids may mimic natural ligands. Mycobacterial lipids are potent transactivators of PPAR-γ mediated transcriptional responses (Mahajan et al. 2012; Sieling et al. 1995), and knockdown of PPAR-γ during *M. tuberculosis* infection of macrophages results in impaired growth of the bacterium that is correlated with increased levels of IL-6, TNF-α, PGE2, and reactive nitrogen species, and decreased levels of IL-10 (Mahajan et al. 2012; Moody 2006; Rajaram et al. 2010). Intriguingly, in PPAR-γ silenced macrophages, a decrease in lipid body formation/foamy macrophage differentiation was also observed. This data support the hypothesis that activation of PPAR-γ by *M. tuberculosis* lipids may promote virulence by modulating macrophage function, both by creating a foamy niche for bacterial persistence and/or by down-regulating harmful inflammatory responses, potentially by influencing the balance of eicosanoids produced during infection.

The ability to regulate immune responses by modulating host lipid sensing nuclear receptors is likely not limited to signaling through PPAR-γ. Indeed, evidence suggests that the crosstalk between mycobacterial lipids and these receptors may be complex, as other receptors such as TR4 seem to play similar roles to PPAR-γ in downregulating macrophage inflammatory responses, LXRα plays a role in upregulating effective immune responses. It is interesting to speculate that the differential requirements for host lipid sensing nuclear receptors have evolved as a mechanism by which *M. tuberculosis* is able to fine-tune immune responses to elicit the optimal balance of immunity that promotes growth, persistence, or transmission.

2.5 CD1 and T Cell Responses to *M. tuberculosis* Lipids

Although much of the work describing the ability of *M. tuberculosis* infection to stimulate and modulate immune responses has focused on the host–pathogen interaction at the level of the infected macrophage, this is only a single component of a multi-dimensional immune response encompassing a large number of cell types with complex and dynamic interactions. The documented effects that *M. tuberculosis* lipids have on macrophage responses likely have ramifications for influencing adaptive immune responses, however it is also likely that *M. tuberculosis* lipids play a more direct role in orchestrating the adaptive response to infection via the activation of CD1 restricted T-cells.

CD1 proteins are MHC like proteins that present lipid antigens to T-cells and NK T-cells. CD1 molecules are classified as group 1 (CD1a, CD1b, CD1c) and group 2 (CD1d) with CD1e considered to be intermediate between the two groups. It is well documented that human T-cells respond to group 1 CD1 presentation of lipid antigens derived from *M. tuberculosis,* including mycolic acids (Beckman et al. 1994; Fischer et al. 2004), lipoarabinomannans (Kasmar et al. 2011; Sieling et al. 1995), and other lipids (Behar et al. 1999; Moody 2006). In addition, there is evidence that mycobacterial phosphatidylinositol mannosides can bind CD1d and activate NK T-cells (Fischer et al. 2004). Considerable effort has gone into identifying the molecular basis by which CD1 molecules acquire and present lipid antigens from mycobacterial species, and a wealth of evidence supports the idea that *M. tuberculosis* lipid antigens presented by CD1 antigens activate T-cells by both direct and indirect mechanisms (Kasmar et al. 2011).

Insights into the functional consequences of CD1 lipid antigen presentation have been limited by the fact that mice lack group 1 CD1 orthologs. Mice do express CD1d molecules, however, CD1d is thought to be less relevant to mycobacterial infection, and knockout mice lacking CD1d lack a discernable phenotype (Behar et al. 1999). Nevertheless, in vitro studies of CD1 function using human cells have demonstrated that human CD1 reactive T-cells express Th1 associated cytokines and can limit the in vitro growth of *M. tuberculosis* through cytolysis and other bactericidal activities, suggesting that these cells may help control infection with *M. tuberculosis* (Stenger et al. 1997, 1998). In addition, clinical studies of human tuberculosis patients have consistently identified lymphocytes that are responsive to mycobacterial lipid antigens (Gilleron et al. 2004; Moody et al. 2000a, b; Ulrichs et al. 2003). These findings demonstrate that *M. tuberculosis* infection results in the activation of CD1 reactive T-cells, however they do not clearly establish the functional consequences of this activation. Several in vitro studies have hypothesized that *M. tuberculosis* may actively downregulate CD1 expression, leading to the hypothesis that CD1 mediated responses may be detrimental to bacterial growth and persistence. However, the hypothesis that *M. tuberculosis* may elicit CD1 mediated responses as a pathogenic strategy has also been proposed (Shinkai and Locksley 2000). The difficulty with which in vitro observations of CD1 function during *M. tuberculosis* infection can be translated into in vivo experimental systems has made a clear understanding of the role of CD1 in *M. tuberculosis* infection difficult to obtain. If CD1 mediated responses are indeed protective, an important question is whether these responses are long lasting and may therefore contribute to memory responses for vaccine development. The area of human immunology describing adaptive immune responses to lipid antigens is therefore an exciting area of research for future investigation.

References

Abdallah AM, Gey van Pittius NC, DiGiuseppe Champion PA et al (2007) Type VII secretion–mycobacteria show the way. Nat Rev Microbiol 5:883–891. doi:10.1038/nrmicro1773

Armstrong JA, Hart PD (1971) Response of cultured macrophages to mycobacterium tuberculosis, with observations on fusion of lysosomes with phagosomes. J Exp Med 134:713–740

Armstrong JA, Hart PD (1975) Phagosome-lysosome interactions in cultured macrophages infected with virulent tubercle bacilli. Reversal of the usual nonfusion pattern and observations on bacterial survival. J Exp Med 142:1–16

Arpaia N, Godec J, Lau L et al (2011) TLR signaling is required for Salmonella typhimurium virulence. Cell 144:675–688. doi:10.1016/j.cell.2011.01.031

Astarie-Dequeker C, Le Guyader L, Malaga W et al (2009) Phthiocerol dimycocerosates of M. tuberculosis participate in macrophage invasion by inducing changes in the organization of plasma membrane lipids. PLoS Pathog 5:e1000289. doi:10.1371/journal.ppat.1000289

Axelrod S, Oschkinat H, Enders J et al (2008) Delay of phagosome maturation by a mycobacterial lipid is reversed by nitric oxide. Cell Microbiol 10:1530–1545. doi:10.1111/j.1462-5822.2008.01147.x

Barber GN (2011) Cytoplasmic DNA innate immune pathways. Immunol Rev 243:99–108. doi:10.1111/j.1600-065X.2011.01051.x

Barkan D, Hedhli D, Yan HG et al (2012) Mycobacterium tuberculosis lacking all mycolic acid cyclopropanation is viable but highly attenuated and hyperinflammatory in mice. Infect Immun 80:1958–1968. doi:10.1128/IAI.00021-12

Beatty WL, Russell DG (2000) Identification of mycobacterial surface proteins released into subcellular compartments of infected macrophages. Infect Immun 68:6997–7002. doi:10.1128/IAI.68.12.6997-7002.2000

Beatty WL, Rhoades ER, Ullrich HJ et al (2000) Trafficking and release of mycobacterial lipids from infected macrophages. Traffic 1:235–247

Beatty WL, Ullrich HJ, Russell DG (2001) Mycobacterial surface moieties are released from infected macrophages by a constitutive exocytic event. Eur J Cell Biol 80:31–40. doi:10.1078/0171-9335-00131

Beckman EM, Porcelli SA, Morita CT et al (1994) Recognition of a lipid antigen by CD1-restricted alpha beta + T cells. Nature 372:691–694. doi:10.1038/372691a0

Behar SM, Dascher CC, Grusby MJ et al (1999) Susceptibility of mice deficient in CD1D or TAP1 to infection with Mycobacterium tuberculosis. J Exp Med 189:1973–1980

Behar SM, Martin CJ, Nunes-Alves C et al (2011) Lipids, apoptosis, and cross-presentation: links in the chain of host defense against Mycobacterium tuberculosis. Microbes Infect 13:749–756. doi:10.1016/j.micinf.2011.03.002

Bekierkunst A, Levij IS, Yarkoni E et al (1969) Granuloma formation induced in mice by chemically defined mycobacterial fractions. J Bacteriol 100:95–102

Berry MPR, Graham CM, Mcnab FW et al (2010) An interferon-inducible neutrophil-driven blood transcriptional signature in human tuberculosis. Nature 466:973–977. doi:10.1038/nature09247

Bloch H (1950) Studies on the virulence of tubercle bacilli; isolation and biological properties of a constituent of virulent organisms. J Exp Med 91:197–218

Bold TD, Ernst JD (2009) Who benefits from granulomas, mycobacteria or host? Cell 136:17–19. doi:10.1016/j.cell.2008.12.032

Camacho LR, Ensergueix D, Perez E et al (1999) Identification of a virulence gene cluster of Mycobacterium tuberculosis by signature-tagged transposon mutagenesis. Mol Microbiol 34:257–267

Capuano SV3, Croix DA, Pawar S et al (2003) Experimental Mycobacterium tuberculosis infection of cynomolgus macaques closely resembles the various manifestations of human M. tuberculosis infection. Infect Immun 71:5831–5844

Clemens DL, Horwitz MA (1995) Characterization of the Mycobacterium tuberculosis phagosome and evidence that phagosomal maturation is inhibited. J Exp Med 181:257–270

Clemens DL, Horwitz MA (1996) The Mycobacterium tuberculosis phagosome interacts with early endosomes and is accessible to exogenously administered transferrin. J Exp Med 184:1349–1355

Clemens DL, Lee B-Y, Horwitz MA (2002) The Mycobacterium tuberculosis phagosome in human macrophages is isolated from the host cell cytoplasm. Infect Immun 70:5800–5807

Cole ST, Brosch R, Parkhill J et al (1998) Deciphering the biology of Mycobacterium tuberculosis from the complete genome sequence. Nature 393:537–544. doi:10.1038/31159

Collins CA, De Mazière A, van Dijk S et al (2009) Atg5-independent sequestration of ubiquitinated mycobacteria. PLoS Pathog 5:e1000430. doi:10.1371/journal.ppat.1000430

Cox JS, Chen B, McNeil M, Jacobs WR (1999) Complex lipid determines tissue-specific replication of Mycobacterium tuberculosis in mice. Nature 402:79–83. doi:10.1038/47042

Crowle AJ, Dahl R, Ross E, May MH (1991) Evidence that vesicles containing living, virulent Mycobacterium tuberculosis or Mycobacterium avium in cultured human macrophages are not acidic. Infect Immun 59:1823–1831

Daffe M, Draper P (1997) The envelope layers of Mycobacteria with reference to their pathogenicity. In: Poole RK (ed) Advances in microbial physiology. Academic Press, London, pp 131–203

Dao DN, Sweeney K, Hsu T et al (2008) Mycolic acid modification by the mmaA4 gene of M. tuberculosis modulates IL-12 production. PLoS Pathog 4:e1000081. doi:10.1371/journal.ppat.1000081

D'Avila H, Melo RCN, Parreira GG et al (2006) Mycobacterium bovis bacillus Calmette-Guérin induces TLR2-mediated formation of lipid bodies: intracellular domains for eicosanoid synthesis in vivo. J Immunol 176:3087–3097

De Jonge M, Pehau-Arnaudet G, Fretz M et al (2007) ESAT-6 from Mycobacterium tuberculosis dissociates from its putative chaperone CFP-10 under acidic conditions and exhibits membrane-lysing activity. J Bacteriol 189:6028. doi:10.1128/JB.00469-07

Deretic V, Levine B (2009) Autophagy, immunity, and microbial adaptations. Cell Host & Microbe 5:527–549. doi:10.1016/j.chom.2009.05.016

Divangahi M, Mostowy S, Coulombe F et al (2008) NOD2-deficient mice have impaired resistance to Mycobacterium tuberculosis infection through defective innate and adaptive immunity. J Immunol 181:7157–7165

Divangahi M, Desjardins D, Nunes-Alves C et al (2010) Eicosanoid pathways regulate adaptive immunity to Mycobacterium tuberculosis. Nat Immunol 11:751–758. doi:10.1038/ni.1904

Fischer K, Scotet E, Niemeyer M et al (2004) Mycobacterial phosphatidylinositol mannoside is a natural antigen for CD1d-restricted T cells. Proc Natl Acad Sci U S A 101:10685–10690. doi:10.1073/pnas.0403787101

Geisel RE, Sakamoto K, Russell DG, Rhoades ER (2005) In vivo activity of released cell wall lipids of Mycobacterium bovis bacillus Calmette-Guérin is due principally to trehalose mycolates. J Immunol 174:5007–5015

George KM, Yuan Y, Sherman DR (1995) Barry, CE3The biosynthesis of cyclopropanated mycolic acids in Mycobacterium tuberculosis. Identification and functional analysis of CMAS-2. J Biol Chem 270:27292–27298

Gey Van Pittius NC, Gamieldien J, Hide W et al (2001) The ESAT-6 gene cluster of Mycobacterium tuberculosis and other high G+C Gram-positive bacteria. Genome Biol 2:research0044

Gill WP, Harik NS, Whiddon MR et al (2009) A replication clock for Mycobacterium tuberculosis. Nat Med 15:211–214. doi:10.1038/nm.1915

Gilleron M, Stenger S, Mazorra Z et al (2004) Diacylated sulfoglycolipids are novel mycobacterial antigens stimulating CD1-restricted T cells during infection with Mycobacterium tuberculosis. J Exp Med 199:649–659

Glickman MS, Cox JS, Jacobs WRJ (2000) A novel mycolic acid cyclopropane synthetase is required for cording, persistence, and virulence of Mycobacterium tuberculosis. Mol Cell 5:717–727

Goren MB, Brokl O, Schaefer WB (1974) Lipids of putative relevance to virulence in Mycobacterium tuberculosis: phthiocerol dimycocerosate and the attenuation indicator lipid. Infect Immun 9:150–158

Guinn KM, Hickey MJ, Mathur SK et al (2004) Individual RD1-region genes are required for export of ESAT-6/CFP-10 and for virulence of Mycobacterium tuberculosis. Mol Microbiol 51:359–370. doi:10.1046/j.1365-2958.2003.03844.x

Gutierrez MG, Master SS, Singh SB et al (2004) Autophagy is a defense mechanism inhibiting BCG and Mycobacterium tuberculosis survival in infected macrophages. Cell 119:753–766. doi:10.1016/j.cell.2004.11.038

Henry T, Monack DM (2007) Activation of the inflammasome upon Francisella tularensis infection: interplay of innate immune pathways and virulence factors. Cell Microbiol 9:2543–2551. doi:10.1111/j.1462-5822.2007.01022.x

Hoffmann C, Leis A, Niederweis M et al (2008) Disclosure of the mycobacterial outer membrane: cryo-electron tomography and vitreous sections reveal the lipid bilayer structure. Proc Natl Acad Sci U S A 105:3963–3967. doi:10.1073/pnas.0709530105

Hornung V, Ablasser A, Charrel-Dennis M et al (2009) AIM2 recognizes cytosolic dsDNA and forms a caspase-1-activating inflammasome with ASC. Nature 458:514–518. doi:10.1038/nature07725

Houben D, Demangel C, van Ingen J et al (2012) ESX-1-mediated translocation to the cytosol controls virulence of mycobacteria. Cell Microbiol 14:1287–1298. doi:10.1111/j.1462-5822.2012.01799.x

Hsu T, Hingley-Wilson S, Chen B et al (2003) The primary mechanism of attenuation of bacillus Calmette-Guerin is a loss of secreted lytic function required for invasion of lung interstitial tissue. Proc Natl Acad Sci U S A 100:12420. doi:10.1073/pnas.1635213100

Hunter RL, Olsen MR, Jagannath C, Actor JK (2006) Multiple roles of cord factor in the pathogenesis of primary, secondary, and cavitary tuberculosis, including a revised description of the pathology of secondary disease. Ann Clin Lab Sci 36:371–386

Indrigo J, Hunter RL, Actor JK (2003) Cord factor trehalose 6,6′-dimycolate (TDM) mediates trafficking events during mycobacterial infection of murine macrophages. Microbiology (Reading, Engl) 149:2049–2059. doi:10.1099/mic.0.26226-0

Ishikawa E, Ishikawa T, Morita YS et al (2009a) Direct recognition of the mycobacterial glycolipid, trehalose dimycolate, by C-type lectin Mincle. J Exp Med 206:2879–2888. doi:10.1084/jem.20091750

Ishikawa H, Ma Z, Barber GN (2009b) STING regulates intracellular DNA-mediated, type I interferon-dependent innate immunity. Nature 461:788–792. doi:10.1038/nature08476

Jagannath C, Lindsey DR, Dhandayuthapani S et al (2009) Autophagy enhances the efficacy of BCG vaccine by increasing peptide presentation in mouse dendritic cells. Nat Med 15:267–276. doi:10.1038/nm.1928

Jayachandran R, Sundaramurthy V, Combaluzier B et al (2007) Survival of mycobacteria in macrophages is mediated by coronin 1-dependent activation of calcineurin. Cell 130:37–50. doi:10.1016/j.cell.2007.04.043

Kasmar AG, van Rhijn I, Cheng T-Y et al (2011) CD1b tetramers bind $\alpha\beta$ T cell receptors to identify a mycobacterial glycolipid-reactive T cell repertoire in humans. J Exp Med 208:1741–1747. doi:10.1084/jem.20110665

Kim M-J, Wainwright HC, Locketz M et al (2010) Caseation of human tuberculosis granulomas correlates with elevated host lipid metabolism. EMBO Mol Med 2:258–274. doi:10.1002/emmm.201000079

Kirkin V, Lamark T, Sou YS et al (2009) A role for NBR1 in autophagosomal degradation of ubiquitinated substrates. Mol Cell 33:505–516. doi:10.1016/j.molcel.2009.01.020

Kirksey MA, Tischler AD, Siméone R et al (2011) Spontaneous phthiocerol dimycocerosate-deficient variants of Mycobacterium tuberculosis are susceptible to gamma interferon-mediated immunity. Infect Immun 79:2829–2838. doi:10.1128/IAI.00097-11

Korf J, Stoltz A, Verschoor J et al (2005) The Mycobacterium tuberculosis cell wall component mycolic acid elicits pathogen-associated host innate immune responses. Eur J Immunol 35:890–900. doi:10.1002/eji.200425332

Kumar D, Nath L, Kamal MA et al (2010) Genome-wide analysis of the host intracellular network that regulates survival of Mycobacterium tuberculosis. Cell 140:731–743. doi:10.1016/j.cell.2010.02.012

Layre E, Sweet L, Hong S et al (2011) A comparative lipidomics platform for chemotaxonomic analysis of Mycobacterium tuberculosis. Chem Biol 18:1537–1549. doi:10.1016/j.chembiol.2011.10.013

Leake ES, Myrvik QN, Wright MJ (1984) Phagosomal membranes of Mycobacterium bovis BCG-immune alveolar macrophages are resistant to disruption by Mycobacterium tuberculosis H37Rv. Infect Immun 45:443–446

Leber JH, Crimmins GT, Raghavan S et al (2008) Distinct TLR- and NLR-mediated transcriptional responses to an intracellular pathogen. PLoS Pathog 4:e6. doi:10.1371/journal.ppat.0040006

Lee B-Y, Clemens DL, Horwitz MA (2008) The metabolic activity of Mycobacterium tuberculosis, assessed by use of a novel inducible GFP expression system, correlates with its capacity to inhibit phagosomal maturation and acidification in human macrophages. Mol Microbiol 68:1047–1060. doi:10.1111/j.1365-2958.2008.06214.x

Lerena MC, Colombo MI (2011) Mycobacterium marinum induces a marked LC3 recruitment to its containing phagosome that depends on a functional ESX-1 secretion system. Cell Microbiol. doi:10.1111/j.1462-5822.2011.01581.x

Lesley R, Ramakrishnan L (2008) Insights into early mycobacterial pathogenesis from the zebrafish. Curr Opin Microbiol 11:277–283. doi:10.1016/j.mib.2008.05.013

Lin PL, Rodgers M, Smith L et al (2009) Quantitative comparison of active and latent tuberculosis in the cynomolgus macaque model. Infect Immun 77:4631–4642. doi:10.1128/IAI.00592-09

Lu Y, Hong S, Tjonahen E, Serhan CN (2005) Mediator-lipidomics: databases and search algorithms for PUFA-derived mediators. J Lipid Res 46:790–802. doi:10.1194/jlr.D400020-JLR200

MacGurn JA, Cox JS (2007) A genetic screen for Mycobacterium tuberculosis mutants defective for phagosome maturation arrest identifies components of the ESX-1 secretion system. Infect Immun 75:2668–2678. doi:10.1128/IAI.01872-06

MacMicking JD, Taylor GA, McKinney JD (2003) Immune control of tuberculosis by IFN-gamma-inducible LRG-47. Science 302:654–659. doi:10.1126/science.1088063

Mahairas GG, Sabo PJ, Hickey MJ et al (1996) Molecular analysis of genetic differences between Mycobacterium bovis BCG and virulent M. bovis. J Bacteriol 178:1274–1282

Mahajan S, Dkhar HK, Chandra V et al (2012) Mycobacterium tuberculosis modulates macrophage lipid-sensing nuclear receptors PPARγ and TR4 for survival. J Immunol 188:5593–5603. doi:10.4049/jimmunol.1103038

Malik ZA, Thompson CR, Hashimi S et al (2003) Cutting edge: Mycobacterium tuberculosis blocks Ca2 + signaling and phagosome maturation in human macrophages via specific inhibition of sphingosine kinase. J Immunol 170:2811–2815

Manca C, Tsenova L, Bergtold A et al (2001) Virulence of a Mycobacterium tuberculosis clinical isolate in mice is determined by failure to induce Th1 type immunity and is associated with induction of IFN-alpha/beta. Proc Natl Acad Sci U S A 98:5752–5757

Manzanillo PS, Shiloh MU, Portnoy DA, Cox JS (2012) Mycobacterium Tuberculosis activates the DNA-dependent cytosolic surveillance pathway within macrophages. Cell Host & Microbe 11:469–480. doi:10.1016/j.chom.2012.03.007

McCaffrey RL, Fawcett P, O'Riordan M et al (2004) A specific gene expression program triggered by Gram-positive bacteria in the cytosol. Proc Natl Acad Sci U S A 101:11386–11391. doi:10.1073/pnas.0403215101

McDonough KA, Kress Y, Bloom BR (1993) Pathogenesis of tuberculosis: interaction of Mycobacterium tuberculosis with macrophages. Infect Immun 61:2763–2773

McFarlane S, Aitken J, Sutherland JS et al (2011) Early induction of autophagy in human fibroblasts after infection with human cytomegalovirus or herpes simplex virus 1. J Virol 85:4212–4221. doi:10.1128/JVI.02435-10

Middlebrook G, Dubos RJ, Pierce C (1947) Virulence and morphological characteristics of mammalian tubercle bacilli. J Exp Med 86:175–184

Monroe KM, McWhirter SM, Vance RE (2010) Induction of type I interferons by bacteria. Cell Microbiol 12:881–890. doi:10.1111/j.1462-5822.2010.01478.x

Moody DB (2006) The surprising diversity of lipid antigens for CD1-restricted T cells. In: Alt FW, Austen KF, Honjo T et al (eds) Advances in immunology. Academic Press, New York, pp 87–139

Moody DB, Guy MR, Grant E et al (2000a) CD1b-mediated T cell recognition of a glycolipid antigen generated from mycobacterial lipid and host carbohydrate during infection. J Exp Med 192:965–976

Moody DB, Ulrichs T, Mühlecker W et al (2000b) CD1c-mediated T cell recognition of isoprenoid glycolipids in Mycobacterium tuberculosis infection. Nature 404:884–888

Murry JP, Pandey AK, Sassetti CM, Rubin EJ (2009) Phthiocerol dimycocerosate transport is required for resisting interferon-gamma-independent immunity. J Infect Dis 200:774–782. doi:10.1086/605128

Myrvik QN, Leake ES, Wright MJ (1984) Disruption of phagosomal membranes of normal alveolar macrophages by the H37Rv strain of Mycobacterium tuberculosis. A correlate of virulence. Am Rev Respir Dis 129:322–328

Nakahira K, Haspel JA, Rathinam VA et al (2011) Autophagy proteins regulate innate immune responses by inhibiting the release of mitochondrial DNA mediated by the NALP3 inflammasome. Nat Immunol 12:222–230. doi:10.1038/ni.1980

Noll H, Bloch H, Asselineau J, Lederer E (1956) The chemical structure of the cord factor of Mycobacterium tuberculosis. Biochim Biophys Acta 20:299–309

Novikov A, Cardone M, Thompson R et al (2011) Mycobacterium tuberculosis triggers host type I IFN signaling to regulate IL-1β production in human macrophages. J Immunol 187:2540–2547. doi:10.4049/jimmunol.1100926

Ohol YM, Goetz DH, Chan K et al (2010) Mycobacterium tuberculosis MycP1 protease plays a dual role in regulation of ESX-1 secretion and virulence. Cell Host & Microbe 7:210–220. doi:10.1016/j.chom.2010.02.006

Pandey AK, Yang Y, Jiang Z et al (2009) NOD2, RIP2 and IRF5 play a critical role in the type I interferon response to Mycobacterium tuberculosis. PLoS Pathog 5:e1000500. doi:10.1371/journal.ppat.1000500

Peyron P, Vaubourgeix J, Poquet Y et al (2008) Foamy macrophages from tuberculous patients' granulomas constitute a nutrient-rich reservoir for M. tuberculosis persistence. PLoS Pathog 4:e1000204. doi:10.1371/journal.ppat.1000204

Philips JA, Ernst JD (2012) Tuberculosis pathogenesis and immunity. Annu Rev Pathol 7:353–384. doi:10.1146/annurev-pathol-011811-132458

Pilli M, Arko-Mensah J, Ponpuak M et al (2012) TBK-1 promotes autophagy-mediated antimicrobial defense by controlling autophagosome maturation. Immunity 37:223–234. doi:10.1016/j.immuni.2012.04.015

Puissegur M-P, Botanch C, Duteyrat J-L et al (2004) An in vitro dual model of mycobacterial granulomas to investigate the molecular interactions between mycobacteria and human host cells. Cell Microbiol 6:423–433. doi:10.1111/j.1462-5822.2004.00371.x

Rajaram MVS, Brooks MN, Morris JD et al (2010) Mycobacterium tuberculosis activates human macrophage peroxisome proliferator-activated receptor gamma linking mannose receptor recognition to regulation of immune responses. J Immunol 185:929–942. doi:10.4049/jimmunol.1000866

Rao V, Fujiwara N, Porcelli SA, Glickman MS (2005) Mycobacterium tuberculosis controls host innate immune activation through cyclopropane modification of a glycolipid effector molecule. J Exp Med 201:535–543. doi:10.1084/jem.20041668

Rao V, Gao F, Chen B et al (2006) Trans-cyclopropanation of mycolic acids on trehalose dimycolate suppresses Mycobacterium tuberculosis -induced inflammation and virulence. J Clin Investig 116:1660–1667. doi:10.1172/JCI27335

Rasmussen SB, Horan KA, Holm CK et al (2011) Activation of autophagy by alpha-herpesviruses in myeloid cells is mediated by cytoplasmic viral DNA through a mechanism dependent on stimulator of IFN genes. J Immunol 187:5268–5276. doi:10.4049/jimmunol.1100949

Rhoades ER, Ullrich HJ (2000) How to establish a lasting relationship with your host: lessons learned from Mycobacterium spp. Immunol Cell Biol 78:301–310. doi:10.1046/j.1440-1711.2000.00938.x

Roberts TL, Idris A, Dunn JA et al (2009) HIN-200 proteins regulate caspase activation in response to foreign cytoplasmic DNA. Science 323:1057–1060. doi:10.1126/science.1169841

Rohde K, Yates RM, Purdy GE, Russell DG (2007) Mycobacterium tuberculosis and the environment within the phagosome. Immunol Rev 219:37–54. doi:10.1111/j.1600-065X.2007.00547.x

Rohde KH, Veiga DFT, Caldwell S et al (2012) Linking the transcriptional profiles and the physiological states of Mycobacterium tuberculosis during an extended intracellular infection. PLoS Pathog 8:e1002769. doi:10.1371/journal.ppat.1002769

Rousseau C, Winter N, Pivert E et al (2004) Production of phthiocerol dimycocerosates protects Mycobacterium tuberculosis from the cidal activity of reactive nitrogen intermediates produced by macrophages and modulates the early immune response to infection. Cell Microbiol 6:277–287. doi:10.1111/j.1462-5822.2004.00368.x

Russell DG (2011) Mycobacterium tuberculosis and the intimate discourse of a chronic infection. Immunol Rev 240:252–268. doi:10.1111/j.1600-065X.2010.00984.x

Russell DG, Cardona P-J, Kim M-J et al (2009) Foamy macrophages and the progression of the human tuberculosis granuloma. Nat Immunol 10:943–948. doi:10.1038/ni.1781

Saitoh T, Fujita N, Jang MH et al (2008) Loss of the autophagy protein Atg16L1 enhances endotoxin-induced IL-1beta production. Nature 456:264–268. doi:10.1038/nature07383

Saitoh T, Fujita N, Hayashi T et al (2009) Atg9a controls dsDNA-driven dynamic translocation of STING and the innate immune response. Proc Natl Acad Sci U S A 106:20842–20846. doi:10.1073/pnas.0911267106

Sani M, Houben ENG, Geurtsen J et al (2010) Direct Visualization by Cryo-EM of the Mycobacterial Capsular Layer: a Labile Structure Containing ESX-1-Secreted Proteins. PLoS Pathog 6:e1000794. doi:10.1371/journal.ppat.1000794

Shi C-S, Shenderov K, Huang N–N et al (2012) Activation of autophagy by inflammatory signals limits IL-1β production by targeting ubiquitinated inflammasomes for destruction. Nat Immunol 13:255–263. doi:10.1038/ni.2215

Shinkai K, Locksley RM (2000) CD1, tuberculosis, and the evolution of major histocompatibility complex molecules. J Exp Med 191:907–914

Sieling PA, Chatterjee D, Porcelli SA et al (1995) CD1-restricted T cell recognition of microbial lipoglycan antigens. Science 269:227–230

Siméone R, Bobard A, Lippmann J et al (2012) Phagosomal rupture by Mycobacterium tuberculosis results in toxicity and host cell death. PLoS Pathog 8:e1002507. doi:10.1371/journal.ppat.1002507

Singh SB, Davis AS, Taylor GA, Deretic V (2006) Human IRGM induces autophagy to eliminate intracellular mycobacteria. Science 313:1438–1441. doi:10.1126/science.1129577

Smith J, Manoranjan J, Pan M et al (2008) Evidence for pore formation in host cell membranes by ESX-1-secreted ESAT-6 and its role in Mycobacterium marinum escape from the vacuole. Infect Immun 76:5478–5487. doi:10.1128/IAI.00614-08

Stamm LM, Morisaki JH, Gao L-Y et al (2003) Mycobacterium marinum escapes from phagosomes and is propelled by actin-based motility. J Exp Med 198:1361–1368. doi:10.1084/jem.20031072

Stanley SA, Raghavan S, Hwang WW, Cox JS (2003) Acute infection and macrophage subversion by Mycobacterium tuberculosis require a specialized secretion system. Proc Natl Acad Sci U S A 100:13001–13006. doi:10.1073/pnas.2235593100

Stanley SA, Johndrow JE, Manzanillo P, Cox JS (2007) The Type I IFN response to infection with Mycobacterium tuberculosis requires ESX-1-mediated secretion and contributes to pathogenesis. J Immunol 178:3143–3152

Stenger S, Mazzaccaro RJ, Uyemura K et al (1997) Differential effects of cytolytic T cell subsets on intracellular infection. Science 276:1684–1687

Stenger S, Hanson DA, Teitelbaum R et al (1998) An antimicrobial activity of cytolytic T cells mediated by granulysin. Science 282:121–125. doi:10.1126/science.282.5386.121

Stetson DB, Ko JS, Heidmann T, Medzhitov R (2008) Trex1 prevents cell-intrinsic initiation of autoimmunity. Cell 134:587–598. doi:10.1016/j.cell.2008.06.032

Sturgill-Koszycki S, Schlesinger PH, Chakraborty P et al (1994) Lack of acidification in Mycobacterium phagosomes produced by exclusion of the vesicular proton-ATPase. Science 263:678–681

Sun L, Wu J, Du F et al (2013) Cyclic GMP-AMP synthase is a cytosolic DNA sensor that activates the type I interferon pathway. Science 339:786–791. doi:10.1126/science.1232458

Takayama K, Wang C, Besra GS (2005) Pathway to synthesis and processing of mycolic acids in Mycobacterium tuberculosis. Clin Microbiol Rev 18:81–101. doi:10.1128/CMR.18.1.81-101.2005

Teles RMB, Graeber TG, Krutzik SR et al (2013) Type I interferon suppresses type II interferon-triggered human anti-mycobacterial responses. Science 339:1448–1453. doi:10.1126/science.1233665

Thurston TLM, Ryzhakov G, Bloor S et al (2009) The TBK1 adaptor and autophagy receptor NDP52 restricts the proliferation of ubiquitin-coated bacteria. Nat Immunol 10:1215–1221. doi:10.1038/ni.1800

Thurston TLM, Wandel MP, von Muhlinen N et al (2012) Galectin 8 targets damaged vesicles for autophagy to defend cells against bacterial invasion. Nature 482:414–418. doi:10.1038/nature10744

Tobin DM, Vary JC, Ray JP et al (2010) The lta4 h locus modulates susceptibility to mycobacterial infection in zebrafish and humans. Cell 140:717–730. doi:10.1016/j.cell.2010.02.013

Tsuchida T, Zou J, Saitoh T et al (2010) The ubiquitin ligase TRIM56 regulates innate immune responses to intracellular double-stranded DNA. Immunity 33:765–776. doi:10.1016/j.immuni.2010.10.013

Ulrichs T, Moody DB, Grant E et al (2003) T-Cell Responses to CD1-Presented Lipid Antigens in Humans with Mycobacterium tuberculosis Infection. Infect Immun 71:3076–3087. doi:10.1128/IAI.71.6.3076-3087.2003

Underhill DM, Ozinsky A, Smith KD, Aderem A (1999) Toll-like receptor-2 mediates mycobacteria-induced proinflammatory signaling in macrophages. Proc Natl Acad Sci U S A 96:14459–14463

van den Elzen P, Garg S, León L et al (2005) Apolipoprotein-mediated pathways of lipid antigen presentation. Nature 437:906–910. doi:10.1038/nature04001

van der Wel N, Hava D, Houben D et al (2007) M. tuberculosis and M. leprae translocate from the phagolysosome to the cytosol in myeloid cells. Cell 129:1287–1298. doi:10.1016/j.cell.2007.05.059

Vance RE, Isberg RR, Portnoy DA (2009) Patterns of pathogenesis: discrimination of pathogenic and nonpathogenic microbes by the innate immune system. Cell Host & Microbe 6:10–21. doi:10.1016/j.chom.2009.06.007

Vandal OH, Pierini LM, Schnappinger D et al (2008) A membrane protein preserves intrabacterial pH in intraphagosomal Mycobacterium tuberculosis. Nat Med 14:849–854. doi:10.1038/nm.1795

Varga T, Czimmerer Z, Nagy L (2011) PPARs are a unique set of fatty acid regulated transcription factors controlling both lipid metabolism and inflammation. Biochim Biophys Acta 1812:1007–1022. doi:10.1016/j.bbadis.2011.02.014

Vergne I, Chua J, Lee H–H et al (2005) Mechanism of phagolysosome biogenesis block by viable Mycobacterium tuberculosis. Proc Natl Acad Sci U S A 102:4033–4038. doi:10.1073/pnas.0409716102

Watson RO, Manzanillo PS, Cox JS (2012) Extracellular M. tuberculosis DNA targets bacteria for autophagy by activating the host DNA-sensing pathway. Cell 150:803–815. doi:10.1016/j.cell.2012.06.040

Welin A, Lerm M (2012) Inside or outside the phagosome? The controversy of the intracellular localization of Mycobacterium tuberculosis. Tuberculosis (Edinb) 92:113–120. doi:10.1016/j.tube.2011.09.009

Wild P, Farhan H, McEwan DG et al (2011) Phosphorylation of the autophagy receptor optineurin restricts Salmonella growth. Science 333:228–233. doi:10.1126/science.1205405

Winter SE, Thiennimitr P, Winter MG et al (2010) Gut inflammation provides a respiratory electron acceptor for Salmonella. Nature 467:426–429. doi:10.1038/nature09415

Wolf AJ, Linas B, Trevejo-Nuñez GJ et al (2007) Mycobacterium tuberculosis infects dendritic cells with high frequency and impairs their function in vivo. J Immunol 179:2509–2519

Wong KW, Jacobs WRJ (2011) Critical role for NLRP3 in necrotic death triggered by Mycobacterium tuberculosis. Cell Microbiol 13:1371–1384. doi:10.1111/j.1462-5822.2011.01625.x

Xu S, Cooper A, Sturgill-Koszycki S et al (1994) Intracellular trafficking in Mycobacterium tuberculosis and Mycobacterium avium-infected macrophages. J Immunol 153:2568–2578

Yoshikawa Y, Ogawa M, Hain T et al (2009) Listeria monocytogenes ActA-mediated escape from autophagic recognition. Nat Cell Biol 11:1233–1240. doi:10.1038/ncb1967

Yuan Y, Barry CE 3rd (1996) A common mechanism for the biosynthesis of methoxy and cyclopropyl mycolic acids in Mycobacterium tuberculosis. Proc Natl Acad Sci U S A 93:12828–12833

Yuan Y, Lee RE, Besra GS et al (1995) Identification of a gene involved in the biosynthesis of cyclopropanated mycolic acids in Mycobacterium tuberculosis. Proc Natl Acad Sci U S A 92:6630–6634

Yuk J-M, Shin D-M, Lee H-M et al (2009) Vitamin D3 induces autophagy in human monocytes/macrophages via cathelicidin. Cell Host & Microbe 6:231–243. doi:10.1016/j.chom.2009.08.004

Zhao Z, Fux B, Goodwin M et al (2008) Autophagosome-independent essential function for the autophagy protein Atg5 in cellular immunity to intracellular pathogens. Cell Host & Microbe 4:458–469. doi:10.1016/j.chom.2008.10.003

Zheng YT, Shahnazari S, Brech A et al (2009) The adaptor protein p62/SQSTM1 targets invading bacteria to the autophagy pathway. J Immunol 183:5909–5916. doi:10.4049/jimmunol.0900441

Index

DOI: 10.1007/978-3-642-40232-6

MIX
Papier aus verantwortungsvollen Quellen
Paper from responsible sources
FSC® C105338

If you have any concerns about our products,
you can contact us on
ProductSafety@springernature.com

In case Publisher is established outside the EU,
the EU authorized representative is:
Springer Nature Customer Service Center GmbH
Europaplatz 3, 69115 Heidelberg, Germany

Printed by Libri Plureos GmbH
in Hamburg, Germany